Textbooks in Electrical and Electronic Engineering

Series Editors

G. Lancaster   E. W. Williams

Electrical Circuits and Systems

An Introduction for Engineers and Physical Scientists

A. M. HOWATSON

Department of Engineering Science
University of Oxford

OXFORD   NEW YORK   TORONTO
OXFORD UNIVERSITY PRESS
1996

Oxford University Press, Walton Street, Oxford OX2 6DP

Oxford   New York
Athens   Auckland   Bangkok   Bombay
Calcutta   Cape Town   Dar es Salaam   Delhi
Florence   Hong Kong   Istanbul   Karachi
Kuala Lumpur   Madras   Madrid   Melbourne
Mexico City   Nairobi   Paris   Singapore
Taipei   Tokyo   Toronto
and associated companies in
Berlin   Ibadan

Oxford is a trade mark of Oxford University Press

Published in the United States
by Oxford University Press Inc., New York

A catalogue record for this book is available from the British Library

Library of Congress Cataloging in Publication Data

Howatson, A. M.
Electrical circuits and systems: an introduction for engineers and
physical scientists / A. M. Howatson.
(Textbooks in electrical and electronic engineering : 5)
Includes index.
1.   Electrical circuits.   2.   Electronic circuits.   I. Title
II. Series.
TK454.H69 1996   621.319'2–dc20   95–42037
ISBN 0 19 856449 X (Hbk)
ISBN 0 19 856448 1 (Pbk)

Typeset by Expo Holdings, Malaysia
Printed in Great Britain by
Bookcraft (Bath) Ltd., Midsomer Norton, Somerset

Preface

This is an introductory book about electrical circuits and those aspects of engineering systems which can be related to circuit theory. It has been aimed at undergraduates in the first two years or so of courses in electrical and electronic engineering, or in engineering science generally; but I hope it may also be useful to other students in the physical sciences and perhaps even, as a refresher, to those who are more advanced in their careers.

There are a great many books on electrical circuits, some of them excellent, and one should have a good reason for presuming to add to their number. The subject is one of the cornerstones of electrical and electronic engineering, and remains no less so after the enormous expansion of that whole profession in recent decades. More circuit theory is now taught to non-specialist undergraduates, and at a higher level, than was once the case; on the other hand, many find it difficult on first acquaintance and have very little by way of introduction before they meet the rigours of a university course. There is a consequent need for books on circuit theory which within a reasonably compact framework go some way beyond, but do not overlook, its elementary reaches.

Accordingly, I have tried to write a book which not only sets out the fundamentals of electrical circuits for beginners but which also goes on to include in later chapters topics outside the usual 'first course' material, such as non-periodic waveforms (including electrical noise) and distributed circuits. Like nearly all comparable books, it is mainly about analysis rather than design: not only is facility in analysis a prerequisite before design can be attempted, it is also invaluable for any engineer who has to assess the behaviour of existing circuits. Design has of course been mentioned whenever appropriate, but no attempt has been made artificially to introduce design procedures. Nor is there any treatment of analysis by computer, which I believe to be of only limited use in acquiring a feel for the subject. Among the appendices there is quite an extensive treatment of the physical sources of the passive parameters resistance, capacitance, and inductance. It is usual to take these for granted in books on circuits, and to regard their physical basis as a question of electromagnetism and the properties of matter, but I have taken the view that in the present context this basis should at least be available.

For the most part the book deals with circuits and systems which can be assumed to be linear in behaviour. It follows that electronic circuits are treated mainly in the small-signal analogue regime and that the whole area of digital applications is mentioned only in passing. An exception to this general constraint of linearity is the final chapter, on diode circuits. These underlie several important applications but from an educational point of view they do not easily fit conventional categories and are quite often omitted from textbooks or dealt with in a piecemeal fashion.

The sequence of chapters is based on the common convention that the steady state should be treated at some length before proceeding to natural response and general driving functions. An alternative approach, also popular, is to take the steady state as a particular case to be addressed after the general foundations are laid. While this has considerable intellectual appeal, it seems to me to be less easily related to our everyday experience of electrical engineering. Having chosen the first of these approaches, however, I have in one or two respects departed from convention. Thus, the chapter on four-terminal networks appears earlier than is usual, immediately after chapters on the d.c. and a.c. steady states, and it not only includes the general relations often found under this heading but also gathers together the most important applications of this class of circuits, ranging from amplifiers to transformers. Operational amplifiers, on the other hand, are deferred until Chapter 8, which is devoted to feedback circuits and follows a chapter on transient behaviour where the idea of stability is introduced.

Chapter 10, on distributed circuits, represents a departure from convention not only because the subject is often omitted from introductory books on circuits but also because the continuous transmission line is approached by way of periodic networks. This choice may be debatable, but it has the advantage of showing immediately how progressive phase shift leads to wave propagation.

I have mostly assumed that readers already have, or are concurrently acquiring, the necessary mathematical background, which hardly goes beyond an elementary knowledge of complex numbers, matrices, and ordinary linear differential equations. In addition, familiarity with the notation of vector analysis has been assumed in the appendix on passive parameters. The worked examples in the text and the sets of problems at the end of each chapter are intended to provide facility and insight while avoiding an excess of mere manipulation.

Much of the book was written during periods of sabbatical leave from Oxford University and Balliol College, for which I am most grateful. It is a pleasure to record my gratitude to colleagues in the Department of Engineering Science, who have taught me a great deal; to generations of engineering undergraduates at Balliol, who have taught me even more; and to the staff and advisers of the Oxford University Press for their ready help, not to say forbearance.

Oxford   A. M. H.
October 1995

Contents

1 Introduction

1.1 General

The electrical circuit is essential to nearly all applications of electricity, because the great majority of these depend on the use of devices in which current flows, and an electrical current can be maintained only in closed paths. A circuit may consist of one such path or of several interconnected paths; in the second case it is often called a *network*, although that term can be formally defined as any set of connected devices whether or not they form a circuit. While it is possible to deal with circuits or networks in a general way, so as to include non-electrical examples such as pipes carrying fluid or a road system carrying vehicles, we consider only the electrical kind in this book.

Any useful electrical circuit must be made up of at least one device which produces some desired effect, and at least one generator or source of energy, such as a battery, to maintain the current through it. By 'device' is meant here an item of electrical apparatus having a particular function: it may be, for example, an electric motor, a radio receiver, or a light bulb. Some devices comprise circuits of considerable complexity, others are rather simple: an ordinary light bulb in essence is nothing more than a length of wire. The simplest physical parts from which a device or a complete circuit can be made up are called, in general, **components**. So far as circuits are concerned, the task of the engineer is **synthesis**, that is choosing and combining components, including sources of energy, to do a specified job. Before that can be done effectively, it is necessary to study circuit **analysis**, in which the task is simpler, namely to predict the behaviour of a given circuit. Analysis is easier because it is purely deductive once the principles are known; nevertheless it is quite extensive, and the greater part of the book is devoted to it.

The fundamental quantities to be found in analysing a circuit are the currents and voltages in its various parts. Although it is common to think of voltage and current as being cause and effect respectively, in that voltage applied to a device will tend to drive current through it, it is equally valid and sometimes more appropriate to consider the reverse, namely that current flowing in a device produces a voltage drop across it.

1.2 Definitions

We need to begin with some knowledge of the electrical quantities encountered in circuits. The definitions given in this section are sufficient for our purpose, but further information can be found in Appendix B.

1.2.1 Current

Electrical current is the flow of electrical charge across any area such as the cross-section of a wire. If positive charge crosses the area from one side to the other then positive current flows in that direction. Negative charge crossing from one side to the other can be taken to be either negative current in that direction or positive current in the other; if both positive and negative charges actually flow, the net current is the resultant of the two effects. In circuits we need rarely be concerned with the physical constitution of the current, it being normally sufficient to know its resultant value in a given direction; further, any given current can always be considered as exactly equivalent to another of identical magnitude and opposite sign, flowing in the opposite direction. The unit of electrical current is the **ampere**[†] (symbol A), equivalent to a flow of one coulomb of charge per second. One ampere is a middling value of current: in practical terms domestic appliances typically require a few amperes; power transmission lines may carry several kiloamperes; and a pocket calculator operates on less than a milliampere. Current is usually given the symbol i, the lower-case letter signifying an instantaneous value in cases where the value may change with time.

1.2.2 Voltage

The name 'voltage' stems from the unit of electrical potential, the volt (symbol V). Potential is one means of describing the forces which act upon charges, in terms of the work done in moving them from point to point: if one joule of work is done in moving one coulomb of positive charge from A to B, then we say that the electrical potential at B exceeds that at A by one volt. In circuits, the **potential difference** (p.d.) between any two points is a measure of the force which tends to move charge, and hence current, from one to the other, and it is often loosely called the voltage or voltage difference; we may speak of the voltage between the points, in which case its direction should be specified, or of the voltage of either one with respect to the other. As with current, any given voltage is equivalent to another of opposite sign acting in the opposite direction; thus, if the voltage of A with respect to B is 1 V, then that of B with respect of A is –1 V. These are sometimes written respectively as v_{AB} and v_{BA}.

[†] A summary of electrical units and their symbols is given in Appendix A.

Current:

flow of electrical charge across a x-sectional area

Voltage:

The measure of potential difference between two points and is a measure of force to move a charge from A to B.

Potential difference in an electrical circuit is produced by **electro-motive force** (e.m.f.), which is also measured in volts and describes a driving force which can maintain current around a closed path. Every source of continuous electrical energy — batteries, generators, thermocouples, and so on — consists essentially of an energy converter which produces e.m.f. Because a source of e.m.f. can maintain a p.d. between its terminals, connecting a source of e.m.f. is often thought of as 'applying a voltage' between two points. Electrical sources are discussed in Section 1.6.

One volt is, in practice, quite a small unit: a battery cell typically provides 1.5 V, but electricity for national power systems is mostly generated at voltages above 10 kV, transmitted at voltages up to 275 kV or more, and typically supplied to domestic consumers at 240 V. Voltage is usually expressed by the symbol v; as for current, the lower-case letter signifies in general an instantaneous value.

1.2.3 Resistance and conductance

When a voltage is maintained between two points in an electrically conducting body, current flows from the more positive point to the other. It is well known that if the physical conditions of the body, for example temperature, are kept constant then the current i which flows is in many cases found to be proportional to the voltage v. This relationship is **Ohm's law**, and may be written

$$v = Ri \qquad (1.1)$$

[handwritten margin note: Ohm's law → assuming constant Temp.]

where R is a constant called the **resistance** between the two points; its unit is the **ohm** (symbol Ω), one ohm being equivalent to one volt per ampere. Ohm's law holds for many materials over a wide range of voltage and current. The heating effect of current means that the value of R tends to change because of the effect of temperature, but for most purposes it is usually justifiable to take R to be a constant for any two points on a given conductor. The inverse of resistance, i.e. the ratio of current to voltage, is **conductance**; it is usually given the symbol G and its unit is the siemens (S) which represents one ampere per volt[†]. A component which is designed to provide a certain resistance to the flow of current through it is known as a resistor and may be made of wire or a solid composite material. The term 'conductor', or conducting material, usually implies as low a resistance as possible.

[handwritten margin note: Resistance: $R = \dfrac{i}{v}$ (ohms Ω) R is constant for a particular material.]

[handwritten margin note: Conductance: The inverse of resistance $G = \dfrac{1}{R}$ (siemens S)]

Values of resistance in practice cover a very large range (as do values for the resistivity of materials). Wires and other conductors used for circuit connections have resistances which are usually a small fraction of an ohm; a heating element might represent around 50 Ω and a lamp filament around 500 Ω; the resistors used as components in electronic circuits commonly have values up to 1 MΩ and more.

[†] The name mho and symbol ℧ for the same unit are still found but are obsolete.

1.2.4 Capacitance

A potential difference is always associated with electric charge, whether or not current flows as in a conductor, and the two quantities are in many circumstances very nearly proportional over a wide range. We define **capacitance** as the charge per unit potential difference, so that it can be expressed as

$$C = q/v \qquad\qquad (1.2)$$

where q is the charge and v the p.d.; we take it to be a constant for a given physical arrangement. The unit of capacitance is the **farad** (symbol F), which is one coulomb per volt.

Capacitance is most easily visualized as a property of two conductors separated by some insulating (or **dielectric**) material, in which case v is the p.d. between conductors which carry equal and opposite charges q; but in fact any arrangement which can sustain a potential difference has some capacitance, whether or not separate conductors can be distinguished. Here we need be concerned only with the essential attribute of capacitance as part of a circuit, namely its current–voltage relationship. Evidently no net current can flow into any body or region which is storing a constant amount of charge; but if the charge is changing with time t then the current flowing on that account must be given by dq/dt. Equation 1.2 then gives by differentiation

$$i = dq/dt = C\, dv/dt \qquad\qquad (1.3)$$

for constant C, and we therefore take C to be that property of a component which shows a rate of change of voltage in proportion to current flowing, however that current may be caused. The same relationship may also be written in the integral form

$$v(t) = \frac{1}{C}\int_0^t i(\tau)\, d\tau \; + \; v(0).$$

One farad is a very large capacitance in practice, and would be unusual. The natural capacitance associated with electrical conductors is typically in the order of picofarads; components designed as capacitors, mostly for electronic circuits, are quite often in the nanofarad to microfarad range.

1.2.5 Inductance

An electric current always produces a magnetic field, and the current i which flows around any closed path produces a certain total magnetic flux Φ which passes through that path. For many materials the two are in proportion over a wide range, and for a given path we may therefore define a constant given by

$$L = \Phi/i \qquad\qquad (1.4)$$

which is known as the **self-inductance** of the path. It is also possible for flux through this path to be produced by current flowing in another, which gives rise to an analogous definition of **mutual inductance**; this we shall consider in Section 3.8. For the present we use the term 'inductance' to mean self-inductance. The unit of both forms of inductance is the **henry** (symbol H), which is one weber per ampere; the weber (Wb) is the unit of magnetic flux.

So far as circuit analysis is concerned, we are again interested mainly in the effect of inductance on voltage–current relations. If the current in a circuit is constant, the only voltage needed to drive it is that to overcome the effects of resistance and capacitance, if any; but if the current changes, so does the flux, and by Faraday's law it is known that this must induce an e.m.f. of magnitude $d\Phi/dt$. It follows by differentiating eqn 1.4 that the e.m.f. can be written

$$e = -L \, di/dt \qquad\qquad (1.5)$$

provided that we may take L to be a constant; the negative sign merely indicates that, according to Lenz's law, the direction of e is such as to oppose the change in current. In circuit analysis, the directions of voltages and currents must be carefully considered (and we return to this below) but it is clear without more ado that the voltage needed to maintain a changing current must be in a direction opposing the induced e.m.f. We therefore rewrite eqn 1.5 in the form

$$v = L \, di/dt \qquad\qquad (1.6)$$

and take inductance to be that property of a circuit which requires an applied voltage in proportion to the rate of change of the current flowing. The same relationship may also be written in the integral form

$$i(t) = \frac{1}{L} \int_0^t v(\tau) \, d\tau \; + \; i(0).$$

In practical terms, one henry is quite a large inductance, typical of devices, such as rotating electrical machines, designed to use high magnetic fields over appreciable volumes; small inductors used as electronic components are more likely to have values in the order of millihenries or less.

1.2.6 Symbols and signs

Resistance, capacitance, and inductance are commonly denoted in circuit diagrams by the standard symbols shown in Fig. 1.1, where for clarity each is shown between a pair of terminals (we shall refer to 'terminals' quite often, meaning in general the end points of components

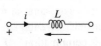

Fig. 1.1

Symbols for resistance, capacitance, and inductance.

or parts of a circuit, by which connections can be made and between which a voltage can be measured). In each case the direction chosen to be that of positive current i is indicated by an arrow, while the voltage v can be defined as the potential of the terminal marked $+$ with respect to that marked $-$. An alternative convention for voltage uses an arrow as shown, with v defined as the potential of the arrowhead with respect to the tail. These signs may not of course give the *actual* directions of current and voltage (as was pointed out above), but their relationship is important: in each case the direction shown for current has been chosen so that it enters by the positive terminal, and eqns 1.1, 1.3, and 1.6 are then correct. The relative directions shown are therefore usually preferred, but if for any reason the specified direction (usually called the **reference** direction) of either i or v is reversed, then a negative sign added to each equation makes it correct. It is clear from eqn 1.3 that the actual direction of current in a capacitance depends not only on the actual direction of voltage but also on whether it is increasing or decreasing, and similarly eqn 1.6 for an inductance shows that the actual direction of v depends on whether i is increasing or decreasing.

Although common sense in the matter of directions and signs will often suffice, in the last resort it is essential to be punctilious: we may choose any reference directions we please; but chosen they must be, and all equations must carry signs to fit our choice.

1.3   Circuit components and ideal elements

From the previous section we can bring together the following relations for resistance, capacitance, and inductance:

$$
\begin{aligned}
v &= Ri \\
i &= C\mathrm{d}v/\mathrm{d}t \\
v &= L\mathrm{d}i/\mathrm{d}t.
\end{aligned}
\tag{1.7}
$$

These describe respectively the physical properties of friction-like resistance to the flow of charge in conductors, the association of electric fields with potential difference, and the association of magnetic fields with current. The quantities R, C, and L are said to be circuit **parameters** and when we take them, as we often shall, to be constants, then eqns 1.7 are **linear**. Since the linearity holds, with the same constant, when the directions of v and i are both reversed, elements having these parameters are sometimes said to be linear and **bilateral**[†]. They are also said to be **passive** because none can generate electrical energy (although, as we shall see, C and L can store energy in their fields). The same terms are applied to circuits which, apart from the sources which energize them (Section 1.6), can be satisfactorily described by combinations of these parameters. Although in practice many circuits include devices such as

---

[†] 'Linear' may be taken to mean bilateral also, unless the context indicates otherwise.

electronic amplifiers and electric motors, which appear to be based on much more complex principles than have so far been mentioned, these devices can often be represented by combinations of passive elements and sources of energy; for this reason we address at some length the behaviour of linear circuits with passive parameters.

It is obvious from the definitions of the previous section that any real component or device in a circuit must to some extent show all of the properties which gave rise to the three parameters, whether by design or not. Thus a length of conducting material which has resistance must also contribute some inductance to its circuit on account of the magnetic flux produced when current flows in it, and some capacitance because of the potential difference which then exists between its ends. However components can be designed to show one or other property predominantly, often to the extent that the others can be neglected. Hence real components can be made in the form of resistors, capacitors, and inductors. Each of these can often be represented with sufficient accuracy by a single parameter, in which case we may refer to a 'perfect' or 'ideal' resistor etc.; this is especially true of the first two, less so for an inductor since it normally needs many turns of wire and has appreciable resistance in consequence. When two or even three parameters are significant in a single component, they can often be legitimately regarded as *separate* entities (however closely they may be mingled in fact) which combine to represent it. Thus an imperfect inductor may be represented by separate L and R carrying the same current, because the current producing the magnetic flux is also that flowing in the resistance of the wire, so it is equivalent to a perfect inductor and a perfect resistor connected in series (see Section 1.4 below). A linear circuit made up in this way with combinations of constant parameters is said to consist of **ideal elements**, although some of these may represent non-ideal components.

It is not always acceptable, however, to represent the parameters of an actual device or component as if they were separate ideal elements; when it is not, the circuit is said to be **distributed** and must be represented rather differently. By contrast, circuits which behave as if they are made up entirely of interconnected but separate elements are sometimes known **lumped**. We shall for the most part deal with lumped circuits; distributed circuits are considered in Chapter 10.

1.4 Circuit connections

It is usual to assume that the components of a circuit (including sources), and the elements in its representation, are connected by conducting paths of negligible resistance which allow current to flow with no potential difference and which likewise have negligible capacitance and inductance. When two or more elements are connected end to end, so that they carry the same current i, as shown in Fig. 1.2(a), they are said to be **series**. It is obvious that the total voltage across all of the elements is then the sum of the voltages across each. When elements are connected between the same pair of points, as in Fig. 1.2(b), so that the

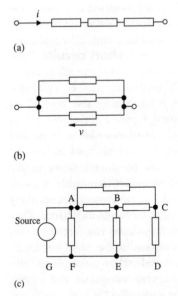

(a)

(b)

(c)

Fig. 1.2
Connected elements: (a) series; (b) parallel; (c) general.

voltage v must be the same for all, they are said to be in **parallel**. (In this case we might suspect that the total current to the combination must be the sum of the individual currents, and in Section 2.2 this will be confirmed by Kirchhoff's first law.)

It is quite possible for a circuit to have none of its passive elements in series or parallel with any other: in Fig. 1.2(c), for example, no two resistances need share either the same voltage or the same current. The same figure illustrates a few definitions which we shall need later. The meeting point of conducting connections which join the ends of two or more elements, as at A and B, is called a **node**; any one path between two nodes, such as BC, containing an element or a series combination of elements, is called a **branch**. Any closed path which passes no node more than once, such as ABEFGA, is a **loop**, and a loop which contains no other loop, such as BCDEB, is a **mesh**. Note that the points D, E, F, G are shown separately for clarity but actually form a single node. The formal relationship between nodes, branches, and meshes without regard for physical behaviour is the concern of **network topology**. A **planar** network is one which can be drawn on a surface with none of its branches crossing any other. There are some formal restrictions in applying circuit theory to non-planar networks, but for practical purposes they are few, and since we shall deal mostly with planar networks we need consider the distinction no further.

1.5 Short circuits and open circuits

Apart from the perfect conductors which are assumed to connect the various elements of a circuit, additional conducting paths can appear, by accident or design, in other places and if these have negligible resistance (as is often the case) they are generally known as **short circuits**. If the resistance of a short circuit is taken to be zero, and if any effect due to inductance and capacitance is similarly neglected, another significant property follows: the voltage across it is necessarily also zero, if we leave aside the possibility of infinite current. Figure 1.3(a) shows a short circuit connected across the terminals of a resistance (that is, in parallel with it) for example. Because the voltage across both must be zero, no current can flow in the resistance; whatever current flows to the combination must be confined to the short circuit (intuitively we may say that it takes the 'easier' path) and the resistance is effectively eliminated; the same would be true for capacitance and inductance. Thus a short circuit has the effect of rendering superfluous any element across which it is connected; the element is then said to be 'short-circuited'. Although a short circuit is usually considered to arise from zero resistance, it follows from eqns 1.7 that zero inductance and infinite capacitance would individually have the same effect.

A non-conducting path, offering infinite resistance (or zero conductance) to current flow, is said to be an **open circuit**. Since this would include any path not actually forming part of a circuit (all of the material surrounding a circuit is assumed to be non-conducting), any number of such paths can be identified; but the significance of the term is

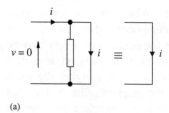

(a)

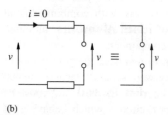

(b)

Fig. 1.3
(a) a short circuit; (b) an open circuit.

seen when a break occurs in an otherwise conducting path, as for example between the two resistances in Fig. 1.3(b). We may say that an infinite resistance has been inserted (that is, connected in series with the others), but the significant consequence is that the current is necessarily zero in the other resistances, no voltage can therefore appear across them, and they are thus effectively eliminated. The same would be true of any other element in series with an open circuit; its effect is therefore to render superfluous any element in series with it. While an open circuit is normally taken to be infinite resistance or zero conductance (since it appears in practice as the absence of a conducting path) it is formally true that infinite inductance and zero capacitance would have the same effect.

The duality between short and open circuits is evident: the former implies zero resistance and zero voltage and removes parallel elements, the latter implies zero conductance and zero current and removes series elements. We shall find that this kind of duality appears very often in circuit analysis. Here it is important to note an ambiguity in using the word 'remove', which in the foregoing simply means to render an element superfluous or ineffective, so that it carries no current and might as well be disconnected and taken away. Any element removed in this sense is really being replaced by an open circuit. On the other hand, in some different context 'removing' a resistance, for example, might mean reducing its value to zero; that would imply that it becomes a *short* circuit. It is an important distinction.

1.6 Sources

In Section 1.2.2 it was stated that continuous current can be supplied only from a source which produces electrical energy from some other form and generates an e.m.f. For circuit analysis we need no knowledge of how the electrical energy is produced; we simply divide sources into two categories, voltage sources and current sources (and even they are not really separate kinds, as we shall see). Strictly a source in either category should also be called *independent* unless it is a *controlled* source as described in Section 1.6.4. Sources are said to be **active** elements, as opposed to the passive elements R, C and L.

1.6.1 Voltage sources

It was mentioned in Section 1.2.2 that a source of e.m.f. produces a potential difference between its terminals; we call this the terminal voltage. If no current is drawn from the source, its terminal voltage must be equal to its e.m.f. An **ideal voltage source** is one which produces a specified terminal voltage v_0 (not necessarily constant in time) whatever the current i drawn from it; it is indicated by the symbol of Fig. 1.4(a). As for passive elements, the directions chosen in specifying v_0 and i are a matter of choice but the relation between them is important: if current is actually leaving by the terminal which is actually at the higher potential, then the source is supplying energy; otherwise it is absorbing energy (and we should then call it a **sink** rather than a source). It follows that with the directions shown v_0 and i

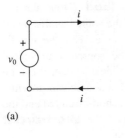

(a)

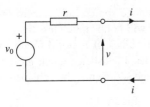

(b)

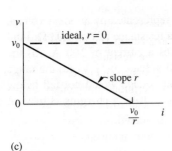

(c)

Fig. 1.4
Voltage sources: (a) ideal;
(b) non-ideal; (c) characteristics.

are both positive (or both negative) if energy is supplied by the source. The idea that a source might actually absorb energy is familiar enough in practice; it arises quite commonly and implies merely that the source is **reversible**. We return to this later.

In practice no voltage source is ideal, because there is always some internal resistance (and possibly also inductance or capacitance or both) associated with a source, which causes the terminal voltage v to fall below its e.m.f. v_0 in proportion to the current it supplies. The effect is easily represented by the appropriate elements in series with an ideal source of voltage v_0 as in Fig. 1.4(b), in which r is the internal resistance. Strictly we need Kirchhoff's voltage law, discussed in Section 2.2, to establish the validity of this, but if we accept that v must differ from v_0 by the voltage across r then we may write

$$v = v_0 - ri \tag{1.8}$$

where the value of i depends on the rest of the circuit to which the source is connected, generally referred to as its **load**. (Note that if i were negative, indicating that energy flows *into* the source as discussed above, then the terminal voltage would have to exceed the e.m.f.) Equation 1.8 gives the **characteristic** of the source, represented graphically in Fig. 1.4(c) together with that of an ideal source. It can be seen that there is a maximum current which such a source can provide, given by v_0/r and attained when the terminals are short-circuited and the terminal voltage therefore zero; it may be called the **short-circuit current**, and for an ideal source would be infinite.

The terminal voltage v of any actual voltage source falls to zero on short-circuit, when $i = v_0/r$; nevertheless v will not be much less than v_0 if the current is restricted to much smaller values. It is not difficult to show that this will be the case if the load has a resistance much greater than r, and under these conditions any voltage source may be considered to approach the ideal.

Example 1.1   A car battery can be represented by an ideal voltage source having $v_0 = 12$ V in series with a resistance of r of 0.03 Ω. Find (a) the terminal voltage when it supplies a current of 100 A; (b) the current which would reduce the terminal voltage to 11 V; (c) the range of load resistance for which the terminal voltage would not fall below 10 V; (d) the terminal voltage required to drive a charging current of 10 A through the battery.

From eqn 1.8 we have

(a) $v = 12 - 0.03 \times 100 = \mathbf{9}$ **V**;

and

$11 = 12 - 0.03\, i,$

from which

(b) $i = \textbf{33.3 A}$.

(c)   The current i which would reduce v to 10 V is given by

$$10 = 12 - 0.03\,i$$

from which

$$i = 66.7 \text{ A}$$

and it follows that the load resistance R_L must have a value of at least

$$v/i = 10/66.7 = \textbf{0.15 } \boldsymbol{\Omega}.$$

(d)   The terminal voltage is

$$v = 12 - 0.03(-10) = \textbf{12.3 V}.$$

(a)

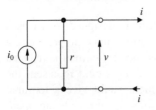

(b)

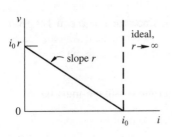

(c)

Fig. 1.5
Current sources: (a) ideal;
(b) non-ideal; (c) characteristics.

1.6.2 Current sources

Although it is physically e.m.f., as represented in a voltage source, which drives current in a circuit, it is possible to design sources which can supply a specified current, whatever the load. This leads to the idea of an **ideal current source**, providing the same current i_0 (not necessarily constant in time) to any load; it is given the symbol shown in Fig. 1.5(a). An actual current source is not ideal: internal resistance (and possibly inductance or capacitance) cause the current flowing from the terminals to fall below the value i_0 in proportion to the terminal voltage v, which depends on the load. This behaviour can be represented by a branch containing appropriate elements in *parallel* with an ideal source, as in Fig. 1.5(b) where r is again an internal resistance. Although the validity of this really depends on Kirchhoff's current law (Section 2.2), we can allege here that, since the current drawn by r must be v/r, the current available for the load must be

$$i = i_0 - v/r. \tag{1.9}$$

This representation of a current source does not mean that in physical fact it comprises a resistance r in parallel with some ideal source, but is only a way of accounting for the observed behaviour of what may be a fairly sophisticated device; it can be called an **equivalent** circuit, a term which we shall often meet in various contexts. The representation of a voltage source in Fig. 1.4(b), on the other hand, is somewhat closer to physical reality; this difference will become clearer in the next section.

Equation 1.9 gives the sloping characteristic of Fig. 1.5(c), which also shows the ideal characteristic as a vertical line. We should note that an ideal current source has *infinite* internal resistance, that is to say zero conductance; this simply means no loss of current to the parallel branch, just as the absence of series resistance in an ideal voltage source means no loss of voltage. (The duality of the two can be emphasized by labelling the parallel branch of a current source as a conductance g, given by $1/r$.) It can be seen from Fig. 1.5(c) that the terminal voltage v reaches a maximum value given by $i_0 r$ when the load current i has fallen to zero, the load being then an open circuit; it may therefore be called the **open-circuit voltage**, and for an ideal current source would be infinite. (Unlikely as it seems to postulate a source which would produce an infinite voltage across an open circuit, it is no more far-fetched in principle than an ideal voltage source which produces an infinite current in a short circuit.)

In the previous section it was stated that a voltage source could be said to approach the ideal, with a terminal voltage close to v_0, if the resistance of its load were always much greater than r. Similarly, a current source is sufficiently close to ideal behaviour, with an output current not much less than i_0, if its load resistance is much *less* than r; an exactly equivalent condition is that its load conductance is much greater than g.

Example 1.2   A certain electronic circuit is designed to act as a current source. When its terminals are short-circuited it produces a current of 10 mA; with a load of 1 kΩ resistance the current is reduced to 9 mA. Find (a) the values of i_0 and g which would represent the current source; (b) the load resistance which would give a terminal voltage of 18 V; (c) the terminal voltage when its terminals are open-circuited.

The short-circuit current gives i_0 directly, since for $v = 0$ eqn 1.9 gives $i = i_0$; thus

(a) $i_0 = \mathbf{10\ mA}$.

Since the 1 kΩ load carries 9 mA, the terminal voltage must be

$$v = 10^3 \times 9 \times 10^{-3} = 9\ \text{V}$$

and eqn 1.9 now gives

$$9 \times 10^{-3} = 10 \times 10^{-3} - 9/r$$

from which

$$g = 1/r = 1/(9 \times 10^3) = \mathbf{1.11 \times 10^{-4}\ S}.$$

(b)   For $v = 18$ V, the same equation becomes

$$i = 10 \times 10^{-3} - 18/(9 \times 10^3) = 8 \times 10^{-3}$$

and the load resistance is therefore

$$R_L = v/i = 18/(8 \times 10^{-3}) = \mathbf{2.25\ k\Omega}.$$

(c)   For an open circuit $i = 0$, and eqn 1.9 becomes

$$0 = 10 \times 10^{-3} - v/(9 \times 10^3)$$

and therefore

$$v = \mathbf{90 \ V}.$$

1.6.3   Equivalent sources

The evident duality between voltage and current sources has been pointed out: a comparison of Figs 1.4(c) and 1.5(c) shows that they are simply different ways of representing the same behaviour. The open-circuit voltage $i_0 r$ of a current source can be regarded as the voltage v_0 of a voltage source having the same internal resistance r (but in series), while similarly the short-circuit current v_0/r of a voltage source can be seen as the current i_0 of a current source with the same resistance r (but in parallel). It follows that any given source of internal resistance r can be easily converted from either form to the other by the relation

$$v_0 = i_0 r. \tag{1.10}$$

Fig. 1.6

Equivalent sources.

The conversion becomes meaningless only in the two limiting cases $r = 0$ (an ideal voltage source) and $r \to \infty$ or $g = 0$ (an ideal current source). The equivalence is illustrated in Fig. 1.6. It is also worth noting here that this equivalence can be applied to convert the combination of an ideal voltage source in series with *any* resistance (whether or not representing an internal resistance) into a corresponding parallel combination of an ideal current source with the same resistance, or vice versa.

At first sight the idea of a current source may seem physically less plausible than that of a voltage source. This is not simply because currents must essentially be driven by e.m.f., but also because in practice electrical energy is commonly obtained from voltage sources which are quite close to ideal: for example the mains power supply and most batteries have terminal voltages which depend very little on current over quite a wide range. Nevertheless good current sources can be made: as a simple example, connecting a high resistance r in series with a good voltage source turns it into a bad voltage source but a good[†] current

---

[†] The description 'good' here is used only in the sense of approaching the ideal characteristic; the efficiency of a source in terms of energy is a different matter, and will be considered in a later chapter.

source (since the source current i_0 is then v_0/r, the voltage v_0 must also be high, enough to produce the required value of i_0 when r is large). It is obvious that such a voltage source will provide a current nearly independent of its load resistance so long as the latter is much smaller than r.

From this point, all sources mentioned are to be taken as ideal unless there is a contrary indication.

1.6.4 Controlled sources

In some circuits there can arise sources of voltage or current having a magnitude which depends on some voltage or current elsewhere in the circuit. Such a source is said to be **controlled** or **dependent**, in contrast to the more familiar independent sources so far considered; unlike an independent source, it cannot be disconnected unless the controlling voltage or current is itself zero; on the other hand it can only take effect if there is at least one independent source to provide the controlling quantity. Sources of this kind do not represent separate physical components but are useful in describing the properties of certain devices: an electronic amplifier, for example, acts like a source which varies with some input voltage or current.

The controlling quantity, for both voltage and current sources, can be a voltage or a current; abbreviations such as CCVS (current-controlled voltage source) are sometimes used for each of the four possible kinds. A circuit containing one or more controlled sources is by its nature active, because it obtains energy from some source additional to the independent sources which are connected to it. In the example mentioned above, an amplifier receives energy from its power supplies as well as its input signal although these supplies, as we shall see, need not be shown in its representation.

Controlled sources can be represented by the same symbols as before, except that the voltage v_0 and the current i_0 depend on some other quantity and that they always appear to be ideal, having no resistance or conductance directly associated with them.

Problems

In the following, t is in seconds.

1.1 A voltage given by $10 \sin 1000t$ V is applied to a capacitance of $1 \ \mu$F. Find the current it takes, and the charge stored.

1.2 A constant current of $1 \ \mu$A flows into a capacitance of 10 nF. Find the voltage across it.

1.3 Find the current taken by an inductance of 1 mH from a voltage of 10 V.

1.4 Find the voltage needed to produce current of value $10 + 10t$ mA in an inductance of 1 mH.

1.5 A certain coil has an inductance of 10 mH and a series resistance of 1 Ω. Find the voltage needed to change its current at the rate of 1 mA μs^{-1}, starting from time $t = 0$.

1.6 A 6 V battery has an internal resistance of 0.1 Ω. Find the load resistance which would reduce its terminal voltage to 5 V.

1.7 Find the voltage needed to charge the battery of Problem 1.6 with a current of 0.5 A.

1.8 A source produces a terminal voltage of 10 V when supplying a current of 1 A. When the current is increased to 2 A the voltage falls to 8 V. Find the e.m.f. and internal resistance of the source.

1.9 A current source is made by connecting a voltage source of e.m.f. 200 V and negligible internal resistance in series with a resistance of 1 kΩ. Find the allowed range of load resistance if the current is to remain constant within 10% of its maximum value.

1.10 Find the range of output voltage obtainable from the source of Problem 1.9.

2 Direct-current circuits

2.1 General

Circuits in which voltages and currents are not meant to be continually changing are known for historical reasons as direct-current or d.c. circuits. (For that reason, the abbreviation d.c. has come to mean 'constant in time' whether applied to current or otherwise.) In this chapter we consider their analysis so as to introduce some of the fundamental principles of circuit theory uncomplicated by time variation, and despite the fact that a minority of actual circuits fall into this class; in later chapters these same principles will be applied more generally.

We first note that constant currents in a circuit imply constant voltages. In a resistance the two are in proportion; in a capacitance eqn 1.3 shows that a constant current must be zero, and the voltage therefore constant, since otherwise the voltage would increase without limit; in an inductance eqn 1.6 shows that for constant current the voltage is zero. It follows not only that a d.c. circuit is one in which all sources have voltages or currents which do not vary in time, but also that *any ideal capacitance becomes an open circuit and any ideal inductance a short circuit*; in other words a d.c. circuit is in effect, if not in physical fact, made up of resistances and d.c. sources only.

One more point should be mentioned here: the d.c. sources do not of themselves guarantee constant currents and voltages in the circuit *at all times*. In practice any circuit gets switched on and off on occasion. The charging of a capacitor when first connected to a battery in series with a resistance is a familiar phenomenon in which the current is plainly not constant: it varies as shown in Fig. 2.1, and it is the *ultimate* current which is constant (and in this case zero). Similarly, the voltage across an inductance cannot be zero while the current in it is growing to its ultimate value after being connected to its circuit. It is this ultimate state of a circuit, after it has been connected (e.g. by closing a switch) with which we are meantime concerned; it is called the **steady state**. Before it is reached the circuit is said to be in the **transient** state, and this we consider further in Chapter 7. Although in theory the steady state is (as we shall see) reached asymptotically, for all practical purposes it is attained in a time which in the majority of electrical circuits is a fraction

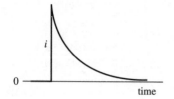

Fig. 2.1
The charging current of a capacitance.

(a)

$$I_3 = I_1 + I_2 + I_4$$

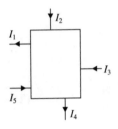

(b)

$$I_2 + I_3 + I_5 = I_1 + I_4$$

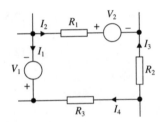

(c)

$$V_1 + V_2 = -I_2 R_1 + I_3 R_2 - I_4 R_3$$

Fig. 2.2
Kirchhoff's laws.

of a second; in circuits which are purely resistive, that is where inductance and capacitance can be taken to be zero, the steady state is reached immediately. In using the term 'd.c. circuit' without qualification we imply the steady state.

It is conventional to designate d.c. currents and voltages by the upper-case letters I and V.

2.2 Kirchhoff's laws

The fundamental laws which enable us to analyse electrical circuits are those of Kirchhoff. They may be stated thus:

> I. The algebraic sum of the currents leaving (or of those entering) any node is always zero.
> II. The algebraic sum of the potential differences across the passive elements around a loop in a given direction is equal to the algebraic sum of the e.m.f.s acting around that loop in the same direction.

These are known as Kirchhoff's first (or current) law and second (or voltage) law. For brevity we shall occasionally refer to them as KCL and KVL. The first, illustrated in Fig. 2.2(a), seems to rest on no more than intuition, but it follows formally from the conservation of charge and the assumption that charge cannot be stored at a node. It can also be applied to any part of a circuit, as shown in (b), provided that that part cannot store a net amount of charge (even on a capacitor the *net* charge is normally zero). The second law, illustrated in Fig. 2.2(c), follows essentially from the fact that electrical potential is single valued: every point in a circuit has a certain potential at any instant, and the potential reached by completing one circuit of a loop must be that of the starting point; in other words, the p.d. provided by the e.m.f. must balance that required by the passive elements. Sometimes this law is couched in the alternative form, that the algebraic sum of the voltages around any loop is zero: in this each e.m.f. counts as a voltage in the opposite direction, relative to current, from that across the passive elements. In practical terms the two definitions are equivalent, but we shall keep to the first.

The analysis of a given circuit requires that all unknown currents and voltages be found from those which are given. Very often the given values are those of the sources, and we have to find the resulting currents in the branches and the voltages between pairs of nodes. In the first instance it is enough to find either the currents or the voltages, since the two are easily related, for d.c. circuits, by Ohm's law. In the case of voltages it is not necessary to find the p.d. between every possible pair of nodes: usually a reference node is chosen (quite often called **earth** or **ground**, in reference to a practical choice for the zero of potential) and it is then sufficient to know the voltage of every other node with respect to that.

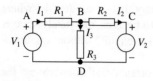

Fig. 2.3
A circuit with unknown branch currents.

To do the analysis we can use Kirchhoff's laws to write equations, in terms of unknown currents or voltages, for as many nodes and loops as are needed for a solution. In the circuit of Fig. 2.3, for example, we may assign three unknown currents to the three branches as shown and then apply the laws thus:

KCL at B:

$$I_1 - I_2 - I_3 = 0 \tag{2.1}$$

KVL around ABDA:

$$I_1 R_1 + I_3 R_3 = V_1 \tag{2.2}$$

KVL around BCDB:

$$I_2 R_2 - I_3 R_3 = -V_2. \tag{2.3}$$

These three equations can be solved to give

$$I_1 = \frac{V_1(R_2 + R_3) - V_2 R_3}{S} \tag{2.4}$$

$$I_2 = \frac{V_1 R_3 - V_2(R_1 + R_3)}{S} \tag{2.5}$$

$$I_3 = \frac{V_1 R_2 + V_2 R_1}{S} \tag{2.6}$$

where

$$S = R_1 R_2 + R_2 R_3 + R_3 R_1. \tag{2.7}$$

Since we now know all of the branch currents (and can if we wish use these to find the node voltages from Ohm's law) it is clear that three equations have sufficed to analyse the circuit. But there is a snag in this approach: we cannot use *any* three, for only three independent equations will yield a solution. It is easy to write two more equations in the same three unknowns, by using the voltage law for loop ABCDA and the current law for node D; but mathematically all five equations cannot be independent, and we could inadvertently choose three which are interdependent and cannot lead to a solution. It can be seen, for example, that the current law at D simply repeats eqn 2.1 for B, so there can be no point in trying to use both of these with one other equation. In this simple circuit it is not difficult to choose independent equations, but in general the choice may not be obvious.

The second drawback of the approach above is that allotting one unknown to every branch makes for a cumbersome set of equations in all but the simplest circuits. Fortunately there are systematic ways of applying Kirchhoff's laws, which overcome the first problem and relieve the second; that is, they ensure that the equations obtained are independent and as few in number as is consistent with a rapid solution (once this number of unknowns are found, others follow very easily). We deal with two such systematic approaches in the following sections.

2.3 Mesh analysis

In the example of the previous section we could have reduced the number of unknown currents from three to two by writing I_3 as $I_1 - I_2$, thereby invoking eqn 2.1 and eliminating it from the formal solution. We may note that the number of unknowns, two, is now the same as the number of meshes in the circuit.

The basis of **mesh analysis** is to assign an unknown current to every mesh in the circuit, as shown in Fig. 2.4 for the same example as before, and then to apply the voltage law to every mesh. Every branch current can be written in terms of one or two mesh currents (for in a planar network no branch is common to more than two meshes) and this process automatically satisfies the current law at every node. Thus, in Fig. 2.4 the current in R_3 is $I_1 - I_2$ (downwards) and the voltage law for the two meshes then gives, if we arbitrarily choose the clockwise direction,

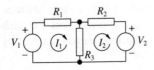

Fig. 2.4
A circuit with unknown mesh currents.

$$I_1 R_1 + (I_1 - I_2)R_3 = V_1 \qquad (2.8)$$

$$(I_2 - I_1)R_3 + I_2 R_2 = -V_2 \qquad (2.9)$$

which are independent equations yielding I_1 and I_2 exactly as in eqns 2.4 and 2.5; the remaining branch current, formerly I_3 as in eqn 2.7, follows immediately as $I_1 - I_2$.

By solving only for mesh currents, therefore, we reduce the number of unknown currents to a minimum; by writing the branch currents in terms of mesh currents we automatically satisfy the current law; and by applying the voltage law to meshes only we ensure independent equations. It should be noted, however, that (as we shall see later) mesh analysis for unknown currents is not necessarily the most efficient way to solve a given circuit.

In the example above, we chose to apply the voltage law in the clockwise direction, having taken the mesh currents clockwise also. Neither choice is essential, provided that signs are made consistent with assumed directions; nor need the same direction be taken for every mesh, although doing so makes for a more methodical analysis.

Example 2.1 Find the currents I_1 and I_2 in the circuit shown in Fig. 2.5, and hence find the voltage at the node B if node D is taken as the 'earth' or reference node, that is as the point of zero potential.

The mesh equations are

$$2I_1 + 3(I_1 - I_2) = 10$$

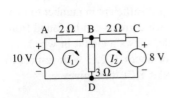

Fig. 2.5
The circuit for Example 2.1.

and

$$3(I_2 - I_1) + 2I_2 = -8$$

from which we find $I_1 = 1.625$ A and $I_2 = -0.625$ A. The voltage at B is then given by

$$V_B = 3(I_1 - I_2) = 6.75 \text{ V}.$$

The examples considered so far have involved only voltage sources. While mesh analysis happens to be particularly convenient for such circuits, it is quite possible to apply it to those with current sources by taking the following steps:

1. The mesh currents should now include both unknown currents and given source currents. Thus, in Fig. 2.6 (our earlier example with a current source added) the current in the mesh containing the current source must be the given I_S, and the branch currents in R_1 and R_3 are $I_1 - I_S$ and $I_2 - I_S$ respectively.

2. The voltage law should be applied only to those meshes which do not contain a current source; so in Fig. 2.6 equations are written for the two lower meshes as before but not for the top one.

Alternatively, it is of course possible, and may be convenient, to change current sources to voltage sources (or vice versa) before carrying out the analysis.

For more than two or three meshes, the solution of the simultaneous equations of mesh analysis is best carried out by a systematic method; in practice it would usually be done by computer with a suitably efficient routine. Some formal aspects of solution, and the notation used, are discussed further in Section 2.6.

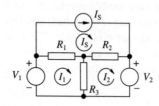

Fig. 2.6

Mesh currents in a circuit with a current source.

2.4 Nodal analysis

In mesh analysis the mesh currents were taken as the unknowns in the first instance, to be found from the application of Kirchhoff's voltage law. It is equally possible to choose instead the node voltages (with respect to some reference) as unknowns, and this is the basis of **nodal analysis**. In terms of these unknown voltages the potential difference across each branch, and hence (by Ohm's law) the current through each branch, can be expressed. The application of Kirchhoff's current law at every node, other than the chosen reference, then yields an independent set of simultaneous equations which are just sufficient in number to give a solution for the node voltages; the branch currents can be found from these if required. Figure 2.7, for example, shows a circuit in which V_1, V_2, V_3 signify the (unknown) voltages of nodes 1, 2, 3 with respect to the reference, node 0. At these three nodes KCL then gives

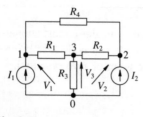

Fig 2.7

A circuit with unknown node voltages.

$$(V_1 - V_2)/R_4 + (V_1 - V_3)/R_1 = I_1 \tag{2.10}$$

$$(V_2 - V_1)/R_4 + (V_2 - V_3)/R_2 = I_2 \tag{2.11}$$

$$(V_3 - V_1)/R_1 + (V_3 - V_2)/R_2 + V_3/R_3 = 0 \tag{2.12}$$

which can be solved for V_1, V_2 and V_3.

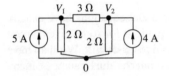

Fig. 2.8

The circuit for Example 2.2.

Example 2.2   Find the voltages V_1 and V_2 in the circuit of Fig. 2.8, and hence find the current in the 3 Ω resistor.

The node equations are

$$V_1/2 + (V_1 - V_2)/3 = 5$$

and

$$V_2/2 + (V_2 - V_1)/3 = 4$$

from which we obtain $V_1 = \mathbf{9.43}$ **V** and $V_2 = \mathbf{8.57}$ **V**. The current in the 3 Ω resistor is then $(V_1 - V_2)/3 = \mathbf{0.286}$ **A**.

In the above cases nodal analysis has been applied to circuits containing current sources only, and it is true that the method is well adapted to such circuits. However, it can readily be used with voltage sources also by noting that the current law need not be applied to nodes whose voltages are fixed by such sources (and are therefore known). In Fig. 2.5, for example, only the equation for node B is required for a nodal solution since we already know the voltages of A and C relative to D; if V_B is the voltage of B relative to D then we have

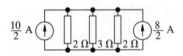

Fig. 2.9

The current-source equivalent of Fig. 2.4.

$$(V_B - 10)/2 + (V_B - 8)/2 + V_B/3 = 0 \tag{2.13}$$

from which $V_B = 6.75$ V as before. Alternatively, it may be convenient to convert voltage sources to current sources: the circuit of Fig. 2.5 would then appear as in Fig. 2.9, for which there is evidently only one node equation; it is easily confirmed to be exactly equivalent to eqn 2.13.

It must be remembered that an *ideal* source cannot usefully be converted unless a finite external resistance can be associated with it, like the 2 Ω resistor in series with each source in Fig. 2.5; thus, no useful conversion can be made of the sources in Figs 2.6 and 2.7.

There is an evident duality between mesh analysis and nodal analysis: although either can be used for any given circuit, in essence the former relates unknown currents to source voltages by Kirchhoff's voltage law, while the latter relates unknown voltages to source currents by the current law. The duality is made more complete when conductance instead of resistance (Section 1.2.3) is used in nodal analysis, as is clear when the form of eqn 2.10 is compared with that of

2.8, for example. But it is important to note that the two methods need not yield the same number of equations: we have already found, for example, that the circuit of Fig. 2.5 yields two mesh equations but only one node equation. In general the numbers are close if not equal, but either may exceed the other. It follows that, because the labour of solution increases rapidly with the number of equations, a careful choice may be profitable.

It is possible to devise a circuit whose nodal equations have exactly the same number and form as the mesh equations for some given circuit (or vice versa); the two circuits are then said to be **duals**. We shall have no need to consider such pairs of circuits, but the dual nature of many circuit relations will quite often be evident.

2.5 Circuit reduction

Any network of resistances and sources can be analysed by setting up and solving mesh or nodal equations as in the foregoing examples; both methods fully account for both of Kirchhoff's laws, which govern all circuit behaviour. But there are several ways in which circuits can be simplified before the equations are set up, and the labour of solution reduced; some of them are described in what follows.

2.5.1 Series and parallel combination

The familiar expressions for the combination of resistances in series and in parallel follow immediately from Kirchhoff's laws. For resistances R_1, R_2, ... in series carrying current I, the total voltage V across the combination must be the sum of their individual voltages, so

$$V = I(R_1 + R_2 + \ldots)$$

and the overall resistance is therefore

$$R_s = V/I = R_1 + R_2 + \ldots \qquad\qquad (2.14)$$

When the resistances are in parallel, with a voltage V common to all, the total current I must be the sum of their individual currents, so

$$I = V/R_1 + V/R_2 + \ldots$$

and the overall resistance is therefore given by

$$1/R_p = I/V = 1/R_1 + 1/R_2 + \ldots \qquad\qquad (2.15)$$

Equation 2.15 may also be written in terms of conductance as

$$G_p = G_1 + G_2 + \dots \tag{2.16}$$

which shows that conductances in parallel, like resistances in series, are directly additive. We might expect from this that conductance should be a more convenient quantity than resistance in dealing with parallel circuits, and in principle this is so. (We saw in the previous section that the use of conductance in nodal analysis corresponded to that of resistance in mesh analysis; this really arises from the fact that parallel elements relate to nodal analysis as do series elements to mesh analysis.) However, in practice engineers tend to use resistance for circuit analysis because by convention it is resistance which is usually specified for individual components and because in any case circuits often include both parallel and series elements. A useful version of eqn 2.15 arises when it is applied to only two resistances in parallel; in that case

$$R_p = \frac{R_1 R_2}{(R_1 + R_2)}. \tag{2.17}$$

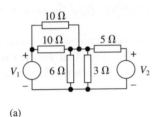

(a)

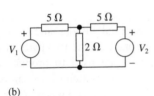

(b)

Fig. 2.10
A circuit (a) before reduction and (b) after reduction.

The first step in simplifying a circuit is to carry out all the combinations possible by the above expressions; even if we wish eventually to know the voltage and current for an individual element of a combination, these are easily retrieved once the values for the whole have been found. For example there is no need to write four mesh equations in order to solve the circuit of Fig. 2.10(a), since it can be quickly reduced to two meshes, as in (b), by the use of eqn 2.17.

Example 2.3 Find the resistance between the points A and B in the circuit of Fig. 2.11, in which the resistances have the values indicated in ohms.

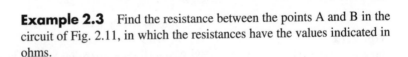

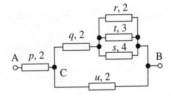

Fig. 2.11
The circuit for Example 2.3.

We must begin by combining elements which are entirely series or entirely parallel, and this points to r, s, and t. By eqn 2.15 their combined resistance is the inverse of $1/2 + 1/3 + 1/4$, which is 12/13; this then can be combined in series with q to give a total of $2 + 12/13$, or 38/13. Combining this in parallel with u gives, by eqn 2.17,

$$R_{BC} = (38/13 \times 2)/(38/13 + 2) = 19/16$$

and to this is added p in series to get, finally,

$$R_{AB} = 19/16 + 2 = \mathbf{51/16 \ \Omega}.$$

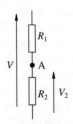

(a)

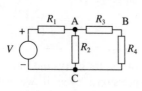

(b)

(c)

Fig. 2.12
(a) a potential divider;
(b) potential division in a
combination; (c) current division.

2.5.2 Voltage and current division

The application of Kirchhoff's laws to series and parallel combinations of resistance, as in the previous section, leads also to standard results for the division of voltage and current which can often speed the processes of circuit analysis in practice.

Since resistances in series must carry the same current, the voltage across each must be in proportion to its value. It follows that, for the pair of resistances shown in Fig. 2.12(a), the voltage V_2 as a fraction of the total voltage V is

$$\frac{V_2}{V} = \frac{R_2}{R_1 + R_2}. \tag{2.18}$$

This is, of course, the principle of the potential divider and of the potentiometer, where two resistances (or, in practice, the two parts of a single resistor with a variable tapping point) are used to produce a diminished voltage V_2 from some input voltage V. But eqn 2.18 can be useful even when this is not the purpose of the circuit, and where R_1 and R_2 may each be the resultant of some combination of resistances. For example, in Fig. 2.12(b) it may be required to know the potential at B relative to C, say, in terms of the source voltage V, and it is easily found as follows. The resistance between A and C due to R_2, R_3 and R_4 is, by eqn 2.17,

$$R_{AC} = R_2(R_3 + R_4)/(R_2 + R_3 + R_4)$$

and eqn 2.18 then gives the voltage at point A with respect to C as

$$V_{AC} = VR_{AC}/(R_1 + R_{AC}).$$

The required potential at B, relative to C, is then, by the same equation,

$$V_{BC} = V_{AC}R_4/(R_3 + R_4).$$

It is very important to note that eqn 2.18 applies *only* when R_1 and R_2 carry the same current, so no other current path must join the node between them: for that reason the voltage V_{AC} in our example must be found by using for R_2 the whole combination between A and C. It follows also that the ratio of a practical potential divider is given correctly by eqn 2.18 only when nothing else is connected to its output — that is, to the point A in Fig. 2.12(a). In that case we may say that the divider output is open-circuited; but if other resistances are connected to A it is said to be **loaded** and account should be taken of them as in our example. In practice the ratio of a potential divider is often taken to be that given by eqn 2.18 so long as any 'output' current flowing from A is relatively small: this is a point which applies to four-terminal networks generally, and more is said about it in Chapter 4.

A similar relationship for the proportional division of current follows from the fact that resistances in parallel share the same voltage. Referring to Fig. 2.12(c) we may write

$$\frac{I_2}{I} = \frac{R_1}{R_1 + R_2} \tag{2.19}$$

noting that now there is inverse proportionality to resistance: the smaller a resistance, the higher its share of current. However, in terms of conductance the proportionality is again direct and the analogy with voltage more precise; thus eqn 2.19 could be written

$$\frac{I_2}{I} = \frac{G_2}{G_1 + G_2}$$

which corresponds to eqn 2.18 for voltage division. As in that case, eqn 2.19 can still be useful when R_1 and R_2 represent combinations of resistances, provided that they share the same potential difference. Neither equation is applicable if the branches themselves contain sources.

2.5.3 Star–delta transformation

Not all interconnections of resistances can be reduced as series and/or parallel combinations. A common instance is that of three resistances connected to three nodes, either in the star (or Y) arrangement of Fig. 2.13(a) or in the delta (or mesh) shown in Fig. 2.13(b). Such a connection can of course exist even when the circuit diagram does not show these precise shapes: for example the circuit of Fig. 2.7 contains one star and one delta, while the bridge circuit of Fig. 2.14(a) contains two of each.

It is often helpful in simplifying circuits to convert one such set of three resistances into its equivalent in the other form; by equivalent is here meant a set having the same resistance between each pair of nodes, when only the three resistances in question are connected to the nodes in each case. From the series and parallel combination rules of the previous section, the resistance between terminals 1 and 2, for example, in Fig. 2.13 is $R_a + R_b$ for the star shown in (a) while in the delta of (b) it is $R_z(R_x + R_y)/(R_x + R_y + R_z)$. For equivalence these two expressions must be equal, and two other equalities follow for the resistances between the terminals 2 and 3, and 3 and 1. By solving for R_a etc. the three equations so obtained, it is readily shown that the two connections in Fig. 2.13 are equivalent when

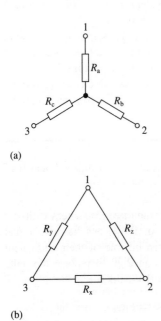

Fig. 2.13

Three resistances connected (a) in star and (b) in delta.

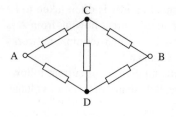

(a)

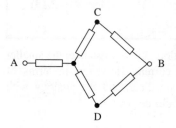

(b)

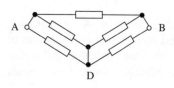

(c)

Fig. 2.14

A circuit (a) before reduction, (b) after a delta–star transformation and (c) after a star–delta transformation.

$$R_a = \frac{R_y R_z}{R_x + R_y + R_z} \qquad (2.20)$$

with similar expressions for R_b and R_c; this set of relations gives the star equivalent to a delta. By solving the same equations instead to give R_x etc. the reverse transformation, giving a delta from a star, can likewise be shown to be

$$R_x = R_b R_c \left(\frac{1}{R_a} + \frac{1}{R_b} + \frac{1}{R_c} \right) \qquad (2.21)$$

with similar expressions for R_y and R_z. When rewritten in terms of conductances, eqn 2.21 becomes

$$G_x = \frac{G_b G_c}{G_a + G_b + G_c}$$

which is exactly analogous to eqn 2.20 for the delta–star case. (The star–delta transformation is actually a special case of a more general transformation between a star network of n elements connecting n terminals to a star point and an equivalent network of $n(n-1)/2$ elements connecting the same n terminals in pairs.)

The resistance between the terminals in Fig. 2.14(a) cannot be found by using series and parallel combination in the circuit as it stands, but eqns 2.20 and 2.21 can be used to give the circuits of Figs 2.14(b) and (c) as two of several possible equivalent versions, which can quickly be reduced further since they now contain only series and parallel combinations.

Example 2.4   Find the resistance between the terminals A and B of the bridge circuit shown in Fig. 2.15(a).

It is sensible to choose to transform either the three resistances with star point D or the delta BCD, since each of these sets has two equal members and by symmetry the same must be true of their equivalent sets. Taking the second choice, we use eqn 2.20 to replace the delta BCD with the three resistances shown in (b), namely

$$R_{CS} = R_{DS} = (5 \times 15)/(5 + 5 + 15) = 3\ \Omega;$$

$$R_{BS} = (5 \times 5)/(5 + 5 + 15) = 1\ \Omega.$$

The resistance between A and S is then given by the parallel combination of the branches ACS and ADS, which is

$$R_{AS} = (10+3)(5+3)/(10+3+5+3) = 4.952 \ \Omega.$$

The resistance between A and B is then given, finally, by

$$R_{AB} = R_{AS} + R_{SB} = 4.952 + 1 = \mathbf{5.95 \ \Omega.}$$

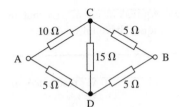

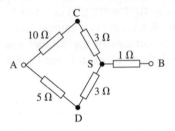

(a)

(b)

Fig. 2.15
The circuit for Example 2.4
(a) before and (b) after
transformation.

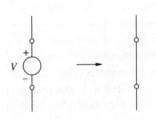

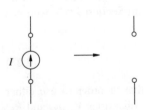

Fig. 2.16
Removing ideal sources.

A particular case to be kept in mind is that of the symmetrical or balanced set in which the three resistances are equal. In this case a set of R_s in star and a set of R_d in delta are equivalent when

$$R_d = 3R_s. \tag{2.22}$$

2.5.4 Thévenin's and Norton's theorems

A powerful technique for circuit reduction is afforded by the theorems of Thévenin and Norton, which are essentially alternative statements of the same principle. They are particularly appropriate where only the state of a circuit at a specified branch is needed by way of solution.

Thévenin's theorem, which we give here without proof, may be stated thus for d.c. circuits:

> With respect to any one passive branch between two nodes of a network containing resistances and d.c. sources, all of the remainder of the network is equivalent to a single d.c. source comprising a voltage V_S in series with a resistance R_S, where V_S is the voltage which would appear between the two nodes were the branch disconnected, and R_S the resistance between the nodes with the branch disconnected and all independent sources removed.

In applying the theorem, we need to bear in mind that 'removing' a source means replacing it with its internal resistance only: thus the ideal voltage source becomes a *short circuit*, and the ideal current source an *open circuit*, as shown in Fig. 2.16.

Consider the network of Fig. 2.17(a) in which we are asked to find the current in branch AB, say. Thévenin's theorem asserts that the whole circuit can be represented as shown in (b). To find V_S we disconnect R_3, in (a), to leave the circuit of (c). Defining V_S as the potential of A with respect to B in order to fix its direction as shown, we have

$$V_S = V_{AC} - V_{BC} = \frac{V_1 R_2}{(R_1 + R_2)} - IR_4.$$

To find R_S, both sources are removed (with R_3 still disconnected) to give circuit (d), in which the resistance between A and B is clearly

$$R_S = \frac{R_1 R_2}{(R_1 + R_2)} + R_4.$$

This completes the specification of the equivalent circuit (b) from which the current in R_3 is easily found. It should be noted that the choice of a

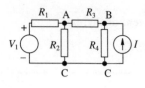

(a)

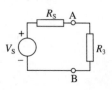

(b)

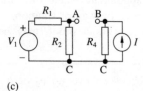

(c)

(d)

(e)

Fig. 2.17
A circuit (a) before reduction,
(b) as a Thévenin equivalent,
(c) with the branch AB removed,
(d) with sources removed and
(e) as a Norton equivalent.

different branch would yield a different equivalent circuit, so that in general a given circuit does not have a unique Thévenin equivalent.

Example 2.5 Find the current I in the $10\,\Omega$ resistance in Fig. 2.18(a).

To find the Thévenin equivalent of the rest of the circuit the $10\,\Omega$ resistance is removed. The voltage V_S from A to B is then given by $V_{AC} - V_{BC}$, which by potential division is $10/2 - 10(2/3)$ or $-5/3$ V. The resistance R_S is that between A and B with the two sources short-circuited as shown in (b), and therefore consists of the parallel pair between A and C in series with the parallel pair between B and C; thus

$$R_S = 4/4 + 2/3 = 5/3\ \Omega.$$

The current I is therefore

$$I = \frac{V_S}{(R_S + 10)} = \frac{-5/3}{35/3} = -\frac{1}{7}\ \mathbf{A}.$$

Because a voltage source with a series resistance can always be converted into an equivalent current source (Section 1.6.3), it follows that there must be a current-source counterpart to Thévenin's theorem. This is Norton's theorem, which for d.c. circuits may be stated thus:

> With respect to any one passive branch between two nodes of a network containing resistances and d.c. sources, all of the remainder of the network is equivalent to a single d.c. source comprising a current I_S in parallel with a resistance R_S, where I_S is the current which would flow between the two nodes were they joined by a short circuit, and R_S is the resistance as defined for Thévenin's theorem.

The Norton equivalent for our earlier example, Fig. 2.17, would therefore take the form shown in (e). A comparison of (b) and (e) shows that, by the equivalence of sources discussed in Section 1.6.3, we must have

$$I_S = V_S/R_S, \tag{2.23}$$

which relates the two theorems. It follows that in order to use either theorem it is sufficient to find any two of the quantities V_S, I_S, and R_S; and that, with respect to any branch, R_S can also be defined as the ratio of open-circuit voltage to short-circuit current at that branch. It is also evident that eqn 2.23 accords with the definitions of V_S and I_S: inspection of Fig. 2.17(b), for example, shows that I_S is indeed the short-

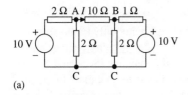

(a)

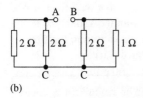

(b)

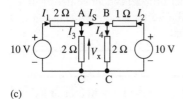

(c)

(d)

Fig. 2.18
The circuit for Examples 2.5 and 2.6: (a) before reduction, (b) with the 10 Ω resistance and the sources removed, (c) with the 10 Ω resistance short-circuited, and (d) the Norton equivalent.

circuit current which would flow between points A and B, exactly as defined for Norton's theorem.

Statements of Norton's theorem often refer to the conductance rather than the resistance of the equivalent source: thus eqn 2.23 could be written

$$I_S = V_S G_S$$

where G_S is the conductance $1/R_S$.

Example 2.6 Find the Norton equivalent of the circuit of Fig. 2.18(a) with respect to the 10 Ω resistance.

The easiest way to find the Norton circuit, since from Example 2.5 we already have the Thévenin circuit, is to use eqn 2.23. However, for illustration, let us find it *ab initio* according to Norton's theorem. The resistance R_S being identical for both theorems, we have that as **5/3 Ω** by the same calculation as in Example 2.5. The current I_S is that which flows in the short circuit between A and B, as shown in Fig. 2.18(c). This appears to be a three-mesh circuit, since we may hesitate to combine the two parallel resistances between A and B (now the same node) and C for fear of losing sight of the current we want. However, the voltage of the node A/B with respect to C, V_x say, is the only one in the circuit which is unknown and it can therefore be found from one nodal equation. From Kirchhoff's current law we may write

$$I_1 + I_2 = I_3 + I_4$$

and hence

$$(10 - V_x)/2 + (10 - V_x) = 2V_x/2$$

which gives $V_x = 6$ V. From this we may find I_S by applying the law again at node A, thus:

$$I_S = (10 - 6)/2 - 6/2 = -1 \text{ A}.$$

The Norton circuit is therefore as shown in Fig. 2.18(d), and we find as expected that the result is exactly what eqn 2.23 would have given. It is worth noting, by comparison of Examples 2.5 and 2.6, that in this particular circuit both V_S and R_S are rather easier to evaluate than I_S.

It is important to recognize that the Thévenin and Norton theorems, while often very useful, do not by any means remove all the labour from circuit analysis: in a circuit of any complexity the evaluation of any of the three basic quantities is not likely to be an insignificant task.

2.5.5 Superposition and reciprocity

There are two important principles which apply not only to the electrical circuits we are now considering but also to other systems, such as elastic structures, which show linear and bilateral behaviour (Section 1.3), that is in which an effect in general is proportional to its cause. In essence these principles assert respectively that the total effect of several causes is the sum of their effects taken one at a time; and that the effect at one point in a system due only to a certain cause at another is identical to the effect at the second point were the same cause transferred to the first. In electrical terms they may be more formally stated thus:

> **Superposition**: The value of any current or voltage in a linear bilateral circuit containing any number of independent sources is the algebraic sum of the contributions due to these sources considered one at a time with the others removed.

> **Reciprocity**: In a linear bilateral circuit, the current in branch p due to a voltage applied in branch q is equal to that which would flow in branch q were the same voltage applied in branch p, with other sources removed and due account taken of relative directions; and the voltage at node r due to a source current entering node s is equal to that which would appear at node s were the same source current to enter node r, with other sources removed.

In these, the 'removal' of a source has the same meaning as in the previous section: short circuit for voltage, open circuit for current in the ideal cases. The usefulness of these two principles in circuit analysis lies in the fact that the labour needed to analyse a circuit containing several sources may well be more than the total needed for a similar circuit with the sources considered one at a time if reciprocity simplifies the latter calculation. Both principles apply to circuits with only linear and bilateral passive elements and independent sources, but not to circuits containing controlled sources because these are in general neither passive nor bilateral. We shall return to this question in Chapter 4.

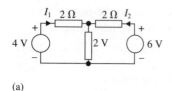

(a)

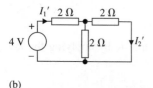

(b)

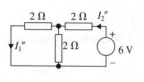

(c)

Fig. 2.19

The circuit for Example 2.7: (a) before reduction, (b) with the 6 V source removed, and (c) with the 4 V source removed.

Example 2.7 Find the currents I_1 and I_2 in the 4 V and 6 V sources shown in Fig. 2.19 (a).

To use superposition, we find the contributions to the currents due to each source on its own. Thus, for the 4 V source we consider the circuit shown in (b). This gives, by simple combination of resistance and by current division,

$$I_1' = 4/(2+1) = 4/3 \text{ A}$$

and

$$I_2' = I_1'/2 = 2/3 \text{ A}.$$

The contributions of the 6 V source can be similarly found from the circuit of (c) to be

$$I_2'' = 6/(2+1) = 2 \text{ A}$$

and

$$I_1'' = I_2''/2 = 1 \text{ A}.$$

The total currents required are then given by

$$I_1 = I_1' - I_1'' = 4/3 - 1 = \mathbf{1/3 \text{ A}}$$

and

$$I_2 = -I_2' + I_2'' = -2/3 + 2 = \mathbf{4/3 \text{ A}},$$

when due account is taken of the directions indicated. Although this is a very simple calculation, we may note that one of the contributions could have been quickly deduced by invoking the reciprocity principle as follows: the current I_1'' due to the 6 V source must be given by the value of I_2' due to the 4 V, increased by the factor 6/4 to allow for the different voltage; thus we could have written

$$I_1'' = (6/4)(2/3) = 1 \text{ A}$$

as before.

2.6 Formal notation for circuit analysis

For the systematic solution of the simultaneous equations of mesh and nodal analysis there is some advantage in a formal notation. Equations 2.8 and 2.9, from our first example of mesh analysis, can be written

$$I_1(R_1 + R_3) + I_2(-R_3) = V_1$$

$$I_1(-R_3) + I_2(R_2 + R_3) = -V_2$$

or, in matrix form,

$$[R][I] = [V] \qquad\qquad (2.24)$$

in which $[I]$ and $[V]$ are column matrices containing the mesh currents and (with signs as appropriate) e.m.f.s., and $[R]$ is the **resistance matrix** of the network, given by

$$[R] = \begin{bmatrix} R_1 + R_3 & -R_3 \\ -R_3 & R_2 + R_3 \end{bmatrix}.$$

We should note that this is a square symmetric matrix; that the elements on the main diagonal represent the total resistance included in each mesh; and that the off-diagonal elements represent, sign apart, the resistance which is common, or mutual, to the two meshes (the negative sign signifying simply that two mesh currents flow in opposite directions in the mutual resistance when both are taken to be clockwise, as here, or anticlockwise). For circuits which contain only linear passive elements and independent sources such as we have so far considered, these properties of eqn 2.24 are quite general. For a circuit with n meshes (not counting any which contain ideal current sources: these, when the mesh equations are written, appear effectively as equivalent voltage sources) the resistance matrix can be written

$$[R] = \begin{bmatrix} R_{11} & R_{12} & \dots & R_{1n} \\ R_{21} & R_{22} & \dots & R_{2n} \\ \dots & \dots & \dots & \dots \\ R_{n1} & R_{n2} & \dots & R_{nn} \end{bmatrix}.$$

In this, R_{jj} is then the total resistance around the jth mesh while R_{jk} and R_{kj} are both the negative of the mutual resistance shared by the jth and kth meshes. In this way, for many circuits, the resistance matrix can be rapidly set up. The condition that $R_{kj} = R_{jk}$ is sometimes known as the **reciprocity condition**, since it is true of any circuit which obeys the reciprocity principle (Section 2.5.5).

Similarly, nodal equations can be compactly expressed in terms of conductance. Equations 2.10 to 2.12 can be written

$$V_1(G_1 + G_4) + V_2(-G_4) + V_3(-G_1) = I_1$$
$$V_1(-G_4) + V_2(G_2 + G_4) + V_3(-G_2) = I_2$$
$$V_1(-G_1) + V_2(-G_2) + V_3(G_1 + G_2 + G_3) = 0$$

in which G_1 signifies $1/R_1$ and so on. These can be simply expressed in matrix form as

$$[G][V] = [I] \tag{2.25}$$

where $[V]$ and $[I]$ are again column matrices, this time of node voltages and source currents respectively, and

$$[G] = \begin{bmatrix} G_1 + G_4 & -G_4 & -G_1 \\ -G_4 & G_2 + G_4 & -G_2 \\ -G_1 & -G_2 & G_1 + G_2 + G_3 \end{bmatrix}$$

which is the **conductance matrix** for the network; like the resistance matrix above, it is square and symmetric. Here reference to Fig. 2.7 shows that the elements on the main diagonal each represent the total conductance connected to a node, while the others each represent the negative of the conductance of a branch joining two nodes. Corresponding to the notation introduced for the resistance matrix, the elements of $[G]$ can be written $G_{11} \ldots G_{nn}$; then G_{jj} is the total conductance terminating on the jth node and $-G_{jk}$, which is also $-G_{kj}$, is that joining the jth and kth nodes.

The analogy between the conductance and resistance matrices is clear, and from eqns 2.24 and 2.25 it would be tempting to conclude that for a given circuit $[G]$ should be simply the inverse of $[R]$; but this is *not* generally the case. For one thing, each of the matrices $[V]$ and $[I]$ has a different meaning in the two equations; for another, we saw earlier that the numbers of mesh and of nodal equations for a given circuit are not generally equal and it follows that its $[R]$ and $[G]$ matrices as defined above may differ in rank and cannot therefore be inverse. However there are important classes of circuits for which a resistance matrix and a conductance matrix can be so defined that each is the inverse of the other, and we return to these in Chapter 4.

In solving eqns 2.24 or 2.25 one is simply inverting the $[R]$ or $[G]$ matrix; thus, for example, eqn 2.24 is in effect an equation for a set of unknown currents $[I]$ in terms of source voltages $[V]$, and its solution can be written

$$[I] = [R]^{-1}[V] \tag{2.26}$$

The systematic solution of such simultaneous equations can therefore be seen as essentially a matter of inverting the appropriate matrix, and this is the basis of what is known as Cramer's method of solution. In practice, however, solutions are usually obtained by a process of Gaussian elimination of unknowns.

2.7 Infinite networks

It is quite possible to postulate networks containing an infinite number of branches and meshes, and some approximations to this idea are of practical importance. Evidently it is not possible to analyse such a network by the conventional application of mesh or nodal analysis, but their behaviour can nevertheless be investigated on the basis of Kirchhoff's laws.

A network like that shown in Fig. 2.20(a) is known as a **ladder network**. When it contains an infinite number of identical sections, as indicated, we may wish to know the current it takes from a given applied voltage V or, more generally, the ratio of the voltage to the current, which may be called its **input resistance**. We may also wish to know the relation between the voltages at any two successive nodes such as P and Q, say V_r and V_{r+1}, and successive currents such as I_r and I_{r+1}, as shown. To find the input resistance, R_{in} say, we suppose one additional section comprising R_1 and R_2 to be inserted between the source and the passive network, as shown in (b); the network being infinite, the addition can make no difference to its behaviour and the input resistance must therefore be unchanged. The whole must then be equivalent to

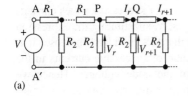

(a)

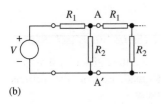

(b)

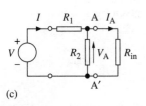

(c)

Fig. 2.20

(a) A ladder network; (b) the same network with a section added; (c) its equivalent circuit.

the combination shown in (c), and by the rules for combining resistances we find that

$$R_{in} = R_1 + \frac{R_2 R_{in}}{R_2 + R_{in}} \tag{2.27}$$

which is easily solved to give R_{in}.

From (c) it can also be deduced by voltage division that

$$\frac{V_A}{V} = \frac{R_2 R_{in}/(R_2 + R_{in})}{R_1 + R_2 R_{in}/(R_2 + R_{in})}$$
$$= \frac{R_2 R_{in}}{R_1 R_2 + R_2 R_{in} + R_{in} R_1} \tag{2.28}$$

and this must also give the general ratio V_{r+1}/V_r for any two successive nodes. Similarly, from (c) it follows by current division that

$$\frac{I_A}{I} = \frac{R_2}{R_{in} + R_2} \tag{2.29}$$

and this is likewise the general ratio I_{r+1}/I_r for any two successive series branches R_1. However, because the input resistance is unchanged by additional sections, it follows that the ratio of successive series currents must be the same as that of node voltages; by substituting the expression for R_{in} in both it could be confirmed that this is indeed the case. It is obvious too, by Ohm's law, that the ratio of successive currents in the shunt branches R_2 must equal that of the node voltages. Hence we deduce that in an infinite ladder corresponding currents and voltages change in identical fashion from section to section. We shall meet this property more generally in Chapter 10.

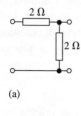

(a)

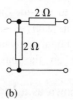

(b)

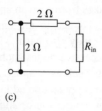

(c)

Fig. 2.21

(a) and (b) Ladder sections for Example 2.8; (c) an equivalent for a ladder of (b).

Example 2.8   Find the input resistances and the ratios of voltages at successive nodes for infinite ladder networks made up from the sections shown in Figs 2.21(a) and 2.21(b) respectively.

From eqn 2.27 the input resistance for (a) is given by

$$R_{in} = 2 + \frac{2 R_{in}}{2 + R_{in}}$$

which gives

$$R_{in}{}^2 - 2R_{in} - 4 = 0$$

with the solution $R_{in} = (\sqrt{5} + 1)\ \Omega$. The ratio of voltages at successive nodes is then, by eqn 2.28,

$$\frac{V_{r+1}}{V_r} = \frac{2(1 + \sqrt{5})}{4 + 4(1 + \sqrt{5})}$$

which simplifies to give the ratio $(3 - \sqrt{5})/2$. It can readily be confirmed that the ratio of series currents according to eqn 2.29 has the same value.

For (b) the argument which led to eqn 2.27 can be applied by redrawing the network as in (c); this gives

$$R_{in} = \frac{2(2 + R_{in})}{4 + R_{in}}$$

which yields

$$R_{in}^2 + 2R_{in} - 4 = 0$$

with the solution $R_{in} = (\sqrt{5} - 1)$ Ω. The ratio of voltages at successive nodes is readily found from the circuit of (c) by voltage division, which gives

$$\frac{V_{r+1}}{V_r} = \frac{R_{in}}{2 + R_{in}}$$

yielding the ratio $(3 - \sqrt{5})/2$. Again, the ratio of series currents can be shown, by current division, to have the same value.

These results are quite illuminating. We may first note that the ratios of voltages are identical in both networks: the reason is that infinite networks made of the two different sections shown in (a) and (b) differ only in their input configurations; a long way from the input they appear to be identical, as Fig. 2.20 shows. Further, the difference in the input is such that the addition of one 2 Ω resistance in series with the ladder made from section (b) yields that made from (a); we should therefore expect the difference between the two results for R_{in} to be 2 Ω, which is indeed the case.

Figure 2.22(a) shows a planar network of resistances which can be said to be 'twice infinite'. Here, whatever property of the network may be required, the approach used for the ladder network would not be very helpful, and the repeated application of Kirchhoff's laws is even less likely to succeed. However, it is still possible to find, for example, the resistance between adjacent nodes when all the resistances are identical, R say. Suppose that in response to some source connected between A and B a total current I flows into the network at A and out at B. We may regard this as the superposition of two states caused by imagined sources: first, a current I flowing into A, where it is divided symmetrically between the four branches and in due course flows out at infinity; and, second, an identical current which flows out at B and is returned to the network, again symmetrically, at infinity. These two

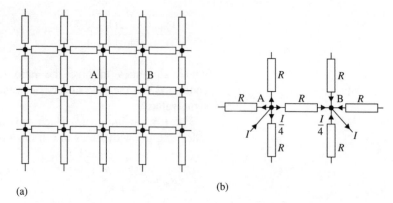

Fig. 2.22
(a) An infinite network and (b) its
input and output currents.

(a)   (b)

states are shown together in (b); clearly if they are superimposed the
resultant current in AB is $I/2$. It follows that the voltage between A and
B is $IR/2$, and this must be the voltage needed to drive I in at A and out
at B (note that no current actually flows in or out of the network at
infinity; the two postulated current systems cancel there). Hence the
overall resistance between A and B is $R/2$.

We return to ladder networks in Chapter 10.

2.8 Power

It is well known that, from the definition of voltage in terms of work
done per unit charge, the flow of current i between points of potential
difference v represents a rate of work, i.e. a power, given by vi. This is
true at every instant and in d.c. circuits power is therefore, like V and I, a
constant quantity in time. Power is measured in joules per second or
watts (symbol W); these units are appropriate for power in any
context, electrical or otherwise.

It is also well known that in practice electrical power must be
produced by conversion from some other form: from mechanical power
in a rotating generator and from chemical power in a battery, for
example. The conversion process need not be considered here; as in
Chapter 1 we take voltage and current sources as being able to provide
electrical energy, and therefore power, without reference to its origin.
But the idea of power flow in a circuit is important. In resistance,
electrical power is converted into heat: we may say that the resistance
receives, or absorbs, electrical power and dissipates it as heat. We also
know that when current flows through resistance its point of entry has
positive potential with respect to its point of exit (Fig. 1.1), and it is
useful to regard this relationship as a criterion for a device to be
receiving electrical power. On the other hand, we know that a source
tends to drive current out of its positive terminal, so by the same token
we may regard current leaving a positive terminal as a general criterion
for a device to be acting as a source, that is *supplying* electrical power.
But, as we saw in Section 1.6.1, a source can be reversible so that, for
example, current flows through a voltage source in the 'wrong'
direction, entering at the positive terminal; in that case, the device,

whatever it is, must be receiving power (but not necessarily converting it into heat as in a resistance). This possibility of reversing the power flow so that a source becomes a sink is common enough in practice: it happens when a battery is charged, the input power being converted into stored chemical energy, and when a rotating generator is operated as a motor instead (quite easy to do in many cases, at least in principle, and a matter of routine in pumped storage systems for peak power supply, which are based on precisely this change of operation).

The direction of power flow in a reversible source is determined, at least in part, by the circuit to which it is connected, according to the laws we have already met. Figure 2.23 shows a simple example, in which V_1 and V_2 are voltage sources such that $V_1 > V_2$. The current therefore actually flows clockwise as shown and is given by

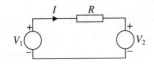

Fig. 2.23
Power flow in a load and two sources.

$$I = (V_1 - V_2)/R.$$

Power of amount $V_1 I$ is supplied by V_1; of this, $V_2 I$ is received by the 'source' V_2 and $(V_1 - V_2)I$ is received and dissipated by R. The last amount is simply the product of the voltage across R and the current in it, but we should note that it follows also from the conservation of energy, which requires that in the steady state power supplied must always balance power received. For the same reason, power contributions from several components are always directly additive (with due signs according as they are receiving or supplying) however the components may be connected to form the circuit.

From Ohm's law it follows that the power received by a resistance R, carrying current I and having voltage V across it, can be expressed not only as VI but also as

$$P = I^2 R = V^2/R. \tag{2.30}$$

The power is therefore proportional to R for given I, inversely so for given V. It should be carefully noted that these simple expressions are correct *only* when both V and I are values for the resistance itself; thus in Fig. 2.23 the power received by R is $I^2 R$ or $(V_1 - V_2)^2/R$ but *not* V_1^2/R, nor V_2^2/R, neither of which expressions has any physical significance in this case.

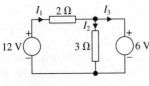

Fig. 2.24
The circuit for Example 2.9.

Example 2.9. For the circuit shown in Fig. 2.24, find the power supplied or received by each component and confirm that the total power received is equal to that supplied.

The currents flowing must first be found, which in this case is very easily done: the voltage across the 3 Ω resistance is clearly that of the 6 V source and the current I_2 in it is thus 2 A; the voltage across the 2 Ω resistance is by inspection $(12 - 6)$ V and the current I_1 is therefore 3 A.

By Kirchhoff's current law I_3 must then be 1 A in the direction shown. Noting that the directions of I_1 and I_3 relative to the source voltages indicate that the 12 V source is supplying power but the 6 V is receiving, we can evaluate all the power components as follows.

From the 12 V source: $12I_1 = \mathbf{36\ W}$;
to the 6 V source: $6I_3 = \mathbf{6\ W}$;
to the 2 Ω resistance: $I_1{}^2 \times 2 = \mathbf{18\ W}$;
to the 3 Ω resistance: $I_2{}^2 \times 3 = \mathbf{12\ W}$.

As required, the last three of these add to the value of the first, confirming that the power supplied balances the total received.

2.9 Matching

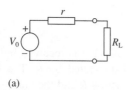

(a)

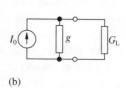

(b)

Fig. 2.25
A load supplied from (a) a voltage source and (b) a current source.

There is a limit to the power which can be delivered to a load from a given source of voltage or current which falls short of the ideal, i.e. which has a finite resistance or conductance. Suppose a voltage source of voltage V_0 and resistance r is connected to a load resistance R_L as shown in Fig. 2.25(a). The current flowing is

$$I = \frac{V_0}{R_L + r}$$

and the power delivered to the load is therefore

$$P = I^2 R_L = \left(\frac{V_0}{R_L + r}\right)^2 R_L. \tag{2.31}$$

Setting the derivative of this, with respect to R_L, to zero shows that P is a maximum when $R_L = r$ and has then the value $V_0{}^2/4r$. In this event the load and source are said to be **matched**.

Since the same current flows in R_L and r, it follows that when the two resistances are equal they dissipate equal amounts of power; that is, as much power is dissipated (uselessly) in the source itself as is delivered to the load so that the efficiency of the arrangement is only 50%. Matching is therefore important in engineering systems where maximizing power takes precedence over efficiency, for example in communications circuits where the 'sources' comprise weak signals; it is irrelevant in any system which handles appreciable amounts of power, since efficiency is then a prime consideration. In any case, efficiency apart, it must not be assumed that either source or load is designed to withstand the matched condition, particularly when the former has a relatively low internal resistance r which makes its maximum output $V_0{}^2/4r$ a large quantity: this is true of a car battery, for example, in which typically r is a small fraction of an ohm, and in the event of its being connected to a load of similar value neither would be likely to withstand for long the consequent levels of current and power dissipation. (Although we have yet to consider a.c. circuits, it can be remarked here that this is also true of the public supply of electricity, which has in effect a very small internal resistance and could not possibly sustain the matched condition.) By contrast, the kind of sources which are applied to low-power electronic circuits have relatively high resistance; not only are they often designed for a matched load, they can also in many cases be connected to a load of lower resistance, even a short circuit, without harm.

The matching of a current source follows the same pattern as that of a voltage source. If the source of current I_0 with parallel conductance g shown in Fig. 2.25(b) is connected to a load of conductance G_L the voltage across the latter is

$$V = \frac{I_0}{g + G_L}$$

and the load power is then

$$P = V^2 G_L = \left(\frac{I_0}{g + G_L}\right)^2 G_L. \tag{2.32}$$

By differentiation, or by analogy with eqn 2.31, it is easily shown that P is in this case a maximum when $g = G_L$ and has then the value $I_0^2/4g$, a result which also follows immediately if I_0 and g are converted to an equivalent voltage source and our previous result applied to that.

The foregoing results apply not only to actual sources but also to the Thévenin and Norton equivalents of any circuit with respect to a specified pair of terminals to which the load R_L may be connected. The Thévenin voltage V_S and resistance R_S correspond to V_0 and r, so it follows that a load can always be chosen to draw maximum power from any pair of terminals in a circuit by setting $R_L = R_S$ for these terminals; the maximum is given by $V_S^2/4R_S$. In the same way I_S and G_S for any pair of terminals, according to Norton's theorem, represent I_0 and g for the equivalent current source.

Problems

2.1 Four branches A, B, C, and D of a certain circuit meet at a node. Currents of 2 A and 3 A enter the node from A and C respectively, while a current of 4 A leaves the node through D. How much current enters the node from B?

2.2 A battery charger, with an effective e.m.f. of 13 V and an internal resistance of 0.5 Ω, charges a battery of internal resistance 0.1 Ω. Find the charging current when the e.m.f. of the battery is 11.5 V.

2.3 The battery in Problem 2.2 is disconnected when its e.m.f. is still 11.5 V and connected in parallel with an

identical battery fully charged to 12 V. A resistance of 1 Ω is connected across the two. Find what current each battery provides to the resistance.

2.4 The circuit in Problem 2.3 remains connected while a third source supplies 5 A (in the same direction) to the resistance. What currents are now supplied by each battery?

2.5 Find the voltages across each current source in the circuit shown.

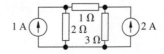

2.6 Find the voltages, relative to C, in the circuit shown.

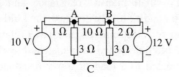

2.7 A potential divider of nominal ratio 1:10 is made from resistances 180 kΩ and 20 kΩ. Find the resistance which, connected across the output, would cause the ratio to fall by 5%.

2.8 Find the resistances needed to make a potential divider for a 200 V supply which will give a ratio of 1:10 for no output current and a ratio of 1:11 for 10 mA output current.

2.9 Find the three resistances which, connected in star, could replace the 10 Ω and 3 Ω resistances in Problem 2.6.

2.10 Find the Thévenin equivalent of the circuit in Problem 2.6 with respect to the 10 Ω resistance.

2.11 Find the Thévenin equivalent of the two batteries in Problem 2.3 with respect to the 1 Ω resistance.

2.12 Find the Norton equivalent of the circuit in Problem 2.5 with respect to the 2 Ω resistance.

2.13 Find the contributions which each source makes to the voltage at node A in Problem 2.6.

2.14 Find the contributions of each of the sources to the current I_S in Fig. 2.18(c).

2.15 With the meshes of the circuit in Problem 2.6 numbered from left to right, find the elements of the resistance matrix in ohms.

2.16 Find how many sections like that of Fig. 2.21(a), connected as a ladder, are needed to reduce the node voltage by a factor of at least 100.

2.17 Find the total power dissipated in the circuit of Problem 2.6.

2.18 Find the power supplied by each source, and dissipated by each resistance, in the circuit of Problem 2.5.

2.19 Find the value of resistance which, replacing the 1 Ω resistance in Problem 2.5, would dissipate most power; and evaluate that power.

2.20 Show that the maximum power available from two sources, of e.m.f.s. V_1 and V_2 and internal resistances R_1 and R_2, connected in parallel is given by

$$\frac{(V_1 R_2 + V_2 R_1)^2}{4R_1 R_2 (R_1 + R_2)}.$$

[Hint: consider the Thévenin equivalent.]

3 Alternating-current circuits

3.1 General

In the last chapter the principles of circuit analysis were set out in terms of d.c. circuits, in which currents and voltages remained constant. We now seek to apply these principles to circuits in which both currents and voltages vary with time, because in practice such circuits are much more common than the d.c. kind. The form of the variation with time of an electrical or other physical quantity is often known as its **waveform**. In this chapter we shall concentrate on circuits in which the waveforms of current and voltage are **periodic** (that is, regularly repeated) and in particular on those in which the variation of both is sinusoidal; the latter kind are known by tradition as **alternating-current**, or a.c., circuits although the same term is occasionally applied more generally to circuits with other waveforms. By a sinusoidal variation we mean having the form of a sine function or a cosine function of time. No distinction between these two functions is necessary at this point: both have the same shape, differ only in their time relationship (or **phase**, a term to which we return later) and may be generally referred to as **sinusoids**. By including an appropriate phase angle any given sinusoid can be expressed as either a sine or a cosine. Figure 3.1(a) shows a waveform which it is natural to regard as a cosine, because it has its maximum at time $t = 0$, but it is equally correct to view it as a sine having its phase shifted by $\pi/2$; we have the same kind of choice wherever the origin of t may be.

3.2 Periodic functions

Before proceeding further we shall need to recognize certain properties which can be defined for any periodic waveform such as the examples shown in Fig. 3.1, all of which are found in electrical engineering. The duration of one complete cycle is the **period** T and its reciprocal, the number of cycles in unit time, is the **frequency** f, measured in **hertz**

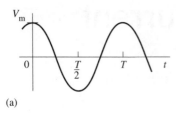

(a)

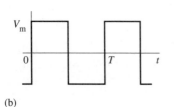

(b)

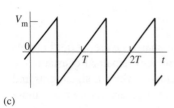

(c)

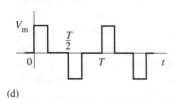

(d)

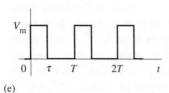

(e)

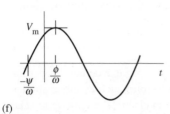

(f)

Fig. 3.1
Periodic waveforms.

when the period is in seconds. Strictly, f should be called the **cyclic frequency**, to distinguish it from the angular frequency defined below; or, in the case of non-sinusoids, the (cyclic) **fundamental frequency**, to distinguish it from harmonic frequencies (see Chapter 9); but by convention 'frequency' is taken to mean f, the reciprocal of T, unless the context indicates otherwise. In many cases the maximum positive and negative values of the waveform are equal; they are then known as the **peak** value, shown in our examples as V_m. The average value over one whole period (or over an infinite time) is often zero; if not, it is sometimes called the **d.c. component**. In Fig. 3.1 it can be seen by inspection that only the waveform of (e) has a d.c. component, and in this case its value is easily seen to be $(\tau/T)V_m$. For those waveforms, including sinusoids, in which positive and negative values occupy one half-period each and take the same form, the average over the positive half-period can be useful: for a sinusoid it is readily shown by integration to be a fraction $2/\pi$ of the peak value. In referring to average values, it is therefore important to specify precisely which is meant.

A particularly useful measure of any periodic quantity is its **root-mean-square**, or r.m.s., value. It is found by averaging the square of the quantity over a period and then taking the square root of this average, and it has the advantage of yielding immediately the true time average of any quantity which depends on the square of another, as for example electrical power depends on the square of current or voltage. The r.m.s. value of any quantity is always taken as positive, being unaffected by the sign of the quantity itself. Because of its relevance to electrical power, r.m.s. value is sometimes defined in terms of the 'heating effect' of current or voltage, but its proper definition is in fact purely mathematical and is not even restricted to functions of time (nor, for that matter, to periodic functions): the r.m.s. value of a function $f(x)$ of any variable x, over the interval x to $x + X$, is

$$\left\{ \frac{1}{X} \int_x^{x+X} f^2(y)\, dy \right\}^{1/2} \tag{3.1}$$

For the purposes of electrical engineering, x is usually time t, X the period T, and the function f a current or voltage. From eqn 3.1 the r.m.s. value of a sinusoid can be shown to be $1/\sqrt{2}$, or about 0.707, of its peak value. The ratio of the r.m.s. value to the half-period average of any symmetrical periodic waveform is known as its **form factor**; for a sinusoid it is given by $\pi/2\sqrt{2}$, or 1.11.

By convention, while the instantaneous value of a time-varying quantity is given a lower-case symbol, thus v or $v(t)$, the r.m.s. value is usually denoted by upper case, V, in the same way as d.c. values; the same letter with a suitable subscript added can signify a peak or other characteristic value, like V_m in Fig. 3.1.

Example 3.1 Find the r.m.s. values of the waveforms shown in Fig. 3.1(c) and (e).

In case (c) the function represented can be expressed as $2V_m t/T$ in the half-period from $t = 0$ to $t = T/2$; this half-period therefore contributes to the integral of eqn 3.1 an amount

$$\int_0^{T/2} \left(\frac{2V_m t}{T}\right)^2 \mathrm{d}t = \frac{V_m^2 T}{6}.$$

It is evident that in this case the second half-period contains the same area as the first, although negative, so its square must produce exactly the same contribution as the first, giving the whole integral the value $V_m^2 T/3$; the r.m.s. value is thus

$$V = \left(\frac{1}{T} \times \frac{V_m^2 T}{3}\right)^{1/2} = \frac{V_m}{\sqrt{3}}.$$

The function in case (e) has the value V_m during an interval τ in each period and is otherwise zero, so in any period the integral of eqn 3.1 is simply

$$\int_0^{\tau} V_m^2 \, \mathrm{d}t = V_m^2 \tau$$

and the r.m.s. value is therefore

$$V = \left(\frac{1}{T} V_m^2 \tau\right)^{1/2} = V_m \sqrt{\frac{\tau}{T}}.$$

The last result of the above example may be compared with the average (or d.c.) value of $V_m \tau/T$ which we previously noted for the same waveform. Since $\tau < T$, the r.m.s. value of this waveform is always larger than its d.c. component; this is true of any waveform (except for a d.c. quantity, in which case both reduce to its actual value).

Non-sinusoidal periodic functions are considered further in Chapter 9.

3.3 Sinusoidal functions

Of all the periodic waveforms, the sinusoidal is by far the most important for several reasons: it is the basic waveform of nature and occurs in every aspect of physical science; it is very easy to generate, electrically and otherwise, to a good approximation; and, as we shall find in Chapter 9, any other waveform can be regarded as the superposition of sinusoids of various frequencies. In addition it has the very convenient properties that adding, subtracting, differentiating and integrating sinusoids at a given frequency produces other sinusoids at that frequency, with the result that in any linear electrical circuit (that is one whose passive elements comprise constant values of R, L, and C) sinusoidal sources at a given frequency produce only currents and voltages which are also sinusoidal at that frequency. Here again, as for d.c. circuits, we refer to the *ultimate* state of affairs. For a.c. circuits too, despite the continuous variation with time, this is often called the steady state to distinguish it from the transient state which usually arises temporarily after any non-periodic change or disturbance (and which is discussed in Chapter 7). In the a.c. steady state, every cycle of a sinusoidal quantity is exactly the same as the preceding cycle. As for the d.c. case, the steady-state currents and voltages which arise from a.c. sources are sometimes known as the **forced response** of a circuit to these sources, as opposed to the transient or natural response.

A sinusoidal variation of voltage, say, like that shown in Fig. 3.1(a) is most conveniently expressed as a cosine function because its origin of time happens (or has been chosen) to occur at a positive maximum, and can be written in the form

$$v = V_m \cos \omega t$$

in which ω is the **angular frequency** and is measured in radians per second, *not* in hertz. Because the angle ωt must change by 2π radians in the course of one period, it follows that ωT must be 2π, so

$$\omega = 2\pi/T = 2\pi f \qquad\qquad (3.2)$$

which merely reflects the fact that there are 2π radians in a revolution (or, as we could say, in a cycle). We evidently have the option of defining the periodicity of any sinusoidal waveform in terms of T or f or ω. The period T is useful for certain calculations, and as an indirect measure for f (for example, from the time base of an oscilloscope). In practice f is very commonly used, for example in quoting 50 Hz as the frequency of the national grid power supply and 92 MHz for a VHF radio frequency. Mathematically ω is often more convenient and is preferred for this chapter, but later we shall sometimes find f more suitable; the vital point, since both are commonly referred to simply as 'frequency', is to avoid confusing them. Waveforms are usually illustrated as functions of time t, as in Fig. 3.1, if only because that is how they are (indirectly) observed on an oscilloscope, but for a sinusoid it is equally reasonable to take the variable to be the angle ωt, as shown for the cosine wave in (a).

In general, the origin $t = 0$ will not occur at a convenient point on the waveform, such as zero or a peak: even if we choose it to be at a zero, say, for one quantity in a circuit, it will be different for any other which, although it has the same frequency, happens to reach zero at different times; we say that the two sinusoidal quantities differ in phase. The waveform shown in Fig. 3.1(f) has an arbitrary relation to the origin $t = 0$, and the voltage, say, which it represents can be written in terms of a **phase angle** ϕ as

$$v = V_m \cos(\omega t - \phi).$$

Alternatively, in terms of a different phase angle ψ the same waveform can be written

$$v = V_m \sin(\omega t + \psi)$$

in which $\psi = \pi/2 - \phi$. The phase difference between two sinusoids is nearly always indicated as an angle rather than a time interval, so we describe the waveforms in (a) and (f) as having a phase difference (or phase shift) ϕ, or as being **out of phase** by ϕ; more precisely we should say that (f) lags (a), or that (a) leads (f), by ϕ radians. It would be equally correct, for a given frequency, to quote a time interval instead of an angle: the phase difference ϕ corresponds to the interval ϕ/ω, and so on. (Note that we are only concerned with time *differences*: there is no physical significance in an absolute value of time.) When two sinusoids have a phase difference of $\pi/2$ they are said to be exactly out of phase or in **quadrature**; if the phase difference is π they are said to be opposite in phase or in **antiphase**, and they can then of course be represented by the same function with different signs. The phase difference between two sinusoidal expressions is immediately obvious if both are written as sines or both as cosines.

In dealing with sinusoidal voltages and currents, it should be remembered that, as for d.c. quantities, the sign given to a voltage or current must be related to its postulated direction in a circuit; the conventions discussed in Section 1.2.6 still apply irrespective of the fact that the *actual* physical directions of voltages and currents are continuously alternating. Thus, Fig. 3.2 shows alternative descriptions of the same voltage source.

This leaves one possible ambiguity to be noted. Although we have said, correctly, that the sinusoid of Fig. 3.1(a) leads that of (f) by the angle ϕ, it would be equally correct to say that it lags by the angle $2\pi - \phi$. In stating that one sinusoid leads or lags another, we refer by convention to the *smaller* of the angles which separate them. In addition, the reference directions for the two quantities can be chosen to keep this angle in the range $\pi/2$ to $-\pi/2$; while this is not essential, it is often done.

Fig. 3.2
Equivalent voltage sources.

3.4 The a.c. behaviour of passive elements

The voltage–current relations for passive elements as given by eqns 1.7 can easily be applied to sinusoids. When a voltage $V_m \cos \omega t$ is applied to a resistance R, then the current is simply given by

$$i = v/R = (V_m/R) \cos \omega t = I_m \cos \omega t \tag{3.3}$$

when I_m is written for V_m/R. This means that current and voltage are in phase in a resistance, and that their peak values (and hence also their r.m.s. values) are related by Ohm's law.

For capacitance, the same voltage gives

$$i = C\,dv/dt = -\omega\,CV_m \sin \omega t = I_m \cos(\omega t + \pi/2) \qquad (3.4)$$

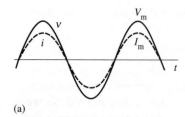

(a)

if I_m is now written for ωCV_m. Here the current is in quadrature with the voltage, leading it by $\pi/2$. The ratio of peak voltage to peak current (or of the r.m.s. values) is now $1/\omega C$; this quantity is known as the **reactance** of the capacitance, and like resistance is measured in ohms although it is physically different.

For inductance, we have from eqns 1.7

$$v = L\,di/dt$$

so we may write

$$i = \frac{1}{L}\int v\,dt = \frac{1}{\omega L}V_m \sin \omega t = I_m \cos(\omega t - \pi/2) \qquad (3.5)$$

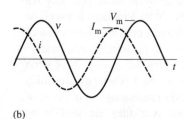

(b)

in which I_m is now $V_m/\omega L$. (The constant of integration has been taken to be zero, because in the steady state there can be no d.c. current from an a.c. source in a purely linear circuit.) In this case the current is again in quadrature with the voltage, but lags it; the ratio of peak, or r.m.s., voltage and current is ωL, which is known as the reactance of the inductance and is again measured in ohms. Reactance in general, whether capacitive ($1/\omega C$) or inductive (ωL) is often given the symbol X: the usefulness of a single symbol for the two will emerge in later sections.

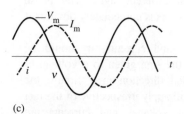

(c)

Fig. 3.3
Sinusoidal voltage and current in (a) resistance, (b) capacitance, and (c) inductance.

We see, then, that a.c. voltage–current relations in the passive elements R, C and L must be described not only by the magnitudes of resistance and reactance R, $1/\omega C$ and ωL, but also by the phase angles 0, $\pi/2$, and $-\pi/2$. By convention it is usual to specify a phase angle as that of current with respect to voltage, so we may say that inductance has a **lagging** phase angle of $\pi/2$ and capacitance a similar **leading** phase angle. A set of sinusoidal waveforms which illustrate the foregoing relationships is shown in Fig. 3.3. Our next problem is how to incorporate these ideas into the laws which govern complete circuits.

3.5 Circuit relationships for sinusoids

3.5.1 Kirchhoff's laws

Kirchhoff's laws are always valid for electrical circuits, and are satisfied at every instant whatever the variation of current and voltage. For a.c. circuits there is therefore no reason of principle why we should not apply them as for d.c. except that every voltage and current must now be expressed as a sinusoid and those which are unknown will in general have phase as well as magnitude unknown. Consider the very simple circuit of Fig. 3.4, with R and L in series, in which the voltage source is

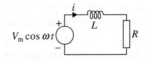

Fig. 3.4

An a.c. circuit.

specified and we wish to find the current i. Let the latter be $I_m \cos(\omega t + \theta)$, in which I_m and θ are unknown. The voltage across the resistance is then, from the previous section,

$$v_R = R I_m \cos(\omega t + \theta)$$

and that across the inductance

$$v_L = \omega L I_m \cos(\omega t + \theta + \pi/2)$$

because in inductance current lags voltage by $\pi/2$, as we saw in the last section, so voltage leads current accordingly. Kirchhoff's second law now gives

$$V_m \cos \omega t = v_R + v_L = I_m \{R \cos(\omega t + \theta) + \omega L \cos(\omega t + \theta + \pi/2)\}$$
$$= I_m(R^2 + \omega^2 L^2)^{1/2} \cos(\omega t + \theta + \phi)$$

where

$$\phi = \tan^{-1}(\omega L/R),$$

and we may equate magnitudes and phases on the two sides to give

$$I_m = \frac{V_m}{\sqrt{R^2 + \omega^2 L^2}}$$

and

$$\theta = -\phi$$

so that the solution for the unknown current i can be written

$$i = \frac{V_m}{\sqrt{R^2 + \omega^2 L^2}} \cos(\omega t - \phi). \tag{3.6}$$

There is nothing new in principle about the above procedure, and it could be applied if necessary to any number of mesh or nodal equations; but the equation for this simple example indicates that writing out the equations alone would be excessively tedious for circuits of even modest complexity. Fortunately there is a shorthand method which allows the equations of a.c. circuits to be written and solved much more compactly; it is set out in the following sections.

3.5.2 Phasors

A very convenient and compact method for handling interrelated sinusoidal quantities, such as are found in a.c. circuits, is based on a theorem in the algebra of complex quantities which enables us to write

$$\cos \theta = \text{Re}[e^{j\theta}]$$
$$\sin \theta = \text{Im}[e^{j\theta}] \tag{3.7}$$

in which j signifies $\sqrt{-1}$. (In some contexts i is traditionally used for this quantity, but j is standard in engineering to avoid confusion with the usual symbol for current). Since any sinusoid can be expressed as a cosine with an appropriate phase angle, we need use only the first of these relations, written for our purposes in the form

$$\cos(\omega t + \phi) = \text{Re}[e^{j(\omega t + \phi)}].$$

We now need to note that, in any system for which superposition holds, a calculation which involves expressions of the form $\text{Re}[e^{j\theta}]$ can be carried out in terms of the whole complex quantity $e^{j\theta}$, the real part of the result being taken to obtain the final correct answer. That is to say, we may legitimately represent the actual (and real) function $\cos(\omega t + \phi)$ by the complex function $e^{j(\omega t + \phi)}$, dropping the 'real part' notation throughout, provided we remember that at every stage of calculation the actual quantities are given by real parts only. We may go further: because every sinusoidal quantity so represented at a given angular frequency ω contains the factor $e^{j\omega t}$, this too can be dropped so that we find ourselves representing a physical sinusoidally-varying quantity $A \cos(\omega t + \phi)$, say, by the quantity $Ae^{j\phi}$ which is complex but constant, and contains only the vital information that the sinusoid has magnitude A and phase angle ϕ relative to cos ωt. (We shall see that its r.m.s. value can equally well be used instead of the peak value A, and that the angle ϕ can be relative to any reference of the correct frequency.) In this lies the justification for the apparent perversity of seeking simplification by using complex quantities to represent real: by removing the time dependence in this way, a.c. circuit calculations can be done with *apparently constant* quantities; the price we pay is that these quantities are in general complex and (what is only a small sacrifice) can be used for only one frequency at a time (which must be separately specified since it too has been removed). The complex quantity $Ae^{j\phi}$, when used to represent a sinusoidal function of time $A \cos(\omega t + \phi)$, is known as a **phasor**. (The term is sometimes also used for complex quantities which retain the factor $e^{j\omega t}$, so-called rotating phasors, but this usage is for practical purposes irrelevant.)

For convenience, the bold symbols commonly used to signify vectors can also be used for phasors; thus, A is used to signify a phasor quantity $Ae^{j\phi}$, and here we use V and I for voltage and current phasors in general. To specify magnitude and phase, the notation $A\underline{|\phi}$ is often used as shorthand for $Ae^{j\phi}$. It follows that $A\underline{|0}$ means simply A and is a real phasor representing, according to the above argument, $A \cos \omega t$. In fact, we can if we wish define $A\underline{|0}$, which may be called the **reference phasor**, to represent a sinusoid of amplitude A and *any* phase, for example $A \cos (\omega t + \theta)$; this is merely equivalent to dropping from our calculations the factor $e^{j(\omega t + \theta)}$, where θ is some constant angle, instead of $e^{j\omega t}$ only, as above. Any phasor for which the frequency ω is known can, of course, be easily converted back to the actual sinusoid when the definition of the reference phasor is known: for example $Be^{j\phi}$ represents

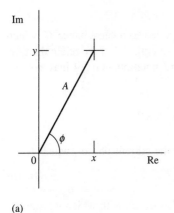

(a)

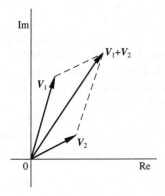

(b)

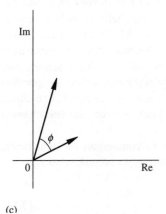

(c)

Fig. 3.5
Phasors on the Argand diagram.

the function $B \cos(\omega t + \theta + \phi)$ when the reference phasor represents $A \cos(\omega t + \theta)$, whatever the values of A and B (amplitude has no particular significance in defining the reference phasor; it is its phase which is important).

Like any other complex quantity, a phasor can be written not only in the form $Ae^{j\phi}$ (usually called the polar form), where A is an amplitude or magnitude and ϕ a phase angle, but also in the Cartesian form $x + jy$; the relation between the two forms is easily confirmed from eqns 3.7 to be given by

$$x = \cos \phi$$
$$y = \sin \phi$$
$$A = \sqrt{(x^2 + y^2)}.$$

Either form is very easy to display on an Argand diagram (that is, on the complex plane with real and imaginary axes) as a line of length A at an angle ϕ to the real axis, the coordinates of its end point being x and y, as shown in Fig. 3.5(a). The rules for the addition and subtraction of phasors are simply those of complex quantities, namely that real and imaginary parts are added and subtracted separately; this can be done graphically in the same way as is common for vector quantities such as force and velocity; an example is shown in (b). When two sinusoids are represented as phasors on the same diagram, their phase difference becomes simply the angle between the two lines as shown in (c); a leading phase angle corresponds to an anticlockwise rotation according to the usual conventions of the Argand diagram (rotation from positive real axis to positive imaginary axis is anticlockwise and positive). A reference phasor (which need not always be shown) is usually taken to lie on the real axis, with phase angle zero according to the above convention.

We should also note here that a phase shift ϕ, which on the diagram is a rotation ϕ, is equivalent to multiplication of a phasor by the factor $e^{j\phi}$. In particular, since (as is easily confirmed from eqns 3.7) we may write

$$e^{j\pi/2} = j \tag{3.8}$$

it follows that a phase shift of $\pi/2$, i.e. an anticlockwise rotation through a right angle, corresponds to multiplication by j; a negative phase shift of $\pi/2$, implying a clockwise rotation, likewise means multiplication by –j.

We have yet to encounter one important consequence of the use of phasors: the operations of differentiation and integration (with respect to time) are no longer necessary. This is hardly surprising, since phasors are not time varying, but it is highly significant for the voltage–current relations in the elements C and L; these are considered in the next section.

3.5.3 Phasor relations in passive elements

Consider again a voltage $V_m \cos \omega t$ applied to a capacitance C, which led to eqn 3.4; in the exponential terms of eqns 3.7, if we retain the time variation but omit as usual the 'real part' notation, eqn 3.4 may now be written

$$i = C \frac{\mathrm{d}v}{\mathrm{d}t} = C \frac{\mathrm{d}}{\mathrm{d}t}(V_m e^{j\omega t}) = j\omega C V_m e^{j\omega t}. \qquad (3.9)$$

Dropping now the factor $e^{j\omega t}$ gives us the current phasor, which is

$$I = j\omega C V \qquad (3.10)$$

because the phasor V for the voltage $V_m \cos \omega t$, with which we started, is simply the real quantity V_m. (We should have arrived at exactly the same result had we chosen a different phase for v to start with.) The factor j in this relation indicates, according to eqn 3.8, that the current is $\pi/2$ ahead of the voltage in phase, which is hardly surprising since eqn 3.4 showed that for capacitance the differential relation between current and voltage gives rise to just such a phase difference. The factor ωC in eqn 3.10 also confirms something we already know from eqn 3.4, that in capacitance the amplitude ratio of voltage to current is $1/\omega C$, which we called the reactance. It follows that the sinusoidal voltage–current relation in capacitance can be represented correctly in both magnitude and phase by eqn 3.10, with no need for differentiation and without reference to any actual function of time.

Two more points should be made about eqn 3.10. At first sight the idea of a.c. current flowing in a capacitance seems to run counter to the fact that a good capacitor relies on dielectric material which ideally cannot carry current; in fact the a.c. current I in the equation, like the instantaneous current i in eqn 1.3, flows in the dielectric only as what is called a displacement current (see Appendix B.2), but it is nevertheless an actual current flowing to and from the capacitance element as it periodically charges and discharges. Secondly, voltage and current phasors, like the actual quantities they represent, relate to specified directions in a circuit (see Section 3.1): eqn 3.10, like eqn 1.3 on which it is ultimately based, is correct as it stands for the directions shown in Fig. 1.1.

As might be expected, eqn 3.10 has its counterpart for inductance, which can be found by applying a comparable procedure to eqn 3.5. This results in the phasor relation

$$V = j\omega L I \qquad (3.11)$$

which shows, as we already know from Section 3.4, that in inductance current lags voltage by the angle $\pi/2$ and that the ratio of peak voltage to peak current is ωL. As before, the equation is correct as it stands for the assumed directions of V and I, which are those of Fig. 1.1.

Equations 3.10 and 3.11 show that the phasor ratios of voltage to current in capacitance and in inductance are given respectively by

$1/j\omega C$, or $-j/\omega C$, and $j\omega L$. This means that they are imaginary quantities, positive and negative respectively, having the magnitude of the reactance in each case. Both are often written in the general form jX, in which case the real quantity X is known as the reactance; on this basis it can be said that inductive reactance is positive and capacitive reactance negative. The term 'reactance' may also be used to describe the relevant imaginary quantity, by including the j in each case, and it then gives us the correct phasor ratio V/I, including phase difference as well as amplitude, for C and L.

Finally, for resistance R we know that sinusoidal voltage and current are in phase and in the ratio R at every instant; it follows that the phasors are in this same real ratio and that we have simply

$$V = RI \tag{3.12}$$

as the a.c. version of Ohm's law.

Let us collect together these last three equations, which give the phasor relations for the three passive elements; they are

$$
\begin{aligned}
V &= RI \\
I &= j\omega CV \\
V &= j\omega LI
\end{aligned}
\tag{3.13}
$$

and they may be compared with the fundamental eqns 1.7. From the comparison it can be deduced that for sinusoids *differentiation is equivalent to multiplication by* $j\omega$, an equivalence which can be useful in applying differential equations to any physical quantities which vary sinusoidally. It is essential to remember that eqns 3.13 apply *only to sinusoids*: for other waveforms phasors cannot be used directly and we must resort to the methods discussed in Chapters 7 and 9. But for a.c. circuits involving only the sinusoidal steady state, these equations lie at the heart of circuit analysis. We apply them to that in the next section.

Before proceeding, we should confirm from eqns 3.13 certain limiting forms of reactive behaviour. The reactance $1/\omega C$ is evidently infinite at zero frequency (that is, d.c.) and an ideal capacitance is therefore then an *open circuit* — which agrees with the familiar idea that a capacitor does not conduct d.c. — while at infinite frequency it would have zero reactance and so approximates to a *short circuit* at sufficiently high frequencies. Actual capacitors are quite close to the ideal, and in electronic circuits especially are often used to take advantage of both properties (the larger the capacitance the closer it approaches to a short circuit at a given frequency). Similarly, inductance becomes a *short circuit* for d.c. but an *open circuit* for infinite frequency; the former property is not very closely attained for actual inductors, because of inevitable resistance, while at a high frequency the second is clearly more closely approached for higher values of inductance.

3.5.4 Phasors in circuit analysis

Let us return to the simple circuit of Fig. 3.4, which was discussed in Section 3.5.1. Now let the phasor current be I. The results of the previous section then give the voltage across the resistance as

$$V_R = RI$$

and that across the inductance as

$$V_L = \mathrm{j}\omega L I$$

and, since phasors add like any other complex quantity, the total voltage across the combination is given by the phasor

$$V = V_R + V_L = I(R + \mathrm{j}\omega L) \tag{3.14}$$

from which the unknown current can be written

$$I = V/(R + \mathrm{j}\omega L),$$

a complex quantity whose magnitude and angle are readily found if V is specified. Taking V to be the (real) reference phasor $V_\mathrm{m}\underline{|0}$ or simply V_m (equivalent to the original voltage $V_\mathrm{m} \cos \omega t$), we find that

$$I = \frac{V_\mathrm{m}}{\sqrt{R^2 + \omega^2 L^2}} \, \underline{|\tan^{-1}(\omega L/R)}$$

which gives exactly the same information as did eqn 3.6, and has plainly been arrived at much more economically. However, its derivation could have been shortened still further: eqn 3.14 shows that the phasor ratio of voltage to current for the series combination of R and L is given by the sum of R and $\mathrm{j}\omega L$, so it appears that reactance, when taken as the appropriate imaginary quantity, obeys the same combination rules as resistance (Section 2.5); eqn 3.14 could therefore have been written immediately by inspection of the circuit. We deduce that a.c. circuit analysis can follow the same pattern as d.c., provided only that the reactive elements C and L are represented, not by resistance, but by the imaginary quantities $1/\mathrm{j}\omega C$ and $\mathrm{j}\omega L$. Given that, *all of the principles, rules, and theorems outlined for d.c. circuits in Chapter 2 can be applied directly to a.c.* (save only those relating to power, which is discussed in Section 10). The resulting equations differ from the d.c. only in these respects: in general they are expressed in terms of complex quantities, and they involve the frequency ω (or f).

So far we have defined a phasor to have magnitude equal to the peak value of the sinusoid it represents, such as V_m, because that followed most easily from the idea of a phasor. In fact, however, r.m.s. values are so commonly used to characterize sinusoids in electrical engineering that phasors are conventionally used with r.m.s. magnitudes. There is no difference in principle, nor is there any need to write the factor $1/\sqrt{2}$ (Section 3.2), because for the linear systems with which we are concerned any factor applied to all sinusoidal sources affects all the derived quantities in proportion; hence the factor $1/\sqrt{2}$ would appear throughout, and it can therefore be dropped from the outset. If source phasors are given r.m.s. magnitudes, all others deduced from them will have the same. From now on we shall assume, as is common in practice, that, unless otherwise made clear, *values given for voltage and current in a.c. circuits are r.m.s.* and that phasors take these magnitudes.

The use of phasors is a highly efficient method for a.c. circuit analysis. One caveat only is needed: although phasors *represent* sinusoidal functions of time, the two do not mix well. For example, whereas $V/j\omega L$ represents a current phasor, it is quite wrong to assign any meaning to $V_m \cos(\omega t)/j\omega L$. As a rule of thumb, one should regard any expression containing both j and t as faulty.

We are now in a position to summarize those features in a.c. circuit analysis which distinguish it from d.c. They are:

1. Reactances must be included as imaginary quantities corresponding to resistances, which are real.
2. All voltages and currents are represented by phasors, which usually have r.m.s. magnitudes, and one is chosen as a reference with phase angle zero.
3. All calculation is carried out in complex notation.

Circuits are commonly specified with values of capacitance and inductance in farads and henries, the frequency being known; alternatively, if none of these is given, reactance values can still be specified directly in imaginary ohms, positive or negative. Thus, j10 would define an inductance having a reactance ωL of 10 Ω at the (perhaps unknown) frequency considered, while –j10 Ω would similarly define a capacitance for which $1/\omega C$ has the value 10 Ω.

Example 3.2   Find the current I in the circuit of Fig. 3.6.

Although superposition could be used here, we shall use straightforward mesh analysis for illustration. The source voltages are already given as phasors in polar form, but in collecting the terms of the mesh equations the Cartesian form will be needed. Thus, V_1 is simply $10 + j0$, while the angle of V_2 has been chosen for simplicity to have cosine 0.8 and sine 0.6, giving the voltage $4 + j3$. The equations for clockwise mesh currents I_1 and I_2 are then

$$10 + j0 = 5I_1 + j10\,(I_1 - I_2)$$

$$4 + j3 = -(-j5)I_2 - j10\,(I_2 - I_1).$$

The first of these gives

$$I_2 = \{-10 + I_1(5 + j10)\}/j10 = j + I_1(1 - j0.5)$$

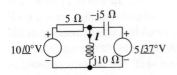

Fig. 3.6
The circuit for Example 3.2.

and inserting this in the second gives

$$4 + j3 = 5 - I_1(2.5 - j5)$$

from which

$$I_1 = (1 - j3)/(2.5 - j5) = (7 - j)/12.5.$$

Inserting this into the equation for I_1 gives

$$I_2 = j + (1 - j0.5)(7 - j)/12.5 = (6.5 + j8)/12.5.$$

The required current is then

$$I = I_1 - I_2 = (0.5 - j9)/12.5 = 0.04 - j0.72$$
$$= \mathbf{0.721 | -86.8°}\ \mathbf{A}\ .$$

Strictly, if we wish to convert this phasor into a function of time we must know what the reference phasor, in this case V_1, represents. If we assume that V_1 represents 14.1 cos ωt V (remembering that the factor $\sqrt{2}$ is needed to get from the r.m.s. phasor value to the peak value) then the actual current function has peak value $0.721\sqrt{2}$ or 1.02 A and can be written

$$i(t) = \mathbf{1.02\ cos\,(\omega t - 86.8°)\ A}$$

in which the value of ω is unknown but is accounted for in the given values of reactance.

3.6 Impedance

Equations 3.13 and 3.14 serve to illustrate one of the most important consequences of using phasor notation: whereas the actual ratio of two sinusoidal quantities would in general be a hopelessly inconvenient function of time, the ratio of their phasors is a constant quantity, complex in general, which tells us how the two sinusoids are related in amplitude and phase. It is important to remember, however, that the ratio of two phasors, although it is in general a complex quantity, is not itself a phasor, for it does not represent a sinusoidal function of time. It would also be as well to point out here that whereas the *product* of two sinusoids at frequency ω includes (as we shall see later) a sinusoid at frequency 2ω, the product of the two corresponding phasors has no physical meaning because the linearity on which the phasor idea, like superposition, depends does not apply to product quantities. For this reason the question of power, which involves the product of voltage and current, needs careful consideration in the case of a.c. circuits when phasors are used: we return to this in Section 3.11.

The ratio of a voltage phasor to a current phasor is in general a complex quantity known as **impedance**. It is measured in ohms, is generally given the symbol Z, and is the a.c. counterpart of resistance; written in the form $Z|\underline{\phi}$, its magnitude Z gives the ratio of r.m.s. (or peak) values and its argument ϕ the phase angle between current and voltage. More precisely, ϕ gives the angle by which the current lags the voltage, because the argument of V/I is given by the angle of V (which has so far been taken as zero) less the angle of I, and is negative if I leads.

We deduced in the previous section that a.c. circuit analysis closely follows d.c. provided that appropriate reactances are used as well as resistance. It is now evident that the idea of impedance offers the simplest generalization of this, and that d.c. circuit analysis (excluding only the question of power) can be directly applied to a.c. simply by substituting impedance for resistance throughout, and, of course, representing sources by phasors.

Equations 3.13 show that resistance R (always real) and the (imaginary) reactances of C and L are special cases of impedance, while eqn 3.14 shows that the impedance of the series combination of R and L is

$$Z = R + j\omega L$$

at the frequency ω. Any branch of a circuit can be represented either by its actual elements, in which case analysis requires the frequency to be specified, or by its impedance, which in general is a frequency-dependent complex quantity. Impedance can be found for any combination of elements by methods analogous to those discussed for d.c. circuits in Chapter 2 — that is, by the rules for series and parallel combination, the various techniques for circuit reduction, or, if all else fails, by the direct application of Kirchhoff's laws. At this point it is possible to view d.c. circuits as merely a special case of a.c., in which all impedances are entirely resistive and therefore real, because even if the circuit includes capacitance and inductance the reactances $1/\omega C$ and ωL become infinite and zero, respectively, in the limit $\omega = 0$. Of course, an a.c. circuit can, in theory at least, comprise resistance only (in practice there would be some capacitance and inductance, however small), in which case all its impedances are real at any frequency and all voltages and currents are in phase.

However many resistances and reactances there may be in a circuit, it is clear that the impedance specified by any required ratio of voltage to current can be reduced for a given frequency ω to one complex quantity, which can be written in the form

$$Z = R_e + jX_e. \tag{3.15}$$

Evidently the overall effect is equivalent to a resistance R_e in series with a reactance X_e (corresponding to an inductance X_e/ω if X_e is positive, and to a capacitance $-1/\omega X_e$ if negative). On this basis we must define resistance generally in a.c. circuits as the real part of impedance, and reactance as the imaginary part. In accepting this definition, we should be aware of its implications: whereas in the simple example of R in series with L, having the impedance $R + j\omega L$, the resistance R_e of the whole is obviously the actual resistance R and similarly X_e is the actual reactance ωL, it turns out that less simple combinations of resistance and reactive elements yield expressions for R_e and X_e which, far from corresponding to particular elements, may *each* involve both resistance and reactive elements; from this it follows that each may be a function of frequency. Thus, although the idea of the equivalent resistance and reactance of a combination can be helpful, and these quantities have the attributes of the corresponding physical entities, they must be distinguished from actual circuit elements. In general a pair of actual components can be equivalent to some other combination only at a certain frequency. A comparable situation can arise with many other kinds of equivalent circuits, for example in the star–delta transformation (Section 2.5.3) applied to reactive elements. Circuit functions which represent what is physically possible are said to be **realizable**. A resistance which is a specified function of frequency is clearly not realizable, nor for example is a negative value of inductance, although both may be perfectly valid for the purpose of analysis. The ideal filter characteristic (Section 5.4.1) is an example of a non-realizable function which is nonetheless useful as an approximation.

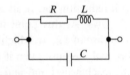

Fig. 3.7
The circuit for Example 3.3.

Example 3.3 Find the equivalent resistance and reactance between the terminals of the circuit shown in Fig. 3.7, at frequency ω.

This deceptively simple-looking circuit is an example of an important class, that of resonant circuits, which we shall meet later. Its impedance $\mathbf{Z}$ can be found straightforwardly by the usual combination rules; it comprises the impedance $R + j\omega L$ in parallel with the reactance $1/j\omega C$. Thus

$$
\begin{aligned}
\mathbf{Z} &= \frac{(R + j\omega L)/j\omega C}{R + j\omega L + 1/j\omega C} \\
&= \frac{R + j\omega L}{1 - \omega^2 LC + j\omega CR} \\
&= \frac{R + j\omega\{L(1 - \omega^2 LC) - CR^2\}}{(1 - \omega^2 LC)^2 + \omega^2 C^2 R^2}
\end{aligned}
$$

in which the quotient of two complex expressions has been rationalized in order to reveal its real and imaginary parts. The result shows that

$$
R_e = \mathrm{Re}[\mathbf{Z}] = \frac{R}{(1 - \omega^2 LC)^2 + \omega^2 C^2 R^2}
$$

and

$$
X_e = \mathrm{Im}[\mathbf{Z}] = \frac{\omega\{L(1 - \omega^2 LC) - CR^2\}}{(1 - \omega^2 LC)^2 + \omega^2 C^2 L^2}
$$

In this case, both components of Z are seen to be functions of all three actual parameters and of ω.

The above example also illustrates one or two general points. First, the equivalent reactance X_e can be positive or negative, representing a net effect which is inductive or capacitive according to the frequency and the circuit parameters; this is characteristic of resonant circuits, as we shall see. More generally, we may note that, as for any impedance, the d.c. behaviour of the circuit is recoverable by setting $\omega = 0$. In that case the expressions give $R_e = R$ and $X_e = 0$, a result which is easily confirmed by inspection of the circuit, since L is a short circuit for d.c. and C an open circuit. It is often also useful to find the limiting behaviour of a circuit at very high frequencies, by setting $\omega \to \infty$. In our example, this gives $R_e = 0$ and $X_e = -1/\omega C$ (X_e would thus also tend to zero, but it is in this case more informative to leave it in a form showing that its high-frequency limit is simply the reactance of C); these can again be confirmed by observing that at infinite frequency L becomes an open circuit and C a short circuit, so that the R, L branch is effectively eliminated and only C is left. Finally, the two expressions show that the frequency ω appears only with the multiplier j, while ω^2 always appears as a real quantity. This is quite general and arises from the fact that ω is introduced in the first instance only in the forms $j\omega L$ and $1/j\omega C$; it is therefore possible, and often desirable, to regard impedance (and other circuit functions still to be encountered) as functions of the imaginary variable $j\omega$ rather than simply of ω.

In the above discussion there has been no need to consider anything other than the Cartesian form of Z. However, sometimes the polar form $Z\underline{|\phi}$ giving magnitude and angle is more helpful; from eqn 3.15 it follows that in general

$$Z = \sqrt{R_e{}^2 + X_e{}^2} \tag{3.16}$$

and

$$\phi = \tan^{-1}(X_e/R_e). \tag{3.17}$$

It is useful to bear in mind that ϕ in this definition is always in the range $\pi/2 > \phi > -\pi/2$, is positive for an inductive circuit, and is the angle by which current lags voltage.

3.7 Susceptance and admittance

It was remarked at several points in Chapter 2 that conductance was occasionally more appropriate than resistance in circuit analysis. There are corresponding alternatives to reactance and impedance. The inverse of reactance is known as **susceptance**; it is given the symbol B and, like conductance, is measured in siemens (symbol S). Like reactance X, it is a measure of the imaginary quantity which relates the phasors V and I for a reactive element. Thus if the ratio I/V for such an element is defined as jB, eqns 3.13 show that for inductance jB is given by $1/j\omega L$ or $-j/\omega L$, so that

$$B = -1/\omega L \tag{3.18}$$

while for capacitance jB is $j\omega C$, so that

$$B = \omega C. \tag{3.19}$$

On this definition we may therefore say that susceptance is positive for capacitance and negative for inductance, which is the reverse of what was deduced for reactance in Section 3.5.3. It also follows that, *for a given reactive element*,

$$B = -1/X.$$

(It will be shown below that this relationship is not true in general for the effective reactance and susceptance of a combination of elements.) Just as reactance, as an imaginary quantity, can be used in circuit analysis in the same way as resistance, so susceptance can be used as an imaginary quantity corresponding to conductance.

The inverse of impedance is known as **admittance**, is given the symbol Y, and is also measured in siemens. Thus we may write in general

$$Y = 1/Z = I/V \tag{3.20}$$

and it follows that the angle of Y is the negative of the angle of Z, so that the admittance corresponding to the impedance $Z\underline{|\phi}$ is

$$Y = (1/Z)\underline{|-\phi}.$$

The change in sign of the angle merely reflects the fact that V leads I if I lags V, and vice versa; the angle of Y, whether positive or negative, is that by which current leads voltage.

Because conductances in parallel, and, for the same reasons, susceptances in parallel, are directly additive, it follows that a conductance G in parallel with a susceptance B produces a resultant admittance

$$Y = G + jB$$

which may be regarded as the dual of the impedance $R + jX$ for the series combination of resistance and reactance. However, it is also true that the overall admittance of *any* combination of elements can always, like their impedance, be reduced for a given frequency to a single complex quantity which we may write in general as

$$Y = G_e + jB_e \tag{3.21}$$

and which corresponds to eqn 3.15 for impedance. Here G_e and B_e are respectively the effective conductance and susceptance of the whole; again we must note that they are by definition the real and imaginary parts of Y, that neither in general represents an actual element, and that, for all but the simplest combinations, each is likely to depend on both resistance and reactive parameters and on frequency (compare Example 3.3 above).

Here we should be aware of a pitfall which was briefly mentioned above. Conductance was introduced (in Section 1.2.3) as the inverse of resistance, and susceptance, above, as the inverse of reactance; both are correct for single elements (and for combinations of resistance alone or reactance alone) but are *not* generally true for combinations. Thus, the effective conductance G_e of any given combination is not in general the inverse of its effective resistance R_e, nor is its susceptance B_e the inverse of its reactance X_e. The question arises in even the simplest combinations: for example, the effective conductance and susceptance of the series combination of R and L, with impedance $R + j\omega L$, are *not* $1/R$ and $-1/\omega L$. Of this more is said in the following section. To avoid trouble it is helpful to remember that admittance and impedance are *always* reciprocal, by definition.

3.8 Combination rules and equivalence

The rules for series and parallel combinations of resistance and conductance, which follow from Kirchhoff's laws, have been discussed in Section 2.5.1. In the foregoing sections it has become clear that the same rules are directly applicable to the corresponding a.c. quantities, but we collect them now for convenience.

3.8.1 Series combination

The general rule is that series impedance, like series resistance, is additive (the addition now being that of complex quantities); thus the total impedance of Z_1, Z_2, ... in series is

$$Z_s = Z_1 + Z_2 + \dots \tag{3.22}$$

If each impedance is purely resistive, this reduces to eqn 2.14; if each is purely reactive, then it follows that

$$X_s = X_1 + X_2 + \dots \tag{3.23}$$

in which each term may be positive or negative. Because inductance L has reactance ωL, it also follows that inductances L_1, L_2, ... in series have a total inductance

$$L_s = L_1 + L_2 + \dots \tag{3.24}$$

exactly as for resistances. On the other hand, capacitance C, having reactance $1/\omega C$, combines in series to give

$$1/C_s = 1/C_1 + 1/C_2 + \dots \tag{3.25}$$

which becomes, for the particular case of two,

$$C_s = C_1 C_2 / (C_1 + C_2) \tag{3.26}$$

corresponding to eqn 2.17 for resistances in parallel.

From the definition of admittance, eqn 3.20, it follows immediately that admittances Y_1, Y_2, ... in series have the resultant admittance

$$1/Y_s = 1/Y_1 + 1/Y_2 + \dots \tag{3.27}$$

while conductance G and susceptance B combine in like fashion, their terms being real but in the latter case carrying either sign.

Voltage division by two series impedances Z_1 and Z_2 carrying the same current can be expressed by the complex relation corresponding exactly to eqn 2.18 for resistances; that is

$$V_2 = VZ_2 / (Z_1 + Z_2). \tag{3.28}$$

Example 3.4   Find the total admittance of the combination shown in Fig. 3.8 at the frequencies (a) 50 Hz and (b) 500 kHz. In each case find also the fraction x of the total circuit voltage which would appear across the capacitance.

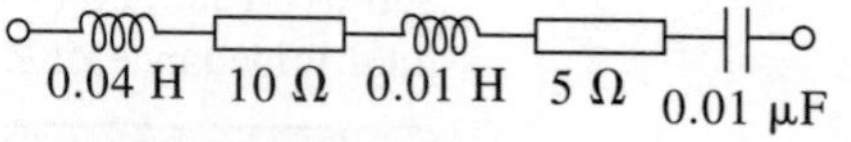

Fig. 3.8
The circuit for Example 3.4.

0.04 H   10 Ω   0.01 H   5 Ω   0.01 μF

Although it is possible to use eqn 3.27 to find admittance for a series combination, it is perhaps simpler to work in terms of impedance first. Here the elements are shown separately, so the individual impedances are identified as resistance and reactance. The two resistances can be added immediately to give 15 Ω, and the two inductances likewise add to a total of 0.05 H. We must now find the reactance values for this inductance and for the capacitance, which at 50 Hz are:

$$jX_L = j\omega L = j2\pi \times 50 \times 0.05 = j15.7 \ \Omega;$$
$$jX_C = -j/(2\pi \times 50 \times 10^{-8}) = -j3.18 \times 10^5 \ \Omega.$$

At 500 kHz the corresponding values are most easily found by proportion, this frequency being 10^4 times greater than the other. Hence:

$$jX_L = j1.57 \times 10^5 \ \Omega;$$
$$jX_C = -j31.8 \ \Omega.$$

The total impedance and admittance can now be found for each frequency.

(a) At 50 Hz the total impedance is, by direct addition,

$$Z = 15 + j15.7 - j3.18 \times 10^5 \approx 15 - j3.18 \times 10^5 \ \Omega$$

and the admittance is therefore

$$Y = 1/Z \approx (15 + j3.18 \times 10^5)/(15^2 + 3.18^2 \times 10^{10})$$
$$= 1.48 \times 10^{-4} + j3.14 \ \mu S$$

which the sign and magnitude of the imaginary part show to be predominantly capacitive.

(b) Similarly at 500 kHz:

$$Z = 15 + j1.57 \times 10^5 - j31.8 \approx 15 + j1.57 \times 10^5 \ \Omega$$

and the admittance is

$$Y = 1/Z \approx (15 - j1.57 \times 10^5)/(15^2 + 1.57^2 \times 10^{10})$$
$$= 6.09 \times 10^{-4} - j6.37 \ \mu S$$

which is predominantly inductive. Because the combination is capacitive at the lower frequency and inductive at the higher, it may be deduced that at some intermediate frequency it is purely resistive; this is characteristic of circuits containing both capacitance and inductance, and is associated with resonance, which is considered in Section 5.3.

The fraction of voltage across the capacitance can be found from eqn 3.28 by taking Z_2 to be its reactance and $(Z_1 + Z_2)$ the total impedance found above, the inverse of which we happen to have already as Y. Thus at 50 Hz

$$x = -j3.18 \times 10^5 \ Y$$
$$= -j3.18 \times 10^5(1.48 \times 10^{-10} + j3.14 \times 10^{-6})$$
$$\approx 1$$

and at 500 kHz

$$x = -j31.8 \ Y = -j31.8(6.09 \times 10^{-10} - j6.37 \times 10^{-6})$$
$$\approx -2.03 \times 10^{-4}.$$

The voltage across the capacitance is a real fraction (to the approximation used here) of the total in both cases because the combination as a whole is predominantly reactive at both frequencies.

3.8.2 Parallel combination

The most convenient rule for parallel connection is that admittances in parallel, being the complex equivalent of conductances, are directly additive; thus the total admittance of Y_1, Y_2, ... in parallel is

$$Y_p = Y_1 + Y_2 + \ldots \qquad (3.29)$$

which reduces to eqn 2.16 if each happens to be a pure conductance. Similarly susceptances B_1, B_2, ... in parallel are additive, giving a total

$$B_p = B_1 + B_2 + \ldots \qquad (3.30)$$

in which the terms can be of either sign. Because the susceptance of capacitance C is ωC, it follows that capacitances in parallel are likewise additive, so that

$$C_p = C_1 + C_2 + \ldots \qquad (3.31)$$

while inductance L, having susceptance $1/\omega L$, combines in parallel with the same reciprocal relation as resistance, thus:

$$1/L_p = 1/L_1 + 1/L_2 + \ldots \ . \qquad (3.32)$$

Equation 3.29 shows immediately that impedances in parallel also combine in the same way as resistance, so that the resultant impedance Z_P is given by

$$1/Z_p = 1/Z_1 + 1/Z_2 + \ldots \qquad (3.33)$$

in which the terms are in general complex; reactance X combines in parallel in the same way, the terms being real and of either sign.

The division of current between two parallel branches can be expressed in terms of impedance or admittance by the complex equivalent of eqn 2.19. Thus, for impedances Z_1 and Z_2 (admittances Y_1 and Y_2) in parallel the current phasor I_2 is related to the total current phasor I by

$$I_2 = I\frac{Z_1}{Z_1 + Z_2} = I\frac{Y_2}{Y_1 + Y_2}. \qquad (3.34)$$

3.8.3   Series and parallel equivalence

Because two resistances in series have a real and constant sum, it is obvious that there must be an infinite number of parallel pairs of resistances having the same combined value. In the case of impedance and admittance, however, the fact that they are complex means that any one impedance, comprising (whether in actuality or as the equivalent of some combination of elements) a resistance R and a reactance X in series, has an equivalent comprising a unique pair of values of conductance G and susceptance B in parallel. The necessary relationship between these parameters is easily found: the admittance of the series pair must be the reciprocal of their impedance, that is

$$Y = 1/(R + jX) = (R - jX)/(R^2 + X^2),$$

and we require this to be also the admittance of the parallel pair, which is simply

$$Y = G + jB.$$

Equating the two expressions yields

$$G = R/(R^2 + X^2) \tag{3.35}$$

and

$$B = -X/(R^2 + X^2). \tag{3.36}$$

Comparable expressions are readily found which give R and X in terms of G and B.

These relations confirm what has already been remarked, first that conductance and susceptance cannot *in general* be defined as the reciprocals of resistance and reactance (although it can be seen that G becomes $1/R$ for small X, and B is $-1/X$ for small R, as expected), and secondly that susceptance and reactance always carry opposite signs for equivalence (as we have seen they do for a single reactive element). This is only to be expected, for if at a given frequency a combination is inductive, say, then any equivalent must also be inductive. There are one or two other points to note. One is that, while G and B are each unique for any given R and X, they do not necessarily imply a unique arrangement of elements, because with more than two elements they can be obtained in an infinite number of ways, even with parallel connections only: the physical significance of eqns 3.35 and 3.36 lies in the definition of series and parallel equivalents having only two elements each. A second important point to note is that, since X and B are frequency dependent, two actual elements in parallel can be the equivalent of another two in series *only at a given frequency.*

Consider, for example, resistance R in series with inductance L and suppose that we wish to find the parallel pair which will be equivalent at frequency ω. Here X is ωL, so eqns 3.35 and 3.36 give

$$G = R/(R^2 + \omega^2 L^2) \tag{3.37}$$

and

$$B = -\omega L/(R^2 + \omega^2 L^2). \tag{3.38}$$

The conductance G would usually in practice be specified by a resistance value, which in this case is

$$R' = 1/G = R + \omega^2 L^2 / R, \tag{3.39}$$

while the susceptance B corresponds to an inductance

$$L' = -1/\omega B = R^2/\omega^2 L + L. \tag{3.40}$$

It is hardly surprising to find that these parallel values are each greater than the original series values, but an interesting case in practice is that of a mainly inductive impedance, such that $R \ll \omega L$. Equations 3.39 and 3.40 show that the equivalent is an almost identical inductance in parallel with a very *large* resistance (i.e. a small conductance). A similar result holds for capacitance: when we consider a good capacitor to be one with a very large resistance (because its dielectric should be a good insulator, allowing little current to pass) it is in effect a *parallel* resistance which we have in mind; the equivalent series resistance should, on the contrary, be small. This, of course, seems perfectly sensible if we remember that to have negligible effect series resistance and parallel conductance should both be very small.

Although eqns 3.35 to 3.40 define only the parallel equivalents of given series values, the reverse expressions are readily found in similar ways.

Example 3.5   Find the two series elements which are equivalent to a resistance R of 1 MΩ in parallel with a capacitance C of 0.01 μF at a frequency of 100 kHz.

Here we need to find the impedance of the parallel combination, which will immediately give the series resistance and reactance. The susceptance of C is given by

$$jB = j\omega C = j6.28 \times 10^{-3} \text{ S}$$

and the conductance of R is 10^{-6} S; the admittance of the parallel combination is therefore

$$Y = 10^{-6} + j6.28 \times 10^{-3} \text{ S}$$

and its impedance

$$\mathbf{Z} = 1/\mathbf{Y} = (10^{-6} - j6.28 \times 10^{-3})/(10^{-12} + 3.94 \times 10^{-5})$$
$$= 0.0254 - j1.59 \times 10^2 \text{ } \Omega.$$

In this, the reactance corresponds at 100 kHz to a capacitance

$$C' = (2\pi \times 10^5 \times 1.59 \times 10^2)^{-1}$$

which, to the accuracy of the calculation, is **0.01 µF**. Hence the equivalent elements are a virtually identical capacitance in series with a resistance of **0.0254 Ω**, in contrast to the parallel resistance of 1 MΩ.

3.9 The Argand diagram

In Section 3.5.2 it was pointed out that phasors, as complex quantities, were easily displayed on the usual Argand diagram, or complex plane, in which horizontal coordinates represent the real components and vertical coordinates the imaginary, of any complex quantity. Figure 3.5 showed examples. In the same way, it can be helpful in describing circuit behaviour to show impedance and admittance on such a diagram, and for that matter any other relevant complex quantity, for example the ratio of two voltage phasors (which will be encountered in the following chapters). When such a quantity is frequency dependent, it can be shown as a locus having frequency as a parameter. A locus of this kind is often known as a **polar diagram**, a term which can be applied to any locus on the complex plane or to any diagram which is based on polar coordinates; we consider this further in Section 3.9.2.

3.9.1 Phasor diagrams

There is no need to confine an Argand diagram to one kind of quantity, and it is often useful to show phasors of current and voltage on the same diagram, each being drawn to indicate (usually) r.m.s. values to some suitable scale. Although it is feasible to use accurate diagrams for measurement, to save calculation, it is much more common to use sketch diagrams which allow rapid assessment of relationships. Figure 3.9(a) is an example which shows current leading voltage by an angle ϕ; the voltage phasor is real and has therefore been chosen as the reference. The angle ϕ must in this case arise from some (capacitive) combination of elements, and a more useful diagram might show voltage and current for each: clarity is the only limit to the number of phasors on a diagram. A reference phasor must be chosen, normally that of an actual (even if unknown) quantity which is assumed to have zero phase angle and is therefore represented by a real phasor. The choice of this phasor may be influenced by the form of the circuit; thus, in a series circuit there is only one current and it is convenient, although not essential, to take that as reference. Figure 3.9(b) shows the complete diagram for the simple series circuit of Fig. 3.4, discussed in Sections 3.5.1 and 3.5.4. Having chosen I as reference, we know that V_R and V_L are respectively in phase with it and leading it by $\pi/2$, and their relative magnitudes are known from R and ωL. Their sum must be V, the source voltage whose magnitude is known. This fixes the scale for the whole diagram. It will be evident that to draw a complete diagram to scale may require as much calculation as a formal analysis; for that reason sketch diagrams are more usual, sometimes labelled with calculated values.

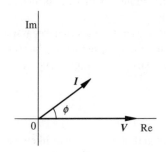

(a)

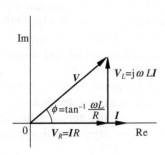

(b)

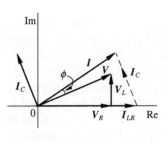

(c)

Fig. 3.9
Phasor diagrams.

In the same way, a phasor diagram for elements in parallel is better constructed by taking the common voltage phasor, whether known or not, as reference, finding the various currents, and summing these to find the total current. When a circuit comprises a mixture of series and parallel elements the choice of reference is more arbitrary and the diagram can be built up as required from any suitable start. For example, Fig. 3.9(c) is the phasor diagram for the circuit of Fig. 3.7 with the current I_L in the inductive branch taken as reference. From this current the total voltage V is obtained as in (b), and the current I_C drawn to lead it by $\pi/2$; the total current I then follows as the sum of I_L and I_C. It can be seen from the diagram that I can lead or lag V by some angle ϕ which depends on the circuit parameters and the frequency; this is exactly equivalent to our earlier observation in Example 3.3 that the effective reactance X_e of this circuit might be positive or negative.

In a more extensive circuit it may be easier to construct separate diagrams for parts of it before superimposing them for the whole; but for more than a few branches and meshes the clarity of a single diagram is greatly reduced.

3.9.2 Impedance and admittance diagrams

Although they are not phasors, impedance and admittance can likewise be shown on an Argand diagram, on which they too add and subtract vectorially. Thus, the impedance Z of the simple R, L series circuit mentioned above is shown in Fig. 3.10(a) as the sum of its components R and $j\omega L$. Its form is exactly that of the phasor voltages in Fig. 3.9(b) because it represents V/I and I here is real. Admittance diagrams are similarly drawn for elements in parallel. If need be, a combined diagram can be drawn for a mixed circuit: Fig. 3.10(b) shows one for the circuit of Fig. 3.7, of which Fig. 3.9(c) is the phasor diagram. Here it is helpful to bear in mind that, whereas the angle of a phasor can be referred to any reference, the angle of impedance is always positive for an inductive circuit; and that the angle of any admittance is the negative of that of the corresponding impedance.

3.9.3 Polar diagrams

In the context of circuit analysis, a polar diagram usually means the locus on the Argand diagram of some frequency-dependent complex quantity. Figure 3.11(a) shows a very simple example, the locus of the impedance Z in Fig. 3.10(a) as the frequency varies; the locus of the voltage V in Fig. 3.9(b) would be identical in form, for fixed real I, because then V and Z are in proportion. The locus of I for fixed real V, however, would correspond to that of the admittance $1/Z$ which can be shown to be a semicircle as in Fig. 3.11(b). (This is an example of mapping on the complex plane: the inverse of a function represented by a straight line is in general circular; more generally, one circle maps into another, a straight line simply implying infinite radius.)

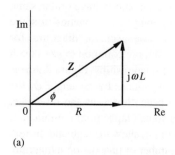

(a)

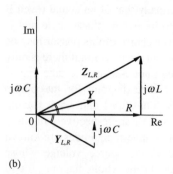

(b)

Fig. 3.10
Impedance and admittance diagrams.

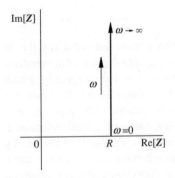

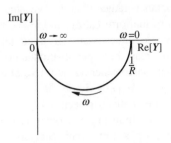

Fig. 3.11
Polar diagrams.

Polar diagrams having frequency as parameter are often useful, and are one way of representing **frequency response**, which is considered further in Chapter 5; an alternative, as we shall see, is to show magnitude and angle as separate functions of frequency.

3.10 Mutual inductance

3.10.1 The nature of mutual inductance

In Section 1.2.5 mutual inductance was mentioned as arising from magnetic flux passing through one closed path when it is produced by current in another, in contrast to self-inductance, which arises from flux passing through the closed path by whose current it is produced; self-inductance is what we normally mean by the term 'inductance' and by the symbol L. However, just as by Faraday's law self-inductance causes an induced e.m.f. given by eqn 1.5, so must mutual inductance, where it exists, cause an e.m.f. in one path to be induced by current in another. This is an effect we have not so far considered, but evidently it must affect the behaviour of any circuit in which it occurs. It can be included in a.c. circuit analysis by adapting the relations which have been used for self-inductance; this we do in Section 3.10.3. First, we consider how to define it to suit our purposes.

Mutual inductance can arise only *with* self-inductance; because if current in one path produces flux through another, that flux must pass through both. Not only that: there is a reciprocal effect which ensures that if one closed path can receive flux produced by current in another, then current in the first will likewise produce flux in the second; this too must pass through both paths, so it follows that the second path must have its own self-inductance. Thus mutual inductance is always associated with two values of self-inductance, although two self-inductances do not necessarily have mutual inductance (it is obvious, for example, that two inductances may be physically so far apart that flux from one cannot reach the other, in which case their mutual inductance must be zero). Whether there actually is current flowing in one or both paths is, of course, immaterial to the existence of mutual inductance, which, like other circuit parameters, depends on geometrical arrangement and material properties (see Appendix B).

By analogy with eqn 1.4, the mutual inductance L_{ab} between two closed paths a and b may be defined as

$$L_{ab} = \Phi_a/i_b \tag{3.41}$$

where Φ_a is that flux linking a which is due to the current i_b in b. (There may be also be flux in a due to its own current, which would depend on its self-inductance, and for that matter the flux in a may include contributions due to other currents i_c, i_d, ... which would depend on mutual inductances L_{ac}, L_{ad}, ...; none of these affects the definition 3.41.) With the same notation it follows that the flux linking b per unit current in a defines the mutual inductance

$$L_{ba} = \Phi_b/i_a. \tag{3.42}$$

On the usual assumption of linearity, which here means that any flux is proportional to the current producing it, L_{ab} and L_{ba} are both constant; but it can then also be shown that, because of the reciprocity which applies to linear systems (Section 2.5.5), they are equal. Thus, we assume in linear circuit theory that in general any pair of self-inductances has associated with it a single value of mutual inductance. The double subscript notation can be used for all three parameters by writing, for example, L_{aa} and L_{bb} for the self-inductances; but it is more usual to write simply L_a and L_b. Also, to avoid confusion, it is common to use the symbol M for mutual inductance rather than L. We shall therefore for the most part use notation of the form L_a, L_b, and M_{ab} (or M_{ba}), dropping the subscripts if they are unnecessary. If a circuit includes several self-inductances there may be mutual inductance between any two; but very often we shall be considering only two at a time, using the notation L_1, L_2, and M_{12} (or M_{21}) or simply M.

In the above discussion, the idea of closed paths has been freely used since it is fundamental to our definition of inductance. However, we have seen that it is legitimate for many purposes of circuit analysis to regard self-inductance as concentrated in lumped elements, which may be interconnected with other elements in any way we please: an inductor, i.e. a component designed to be mainly inductive, is in practice usually made as a coil of wire which concentrates magnetic flux so that its inductance is virtually independent of the precise path by which the circuit is completed (Appendix B.3). By the same token it is legitimate and convenient to regard mutual inductance as a circuit parameter associated with two self-inductances, however these may be interconnected and wherever they may appear in a circuit, provided we remember that the value of their mutual inductance actually depends on their *geometrical* relationship. Sometimes the two self-inductances may exist in separate circuits having no electrical connection whatever but connected magnetically by their mutual inductance (this being essentially how most transformers and electrical machines work), or they may be in different loops of a given circuit, or they may even be different elements in the same loop.

Like other circuit parameters, mutual inductance is present to some extent in all circuits, and the engineer may seek to minimize it for some purposes and maximize it for others: for example, to be efficient a transformer needs the maximum economic value of M; but unwanted mutual inductance in electronic circuits contributes to interference or **crosstalk** and should be kept to a minimum. Circuits having mutual inductance by design are known as coupled circuits; we consider them further in later sections.

In view of the attention given to mutual inductance, it is natural to ask about mutual capacitance. This does indeed exist, although it is rarely introduced by design and does not really justify special treatment. It arises in an analogous way (Appendix B2) and, like mutual inductance, contributes to crosstalk. In a circuit it can be represented when necessary by normal capacitance connected between appropriate points.

3.10.2 The relationship of mutual and self-inductance

If we consider two coils having inductances L_1 and L_2, it is obvious that in general only a part of the flux produced by current in one will pass through the other. Inductances in which all of the flux produced by either passes through both as shown in Fig. 3.12(a), so that all of the flux is mutual whatever the contribution of each coil, are said to be **perfectly coupled**, and it can be shown (Appendix B.3) that their mutual inductance is given by

$$M = \sqrt{(L_1 L_2)} \tag{3.43}$$

At the other extreme, if two coils are so far apart, or are so arranged, that none of the flux produced by either links the other, as indicated in Fig. 3.12(b), then they have zero mutual inductance and may be said to be uncoupled. In general we may write

$$M = k\sqrt{(L_1 L_2)} \tag{3.44}$$

in which k, a constant of value between zero and unity, is the **coefficient of coupling** between L_1 and L_2. We should note here that mutual inductance, unlike self-inductance, can take either sign because the flux due to one coil can pass through the other in either direction according to physical arrangement, whereas its direction through its own coil is uniquely related to the direction of its current: we return to the question of sign in the following sections.

When one inductance is imperfectly coupled to a second, the total flux it produces can be divided as shown in Fig. 3.12(c) into the mutual flux, which passes through both, and the remainder, which does not; the latter is often called the **leakage flux**. Thus, if of the flux produced by coil 1 a fraction k_1 links 2, then the **leakage inductance** of coil 1 is $(1 - k_1)L_1$; the leakage inductance of coil 2 is likewise $(1 - k_2)L_2$. It can be shown that the mutual inductance is then given by the geometric mean of $k_1 L_1$ and $k_2 L_2$, that is by $\sqrt{(k_1 k_2 L_1 L_2)}$. It follows according to eqn 3.44 that the coefficient of coupling can also be written as

$$k = \sqrt{(k_1 k_2)}. \tag{3.45}$$

The idea of leakage inductance, and hence of leakage reactance, can be especially useful in dealing with devices, such as transformers (Section 4.9) and rotating electrical machines, which depend on closely coupled coils.

3.10.3 Mutual inductance in circuit analysis

From the foregoing definitions, it follows from Faraday's law that the e.m.f. induced in coil 1, say, due to a current i_2 in coil 2 is given by

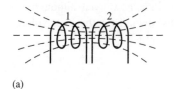

(a)

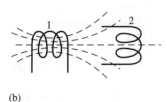

(b)

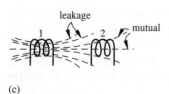

(c)

Fig. 3.12
Mutual inductance: (a) perfectly coupled and (b) uncoupled inductances; (c) mutual and leakage flux.

$$e_1 = -M_{12}\, di_2/dt \qquad\qquad (3.46)$$

corresponding to eqn 1.5 for self-inductance; the negative sign is included for consistency, but now the actual sign of the e.m.f. depends on the physical relationship of the coils as well as the reference directions chosen for both (which may be quite independent). For the time being we retain the double subscript for M. In the case of a.c. circuits in the steady state eqn 3.46 can be written in phasor form as

$$\boldsymbol{E}_1 = -j\omega M_{12}\boldsymbol{I}_2. \qquad\qquad (3.47)$$

In Chapter 1 the negative sign attached to the e.m.f. of self-inductance was eliminated by considering instead the applied voltage needed to overcome that e.m.f.; the same can be done now even although the actual direction of the latter is as yet unknown. Thus the *total* voltage needed for coil 1 may be written as

$$\boldsymbol{V}_1 = j\omega L_1\boldsymbol{I}_1 + j\omega M_{12}\boldsymbol{I}_2. \qquad\qquad (3.48)$$

The uncertainty about the actual direction of the second contribution may now be resolved by allocating one or other sign to M_{12}, as the physical circumstances may dictate: this we discuss again below.

For coil 2, in the same way, the applied voltage may be written

$$\boldsymbol{V}_2 = j\omega L_2\boldsymbol{I}_2 + j\omega M_{21}\boldsymbol{I}_1. \qquad\qquad (3.49)$$

If the two coils are connected to other elements, then the mesh or nodal equations for the circuit can be written in the usual way except that the simple voltage drops of self-inductance are replaced by the voltages $\boldsymbol{V}_1$ and $\boldsymbol{V}_2$ of the above equations. In the event that coil 1, say, is coupled to several other coils, then its voltage drop includes a term from each of these, of the form $j\omega M_{1r}\boldsymbol{I}_r$. With such terms included to account for mutual inductance, a.c. circuit analysis proceeds exactly as described in earlier sections.

3.10.4 The dot convention and the sign of M

We have seen that the actual direction of an induced e.m.f. due to mutual inductance, and hence also the sign to be given to M, depends on the physical arrangement of the coils concerned as well as on the direction of the current causing it. From the form of eqn 3.48 it is readily deduced that positive M_{12} must correspond to an arrangement in which positive currents in both coils produce fluxes in coil 1 which are additive (that is, reinforce each other); if that is so, then the fluxes which positive currents cause in coil 2 are also additive and M_{21} is therefore positive likewise. It follows that M_{12} and M_{21} always carry the same sign and we may therefore revert to the use of M to represent both.

Figure 3.13(a) shows two coils so arranged that for the specified current directions M is clearly positive by the above reasoning; (b) and (c) show two situations in which M is negative. Since it is not easy to

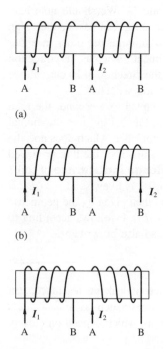

Fig. 3.13
Mutual inductance: (a) positive;
(b), (c) negative.

show on a circuit diagram the precise arrangement of coils in this way, some means is needed of specifying the sign to be used for M for given current reference directions. The **dot convention** is commonly used to do this: each coil is marked with a dot at one end, as shown in Fig. 3.14(a); then M is taken as positive when both assumed currents enter or leave by the dot end, and otherwise negative, as indicated. Of course the dots cannot be *allocated* without detailed knowledge: in using the convention we are accepting the dots in lieu of that knowledge, just as we accept the polarity given for a battery without detailed knowledge of its construction. If we happen to know the physical arrangement, as in Fig. 3.13, then the dot positions are of course easily allocated in either of two equivalent ways: in all three examples of that figure, the dots could either be both at ends A or both at ends B.

An alternative statement of the dot convention gives the direction of the induced e.m.f.s rather than the sign of M, thus: current I entering the dot end of one coil induces in the other an e.m.f. $j\omega|M|I$ acting towards the dot end. This version is illustrated in Fig. 3.14(b); it has the advantage of eliminating any confusion over the significance of negative M, or over the distinction between an induced e.m.f. and the voltage needed to overcome it.

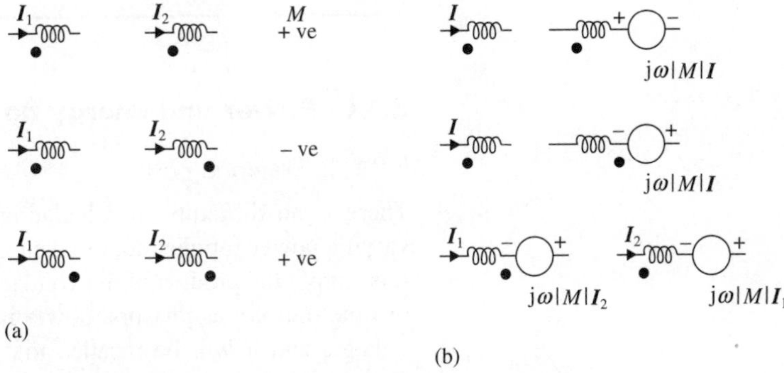

Fig. 3.14
Mutual inductance: the dot convention.

(a)

(b)

The actual direction of an e.m.f. due to mutual inductance is important in circuits where the two inductances are connected electrically as well as magnetically, but often less so if they are electrically separate. In the latter case the sign of M cannot affect the *magnitude* of any quantity unless both circuits include other sources.

Example 3.6 Find the effective inductance between the terminals of each of the circuits shown in Fig. 3.15.

In circuit (a) the current I is common to both coils, so the total voltage drop can be written, from eqns 3.48 and 3.49, as

$$V = j\omega L_1 I + j\omega M I + j\omega L_2 I + j\omega M I$$

from which the effective inductance of the whole is clearly

(a)

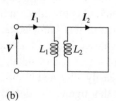

(b)

Fig. 3.15
Circuits for Example 3.6.

$$L' = V/j\omega I = L_1 + L_2 + 2M.$$

The value of this expression depends on the sign of M, which is unknown here since no dots are shown. (To make the dependence explicit, the result is often written as $L_1 + L_2 \pm 2M$.)

For circuit (b), if an applied voltage V produces currents as shown the mesh equations are

$$V = j\omega L_1 I_1 + j\omega M I_2$$
$$0 = j\omega L_2 I_2 + j\omega M I_1$$

from which we may eliminate I_2 to give

$$V/I_1 = j\omega L_1 - j\omega M^2/L_2$$

and the effective inductance of the whole is therefore

$$L' = V/j\omega I_1 = L_1 - M^2/L_2.$$

In this case the sign of M can evidently make no difference to the value of L', as we should expect from the fact that the two loops are electrically separate.

3.11 Power and energy flow in a.c. circuits

3.11.1 Average power

There is no difficulty in calculating the *instantaneous* value of time-varying power supplied to, or taken from, an a.c. circuit or any part of it: it is simply the product of the voltage and current expressed as functions of time (but *not* as phasors; power is not a linear function of current or voltage, and it will be recalled that phasor representation depends on superposition and therefore on linearity). However, the resulting expressions are cumbersome and in any case we are more likely to be interested in the *average* power — averaged, that is, over one period (or any other integral number of periods), and so equivalent to the steady-state average, over many periods, which is what is usually wanted in practice. (For example, it is more helpful to know that a 100 W lamp consumes that much on average, than that its instantaneous power varies in a certain way between two limits.)

Before investigating average power we should consider the nature of power flow in a.c. circuits, because capacitance and inductance, while (ideally) *dissipating* no power, can store energy in their fields and therefore receive or deliver electrical power as this energy fluctuates. It is shown in Appendix B that the amounts of energy they store are respectively

$$u_C = \frac{1}{2} C v^2 \qquad\qquad (3.50)$$

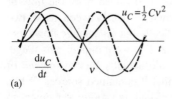

(a)

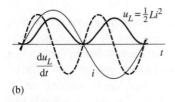

(b)

Fig. 3.16

Power flow in (a) capacitance
and (b) inductance.

and

$$u_L = \frac{1}{2} L i^2 \qquad (3.51)$$

where v and i are the instantaneous values of voltage across C and current in L. In an a.c. circuit, therefore, the energy stored in each reactive element varies as the square of a sinusoid, between zero and a maximum; this variation, and the corresponding flow of power, are illustrated for capacitance and inductance in Fig. 3.16(a) and (b). These elements evidently show a periodic reversal of power flow when they carry alternating voltage and current. In an entire a.c. circuit it is therefore possible for power to flow to and fro between the sources and the passive elements, or between those elements, irrespective of how much is being consumed by dissipation or otherwise. It is also clear from the figure that in the steady state the *average* power flowing to L and C is zero (it must be, since pure reactance cannot dissipate energy). On average, electrical power taken from sources can be consumed only by energy conversion: for passive elements this means the dissipation of heat in resistance, but it may also include energy-converting devices, such as an electric motor or a rechargeable battery.

Now let us consider the power flow in or out of a pair of terminals at which the voltage is

$$v = V_m \cos \omega t$$

and the current, out of phase with the voltage by some angle ϕ, is

$$i = I_m \cos (\omega t - \phi).$$

As in the case of d.c. circuits (Section 2.8) the product vi will yield the power *entering* at the terminals if i is defined to flow into the terminal defined as positive. This power is

$$
\begin{aligned}
vi &= V_m \cos \omega t \times I_m \cos(\omega t - \phi) \\
&= \frac{1}{2} V_m I_m \{ \cos(2\omega t - \phi) + \cos \phi \} \qquad (3.52) \\
&= VI \{ \cos(2\omega t - \phi) + \cos \phi \}
\end{aligned}
$$

where V and I are r.m.s. magnitudes. This expression contains a sinusoidal term at twice the frequency of v and i, together with the term in $\cos \phi$. It is clear that the average power over one period or over a long interval is given by

$$P = VI \cos \phi. \qquad (3.53)$$

Figure 3.17 illustrates this result for three cases. For a purely resistive circuit $\phi = 0$ and the instantaneous power is always positive, varying between zero and $2VI$ as shown in (a); the average power is simply VI, corresponding to the d.c. expression. For a purely reactive circuit $\phi = \pm \pi/2$, the instantaneous power varies between the limits $\pm VI$ and the average is zero, as shown in (b) for the case of inductance (and in

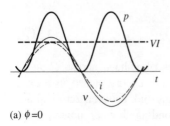

(a) $\phi = 0$

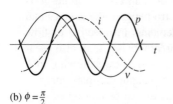

(b) $\phi = \frac{\pi}{2}$

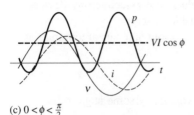

(c) $0 < \phi < \frac{\pi}{2}$

Fig. 3.17
Power flow in a.c. circuits: (a) purely resistive; (b) purely inductive; (c) general case for lagging current.

agreement with Fig. 3.16(b) where power is deduced from the flow of stored energy). In the general case shown in (c), between these two extremes, the power flow reverses periodically but on average flows into the circuit provided that $-\pi/2 < \phi < \pi/2$. If ϕ were outside this range, making the average negative, power would actually be flowing *out* of the circuit at these terminals, implying that it contained sources. When this is likely to be the case, it is conventional and more convenient to reverse the chosen direction for either v or i so that P becomes positive and represents the average output power.

Because of its critical importance in determining average power, $\cos \phi$ is known as the **power factor**. It applies *only* so long as v and i are sinusoidal at a given frequency, but its usefulness lies in the fact that eqn 3.53 then gives the power due to known voltage and current, whatever the nature of the circuit: we need know nothing of its composition. The expression can be evaluated for any identifiable pair of terminals, for example those of each source, and the total power flow into the circuit is then, by conservation of energy, the algebraic sum of all such contributions. The power factor $\cos \phi$, as well as V and I, may be different for different terminal pairs of a given circuit.

If a terminal pair is connected only to passive elements, then the angle ϕ as defined above is simply the phase angle (or argument) of the total impedance; its negative, the angle of the admittance, may equally well be used since $\cos \phi = \cos(-\phi)$. For the same reason, a value of power factor cannot specify the sign of ϕ, so the term 'lagging' or 'leading' is usually added according as current lags the voltage (as assumed in our derivation) or leads it.

Because $\phi = 0$ for a pure resistance, it follows that the average power dissipated in a resistance R, or a conductance G, is VI where V and I are the r.m.s. values measured *at its terminals*; it follows too that this power can also be written

$$P = I^2 R = V^2/R = V^2/G = I^2/G, \tag{3.54}$$

which forms are readily confirmed by noting that V^2 and I^2 are mean square values and hence give the average of v^2 and i^2, on which the instantaneous power depends. At this point we should note especially that the power dissipated in an impedance Z is *not* in general given by $I^2 Z$ or V^2/Z, and that in eqns 3.53 and 3.54 V and I are magnitudes only, *not* phasors.

There is, however, a way of expressing average power from the phasors V and I. We may write eqn 3.53 as

$$P = VI \cos \phi = \text{Re}[VIe^{j\phi}].$$

Now the phasors, written in full, are

$$V = Ve^{j\omega t}; \quad I = Ie^{j(\omega t - \phi)}$$

and clearly their product is a function of time and cannot yield our expression for P; but if we reverse the sign of $j\omega t$ in, say, the phasor I by taking its complex conjugate I^*, then we may write

$$VI^* = Ve^{j\omega t} Ie^{-j(\omega t - \phi)} = VIe^{j\theta}$$

and so

$$P = VI \cos \phi = \text{Re}[VI^*]. \tag{3.55}$$

The two forms for P in eqn 3.55 are equivalent alternatives, to be used as convenient in dealing with sinusoids at a given frequency. It is equally correct to take the conjugate of V and write P as $\text{Re}[IV^*]$, but for reasons which will become clear in the next section it is conventional to conjugate I rather than V.

Example 3.7 In the circuit of Fig. 3.18 find the average power produced by the source and the average power delivered to the impedance Z_L, and confirm that their difference is dissipated in the resistance R.

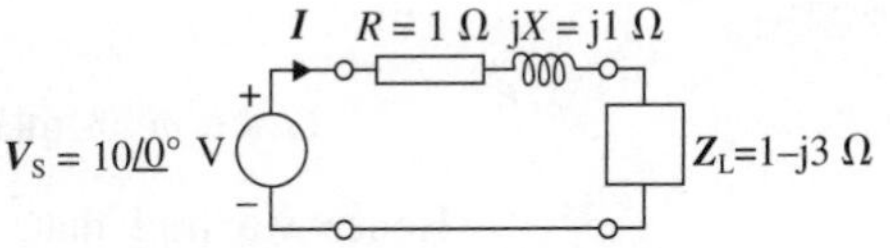

Fig. 3.18
The circuit for Example 3.7.

It is necessary first to find the current which in this case is common to the whole circuit. It is

$$I = \frac{V_S}{R + jX + Z_L} = \frac{10}{2 - j2} = 3.535 \underline{|45°} \text{ A}.$$

The power produced by the source is then

$$P_S = V_S I \cos \phi = 10I \cos 45° = \textbf{25 W}.$$

The power delivered to the impedance Z_L could be found by first finding the voltage across it, but it is quicker to use the fact that its real part is its resistance R_L, so the power dissipated in it is

$$P_L = I^2 R_L = 12.5 \times 1 = \textbf{12.5 W}.$$

The power dissipated in R is similarly

$$P_R = I^2 R = \textbf{12.5 W}.$$

and this is, as expected, the difference between P_S and P_L.

3.11.2 Reactive volt–amperes

In the last section it was shown that in an a.c. circuit a reactance consumes no average power but alternately receives and delivers power as its stored energy rises and falls. By differentiating eqns 3.50 and 3.51 with sinusoidal v and i, it can be shown that the *peak* power which flows to and from a reactance X as its stored energy rises and falls is VI, where V and I are its r.m.s. voltage and current. This peak power can also be written as

$$Q = I^2X = V^2/X = -V^2B = -I^2/B.$$

The significance of the sign here is considered below (it was shown in Section 3.7 that B takes the opposite sign to X for a given reactance). These expressions correspond to eqns 3.54 for the average power in a resistance.

Now suppose a pair of terminals has, as in the last section, current lagging voltage by the angle ϕ. Whatever the circuit may consist of, it is evidently equivalent to some impedance $Z\underline{/\phi}$ or $R + jX$ where $Z = V/I$, $R = Z \cos \phi$, and $X = Z \sin \phi$. We may then write

$$I^2X = I^2Z \sin \phi = VI \sin \phi \qquad (3.56)$$

and by the same reasoning as we used in the last section we may also write

$$VI \sin \phi = \text{Im}[VI^*]. \qquad (3.57)$$

Hence we find that, corresponding to the expressions $VI \cos \phi$ and $\text{Re}[VI^*]$ for the average flow of power, the expressions $VI \sin \phi$ and $\text{Im}[VI^*]$ represent a power which flows back and forth as stored energy rises and falls in different parts of an a.c. circuit. Some engineers call this quantity **reactive power** or even imaginary power, but since, notwithstanding the symmetry of the above relations, it is radically different from the power flow P, having a zero average and representing neither consumption nor production, it is better called **reactive volt-amperes**, or volt-amperes reactive, or VAR. Its units are given this name to distinguish them from watts, with which they are dimensionally identical.

The reactive case also needs careful regard to sign, since VAR may evidently be positive or negative: the reactance X carries a sign (and susceptance B its opposite); so too does ϕ and hence $VI \sin \phi$, if ϕ is measured consistently as the angle of voltage with respect to current, say (that is, as the angle of Z). It can be shown that $\text{Im}[VI^*]$ likewise takes a sign. But whereas the sign of the average power P merely indicates the direction of flow, the sign of VAR indicates the nature of the reactance: the several expressions above are positive for an inductive circuit and negative for a capacitive, when, as in the previous section, the directions of V and I are chosen to make received power positive. We could say with equal truth that an inductive circuit receives positive VAR or that it delivers negative VAR, but for a passive circuit the former is more natural. Although physically neither statement has much meaning, since on average there is no actual flow, the sign of VAR is in fact very important for engineering purposes, as we shall see below.

The significance of different signs for inductive and capacitive VAR can be better appreciated by considering the flow of energy, as in Fig. 3.16, for L and C when both have the same current or the same voltage: we should find that in each case the power flows to the two reactances were exactly opposite in phase. Thus in a circuit containing both L and C, in series or in parallel, energy tends to swing between the two. In general, of course, their peak energies are not equal (when they are, the two are in the state of resonance, which is discussed in Chapter 5) so energy also swings to and from the source, as for a single reactance; nor, of course, are the energy flows in any L and any C of a given circuit necessarily in phase opposition.

It is because Im[VI^*] carries a sign that we opt for VI^* rather than IV^*: the latter would have a positive imaginary part for a capacitive circuit, in contrast to our other conventions.

3.11.3 Resolution of volt-amperes

Having seen that average power and reactive volt–amperes can be expressed either as $VI \cos \phi$ and $VI \sin \phi$ or as Re[VI^*] and Im[VI^*], we may deduce that these two quantities can be regarded as the real and imaginary parts of a complex quantity VI^* whose argument is ϕ and whose magnitude is VI. We may write in general

$$VI^* = S = \text{Re}[VI^*] + j\text{Im}[VI^*]$$
$$= VI \cos \phi + jVI \sin \phi \qquad (3.58)$$
$$= P + jQ$$

in which Q represents VAR, positive or negative as the case may be. The quantity S is called simply the **volt-amperes** or VA, and its units (which are dimensionally the same as watts) are also so-named to distinguish it from its component parts. It is sometimes, perhaps unfortunately, called the **complex power** and its magnitude VI the **apparent power**. It is a very significant quantity, for its magnitude VI comes immediately from the most basic measurements at any pair of terminals (whether of a source, a passive element, or any another device) and its angle is simply the phase of V with respect to I. Although its magnitude may alternatively be expressed as I^2Z or V^2Y (or I^2/Y or V^2/Z) it is *not* in general correct to write it as the complex quantity given by I^2Z etc.: that can attribute the wrong angle to S, and hence imply incorrect components, because it takes no account of the need to conjugate one of the phasors I and V as explained above.

Volt-amperes and its components can also be expressed on the Argand diagram just as easily as voltage and current and impedance or admittance. Figure 3.19(a) shows such a diagram for an inductive circuit, and the similarity with impedance, admittance, and phasor diagrams is clear. If the circuit to which V and I refer is considered as an

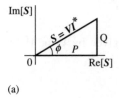

(a)

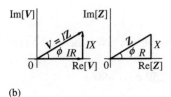

(b)

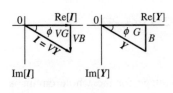

(c)

Fig. 3.19
Argand diagrams (a) for volt–amperes, (b) for the corresponding impedance, and (c) for the admittance.

impedance $R + jX$ in which current is a real reference phasor, then successive division of the volt-ampere diagram by I gives the voltage and impedance diagrams as in (b). Viewed as an admittance $G + jB$ with a real reference voltage, the same inductive circuit gives the current and admittance triangles of (c); these have the same form as a volt-ampere diagram based on IV^*, with the imaginary part negative. All of the diagrams shown would be inverted for a capacitive circuit.

One more important and useful point remains. It is evident that by conservation of energy all of the contributions to the power consumed in a complete circuit can be directly added to give the total power supplied; it is also true that VAR can be likewise added, with due regard to sign, to give a total for the circuit. In other words, volt-amperes can be summed, like any other complex quantity, by the addition of components.

Example 3.8 Find the total average power and VAR input to the circuit shown in Fig. 3.20, and its overall power factor.

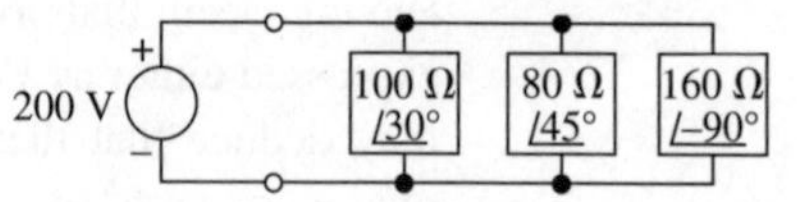

Fig. 3.20
The circuit for Example 3.8.

200 V | 100 Ω /30° | 80 Ω /45° | 160 Ω /−90° |

The magnitude of the VA in each impedance is given by VI or V^2/Z, giving the values

$$200^2/100 = 400,$$

$$200^2/80 = 500,$$

and

$$200^2/160 = 250.$$

The power in each is $VI \cos \phi$, so with the angles given the total power is

$$P = 400 \cos 30° + 500 \cos 45° + 250 \cos (-90°)$$

$$= 346.4 + 353.6 = \textbf{700 W}$$

and the total VAR is similarly

$$Q = 400 \sin 30° + 500 \sin 45° + 250 \sin (-90°)$$

$$= 200 + 353.6 - 250$$

$$= \textbf{304 VAR}.$$

The phase angle between current and voltage for the whole circuit is simply the angle of the total VA S, which is

$$\phi = \tan^{-1}(303.6/700) = 23.45°$$

and the overall power factor is therefore

$$\cos \phi = \cos 23.45° = \textbf{0.917}.$$

3.11.4 Power-factor correction

The idea of power factor has particular significance in any electrical device or circuit or system which consumes sufficient power for efficiency, and the cost of wasted energy, to be of prime importance: this includes, for example, all but the smallest a.c. motors whether considered singly or in their aggregate effect on a power supply system.

Since a.c. power can be expressed generally as $VI \cos \phi$, it follows that any circuit supplies at more or less constant voltage, including any part of a national supply system, will require a minimum current for a given power if its power factor is unity. This minimum is highly desirable, for the current to any load must be provided by generators and transmitted through conductors which inevitably have finite resistance and hence dissipate some power apart from the load they are supplying. A power system is therefore more efficient and less expensive (in all respects) if the power factors of the loads it supplies, and hence also its overall power factor, can be maximized. On the other hand, there is a limit to how much a designer can affect the power factor of a given device: the most common kinds of a.c. motor, for example, inevitably have a lagging power factor which rarely exceeds 0.9 and can be a good deal lower; the power factor on the U.K. national grid as a whole is typically in the region of 0.8–0.9 lagging. The cost of this to the supplier is such that large consumers have to pay not only for the power they use but also a charge which penalizes a poor power factor.

This difficulty can sometimes be answered for any particular consumer by providing additional devices whose sole function is power-factor correction or improvement. Because a typical load receives a lagging current, and therefore lagging VAR, the power factor at its terminals can be brought nearer to unity, while leaving its own behaviour unchanged, by adding to it a device which requires leading VAR but no power. At its simplest this means no more than a capacitance of suitable value added in parallel; a capacitance in series would also raise the power factor and is sometimes used. Some electrical machines can be arranged to take leading VAR (usually, but not invariably, while also taking power for a separate purpose) and hence to improve the overall power factor when connected in parallel with a lagging load. In a system which in total has a lagging power factor, any device which takes leading VAR, i.e. which is capacitive in nature and has itself a leading power factor, is beneficial.

It is not feasible to bring the power factor of an entire national system to unity in such ways; the extent to which it is done for particular parts or for particular consumers is a matter of the economics of providing capital equipment on the one hand and supporting energy losses on the other.

Example 3.9   A certain electric motor consumes 2 kW at a power factor of 0.8 lagging when supplied at 240 V and 50 Hz. Find the value of parallel capacitance which would give unity power factor to the combination.

Since the power consumed is $VI \cos \phi$, in this case $0.8VI$, the volt–ampere product VI is given by $2000/0.8$ and the lagging VAR taken by the motor is therefore $VI \sin \phi = (2000/0.8) \times 0.6 = 1500$ VAR. To provide unity power factor overall a capacitance must take the same value of leading VAR; since capacitance has $\phi = \pi/2$ and $\sin \phi = 1$, its VAR is given simply by its VI product and we therefore need a parallel capacitance to take a current I_C given by $1500/240 = 6.25$ A. But capacitance C takes current of magnitude $V\omega C$ so the value required is

$$C = I_C/V\omega = 6.25/(240 \times 2\pi \times 50) = \textbf{82.9 } \boldsymbol{\mu}\textbf{F}.$$

This is a large value, and unlikely to be economic in practice for such a relatively small motor. Generally, quite large values of capacitance are needed to provide even modest values of leading VAR at a low supply voltage and a typical supply frequency. Other means of providing the same effect would also be relatively expensive, and power-factor correction is therefore economic only in certain cases.

In principle the necessary capacitance for power-factor correction could also be found from a knowledge of the impedance of the original load, by requiring that the angle of the resultant impedance be zero (or some suitably small value); cancelling the lagging VAR, as in the example, is exactly equivalent to cancelling an inductive reactance. In power-system calculations, however, it is common practice, and for a number of reasons generally more convenient, to specify loads in terms of power, or volt-amperes, and power factor (or, for small devices, power only) rather than impedance. One reason is that the power rating of a device at fixed voltage is usually more directly informative: it would hardly be helpful to designate an electric lamp, for example, by its impedance. Another is that a device such as a motor which converts energy, but does not simply dissipate it as heat by virtue of resistance, cannot be represented as a fixed impedance; its effective impedance depends on its operating conditions (supply voltage, speed, and mechanical load).

3.11.5 Matching

The matching of d.c. sources was discussed in Section 2.9, and it was shown that the maximum power which could be delivered to a load by a source of voltage V_0 and resistance r had the value $V_0^2/4r$ and was obtained when the load resistance R_L had the value r; a similar condition applied to a current source. There is a counterpart to this matched condition which applies to a.c. sources. In general an a.c. source, unless it is taken to be ideal, has an internal impedance z which corresponds to the resistance r for the d.c. case and which we may write as $r + jx$ where

x is as usual positive for an inductive reactance and negative for capacitance. For a current source we might use instead an admittance y and write it as $g + jb$, but since its behaviour is exactly analogous (as was seen for d.c. in Section 2.9), it is sufficient to keep to impedance and consider only a voltage source. The internal impedance of an a.c. source is often known as its **output impedance**, a term which we shall encounter again in later chapters. For any pair of terminals in an a.c. circuit to which a load may be connected the corresponding quantity for the equivalent source is simply the Thévenin impedance.

The load which is connected to an a.c. source can be represented in general by an impedance Z_L or $R_L + jX_L$, as shown in Fig. 3.21. The question now is: if R_L and X_L can be independently chosen, what should be their values in order to maximize the power delivered to the load (i.e. to R_L, because X_L cannot consume power)? It is possible to deduce the answer to this by finding the current I in the circuit, hence the load power $I^2 R_L$, and differentiating the latter with respect to R_L and X_L in

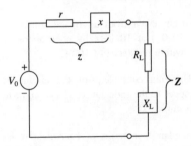

Fig. 3.21

A general a.c. voltage source and its load.

turn. However, it is simpler to use the following argument. Whatever the value of R_L, its power can clearly be maximized by maximizing the current in it. For given r and x this is achieved by using our choice of X_L to cancel the reactance x, i.e. by choosing $X_L = -x$. This done, reactance plays no part in determining current or power, so we have in effect a resistive circuit as in the d.c. case, for which the power in R_L is a maximum for $R_L = r$. It therefore follows that the matched condition for the a.c. case is that $R_L = r$ and $X_L = -x$; in other words a source having impedance z is matched by a load Z_L such that

$$Z_L = z^*. \tag{3.59}$$

The value of the maximum power is, as for the d.c. case, given by $V_0^2/4r$.

It is important to bear in mind that this result applies *only* if R_L and X_L can be separately chosen. If, on the other hand, we need to maximize the power into a load which is restricted to resistance R_L only, then it can be shown that maximum power is delivered to the load when its resistance is equal to the modulus of the source impedance, i.e. when

$R_L = z = \sqrt{(r^2 + x^2)}$. It also follows from this result that if the reactance of the load has some fixed value X_L then maximum power occurs when $R_L = \sqrt{\{r^2 + (x + X_L)^2\}}$.

Problems

3.1 Find the values of capacitance and inductance which would give a reactance of magnitude 100 Ω (a) at 50 Hz and (b) at 500 MHz.

3.2 Two sinusoidal voltages are given by $10 \sin 100\pi t$ V and $20 \cos(100\pi t - \pi/6)$ V respectively. Find (a) their r.m.s. values, (b) their frequency, (c) the phase angle between them, (d) the r.m.s. value of their sum, and (e) the phase angle of their sum relative to the first.

3.3 Find the current taken from a voltage source of (r.m.s.) voltage $20\lfloor 0° \rfloor$ at 10 kHz by a capacitor of 100 nF in series with a resistance of 100 Ω.

3.4 Find the Thévenin equivalent of the circuit of Fig. 3.6 with respect to the inductance. Hence confirm the answer to Example 3.2.

3.5 Find the impedance, in terms of equivalent resistance and reactance, of a capacitance C in parallel with a resistance R at frequency ω.

3.6 Find the impedance of a device which takes current $(2 + j1.5)$ A from an applied voltage $(20 + j10)$ V.

3.7 An impedance $10\lfloor 30° \rfloor$ Ω is connected in parallel with another of $20\lfloor 45° \rfloor$ Ω. Find the impedance of the combination.

3.8 Show that the three elements connected as a star in Fig. 3.6 cannot be realized as passive elements in a delta connection at the same frequency.

3.9 The circuit shown can represent a transmission line supplying current from the voltage source V to the resistance R. Show that the voltage across R is given by

$$V_R = \frac{V}{1 + j\omega L/R - \omega^2 LC}.$$

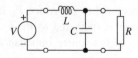

3.10 A current I is supplied to the circuit of Fig. 3.7. Find an expression for the current in the capacitance.

3.11 A certain transformer when unloaded takes a current of 1 A lagging at 75° from a 240 V source. Find its effective susceptance and conductance.

3.12 A coil has inductance 2 mH and resistance 8 Ω. Find the parallel elements which represent it correctly at 10 kHz.

3.13 Two coils have identical self-inductances L, coupling coefficient k and negligible resistance. One is short-circuited. Show that the impedance between the terminals of the other at frequency ω is $j\omega L(1 - k^2)$.

3.14 The two coils of Problem 3.13 are connected in parallel. Show that the effective inductance of the pair is $\frac{1}{2}L(1 \pm k)$.

3.15 Find the power dissipated in the coil of Problem 3.12 when it is connected to a 24 V source, at d.c., at 50 Hz, and at 10 kHz.

3.16 Find the instantaneous maximum power delivered to the coil in Problem 3.15, and its power factor, at 10 kHz.

3.17 Find the power dissipated in the circuit of Fig. 3.8 when the applied voltage is 240 V at 50 Hz and at 500 kHz. [Hint: use the results of Example 3.4.]

3.18 Find the power and reactive volt-amperes absorbed by the device described in Problem 3.6.

3.19 Find the capacitive reactance which in parallel with the impedances shown in Fig. 3.20 would raise the power factor of the circuit to 0.95. [Hint: use the results of Example 3.8.]

3.20 If the circuit in Fig. 3.18 refers to frequency ω, at what frequency would the load Z_L receive maximum power and what would be the value of that maximum?

4 Four-terminal networks

4.1 General

The previous two chapters have outline the principles and methods by which linear circuits may be analysed in the steady state, whether d.c. or a.c. In this chapter we consider in some detail the steady-state behaviour of a very important class of circuits which are often called four-terminal networks, and sometimes **two-ports**, the latter name referring to the fact that a pair of terminals may be regarded as a port, or access, to a circuit by which other things may be connected to it. As we shall see, in these circuits it is very often the case that one terminal pair can be regarded as an input and the other as an output.

Until now we have taken circuits to be essentially collections of branches and meshes containing sources and passive elements, with no attempt to define a number of terminals. Indeed, since we have referred to terminals variously as the end points of components in general, and sometimes of sources in particular, or of any passive branch or combination of elements, it might seem that a pair of terminals, in the abstract sense, could be *any* two points of a circuit; but in fact so wide a meaning would be almost useless, for the essential property of what we mean by a terminal pair is that the same current enters one terminal as leaves the other. Thus the two ends of any passive element or of any source can be considered a terminal pair, as can two points by which current may enter and leave a combination of connected elements. If a combination has only two such points available for connection it might be known as a 'one-terminal-pair' circuit or one-port.

In practice the essential parts of any electrical circuit are first a source, and secondly a 'load' which is the ultimate destination of the energy or information and which might be, for example, an electric lamp or a motor or a loudspeaker or a printer. The two may be connected by no more than the switch necessary to start and stop the current, but more often there is an intervening circuit: a transformer perhaps, or an amplifier, or simply a distribution system. This circuit must evidently have at least two terminal pairs, one, the **input**, for the source and another, the **output**, for the load. This chapter deals with such circuits,

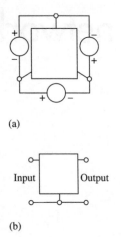

(a)

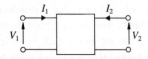

(b)

Fig. 4.1

Three-terminal networks (a) with three inputs, and (b) with input and output.

although for generality we shall also allow the possibility that both terminal pairs may be connected to sources. Not all of the circuits to be considered are entirely passive between input and output; in Section 4.4, we consider those which contain controlled sources (Section 1.6.4) and are used to represent transistors.

In some circuits, for example the three-terminal networks of Fig. 2.12, it is not obvious that terminals can be paired. However, it is legitimate to regard any terminal as common to two pairs. In that figure we could therefore imagine three sources connected to the circuit, such that each terminal was common to two sources as in Fig. 4.1(a), and this is in fact an arrangement used in three-phase circuits (Chapter 6). Equally well, we may imagine an input pair and an output pair sharing a common terminal as in (b); this is how we are able in later sections to treat the important class of three-terminal electronic devices, in particular transistors, as four-terminal networks.

4.2 Network parameters

4.2.1 Impedance parameters

Consider the very general picture of a four-terminal network shown in Fig. 4.2, where subscript 1 refers to the input and subscript 2 to the output. For the sake of symmetry the current I_2 is shown entering the network by the positive terminal, as if the output were connected to a source of voltage V_2; this is quite a common convention, but has no physical significance: the output is more usually connected to a load, which may be a two-terminal impedance or another four-terminal network. Choosing the opposite reference direction for I_2, which is sometimes done, would simply change the sign of every coefficient of I_2 in the equations which follow.

If the actual circuit were known, we could analyse it by the methods of previous chapters. With mesh analysis, for example, the equations could be quickly reduced to the following two:

$$V_1 = z_{11}I_1 + z_{12}I_2$$
$$V_2 = z_{21}I_1 + z_{22}I_2$$

(4.1)

Fig. 4.2

The general four-terminal network.

or, in matrix form, $[V] = [z][I]$. Here by a convention often used in this context, the lower-case symbol z has been used for impedance; we shall also in this chapter drop the bold notation for complex quantities, on the understanding that in general these are implied throughout. The subscript notation is that of Section 2.6, and if the actual circuit contained only two meshes the values of z could be found by inspection as described in that section.

The significance of eqns 4.1 is not simply that the two currents can be found if the voltages are given, or vice versa; it is that these same equations remain true *whatever* may be connected to the input and output terminals. The coefficients z are constants of the network and are called its **impedance parameters** or z parameters. Given any two of the four variables V_1, V_2, I_1 and I_2, the other two can be found from the equations. Alternatively, additional equations can take the place of one or both of the given quantities: for example, if it is known that the input is connected to a source of voltage V_S and impedance Z_S then

$$V_1 = V_S - I_1 Z_S,$$

and a similar equation would apply to the output were that too connected to a source. But if the output is known to be connected to a load impedance Z_L then

$$V_2 = -I_2 Z_L,$$

in which the negative sign arises from the chosen directions. (The case of both terminal pairs connected to passive loads, giving two such equations, does not arise since we should naturally then find that only the trivial solution $V_1 = V_2 = 0$, $I_1 = I_2 = 0$ would satisfy the network equations.)

In eqns 4.1 each of the z parameters has its own significance. Let us consider them in turn.

The parameter z_{11} is the ratio V_1/I_1 when I_2 is zero; it is therefore the impedance at the input terminals when the output is open circuit, and it is called the **open-circuit input impedance**.

The parameter z_{12} is the ratio V_1/I_2 when I_1 is zero, and so represents the voltage which would appear between the open-circuit input terminals due to unit current (however it may be produced) at the output. It is called the **open-circuit transfer impedance**.

The parameter z_{21} is the counterpart of z_{12} with the terminals interchanged, and is also therefore a transfer impedance. If the network consists only of linear passive elements (mutual inductance included) then it always turns out that $z_{12} = z_{21}$, which corresponds to the reciprocity condition mentioned in Section 2.6 and confirms that the circuit as a whole shows reciprocity (Section 2.5.5). Four-terminal networks representing electronic devices do *not* produce this result: as we shall see, even their linearized representations do not satisfy reciprocity because they must include controlled sources.

The parameter z_{22} is the ratio V_2/I_2 when I_1 is zero and is therefore the counterpart of z_{11}; it is called the **open-circuit output impedance** (the 'open-circuit' now referring to the input). We should note however that, while z_{11} and z_{22} have perfect mathematical symmetry, the fact that the output is often a load and the input a source gives them different practical significance.

The foregoing definitions can sometimes be used to evaluate the z parameters of a network, if not by inspection, at least more rapidly than formal analysis.

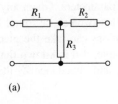

(a)

(b)

Fig. 4.3
Circuits for Examples 4.1 and
4.2.

Example 4.1 Find the four impedance parameters for the circuits of
Fig. 4.3.

(a) Here there are only two meshes and the parameters can be written by
inspection, from the above definitions, as

$$z_{11} = R_1 + R_3$$
$$z_{12} = z_{21} = R_3$$
$$z_{22} = R_2 + R_3.$$

These agree with the example discussed in Section 2.6, except for the
sign of the transfer impedances: this arises simply from the chosen
current directions.

(b) There are now three meshes. The three equations could be written
and combined into two, or the circuit reduced to two meshes by the
delta–star transformation (Section 2.4.3); alternatively the above
definitions can be used together with series and parallel combination
to write

$$z_{11} = \frac{R_1(R_2 + 1/j\omega C)}{R_1 + R_2 + 1/j\omega C} = \frac{R_1(1 + j\omega CR_2)}{1 + j\omega C(R_1 + R_2)}$$

and similarly

$$z_{22} = \frac{R_2(1 + j\omega CR_1)}{1 + j\omega C(R_1 + R_2)}.$$

To find the transfer impedance we note that z_{12} is the open-circuit
voltage at the input when unit current flows at the output; in that case
current division gives the current in R_1 and the voltage across R_1 follows
to give

$$z_{12} = \frac{R_1 R_2}{R_1 + R_2 + 1/j\omega C} = \frac{j\omega CR_1 R_2}{1 + j\omega C(R_1 + R_2)}$$

which is also z_{21}.

4.2.2 Admittance parameters

A circuit with the form of Fig. 4.2 could also yield, either from nodal
analysis or merely by algebraically manipulating eqns 4.1, two
equations in the form

$$I_1 = y_{11}V_1 + y_{12}V_2$$
$$I_2 = y_{21}V_1 + y_{22}V_2$$

(4.2)

of which the coefficients have the dimensions of admittance and are
known as the **admittance** or **y parameters**. They correspond to
the elements of a conductance matrix as in Section 2.6. Here, however,

the form of the equations shows that the matrix of y parameters is the inverse of the z matrix, and vice versa.

Like the impedance parameters, each of the admittance parameters has physical significance. Thus y_{11} is the ratio I_1/V_1 when $V_2 = 0$, that is when the output is short-circuited, and so is known as the **short-circuit input admittance**; y_{22} similarly is the ratio I_2/V_2 for a short-circuited input, known as the **short-circuit output admittance**. In a like way y_{12} is the ratio I_1/V_2 for $V_1 = 0$, that is the current which flows in the short-circuited input due to unit voltage at the output; it is the **short-circuit transfer admittance**, and, for a circuit of linear passive elements only, y_{21} is identical.

Example 4.2 Find the admittance parameters for the circuit of Fig. 4.3(a).

With the output short-circuited, the input admittance is that of R_1 in series with the parallel combination of R_2 and R_3. With G_1 for $1/R_1$, etc., this becomes G_1 in series with $G_2 + G_3$; we therefore have

$$y_{11} = \frac{G_1(G_2 + G_3)}{G_1 + G_2 + G_3}.$$

The output admittance, by symmetry, is then

$$y_{22} = \frac{G_2(G_1 + G_3)}{G_1 + G_2 + G_3}.$$

With the input short-circuited, the current I_1 due to V_2 is

$$I_1 = -I_2 \frac{R_3}{R_1 + R_3} = -y_{22}V_2 \frac{R_3}{R_1 + R_3}$$

which gives

$$y_{12} = \frac{I_1}{V_2} = -y_{22}\frac{R_3}{R_1 + R_3} = -y_{22}\frac{G_1}{G_1 + G_3} = \frac{G_1 G_2}{G_1 + G_2 + G_3}$$

and y_{21} is obviously given, as we expect, by the same expression.

The duality between admittance parameters and impedance parameters is clear from the descriptive definitions given for them above and in the previous section; in particular, we may note that in the definition of z parameters one or other terminal pair is open-circuited, while for the y parameters one pair is short-circuited; and that in both a transfer parameter relates a voltage at one terminal pair to current at another for the appropriate condition. Because of this dual relationship, the word **immittance** has been coined to signify impedance or admittance. Thus we may in general use the term **transfer immittance** to mean

z_{12}, z_{21}, y_{12}, or y_{21}. By contrast, the parameters z_{11}, z_{22}, y_{11}, and y_{22} which relate current and voltage at the same terminals, whether input or output, are often known generally as **driving-point immittances**.

4.2.3 Transmission parameters

In the foregoing sections we have seen that a four-terminal network can be described equally well by impedance parameters which allow the two voltages to be expressed in terms of the two currents, and by admittance parameters for which the reverse is true. Parameters to describe the network are not restricted to these two forms, however, for any two of the four quantities V_1, V_2, I_1, I_2 can be chosen as the independent variables on which the other two depend; since there are six such choices, we may consider four more, for each of which a new set of parameters can be defined. A choice which is often useful is that which expresses the input variables V_1 and I_1 in terms of the output, so that the circuit equations are cast in the form

$$V_1 = a_{11}V_2 + a_{12}I_2$$
$$I_1 = a_{21}V_2 + a_{22}I_2. \qquad (4.3)$$

The coefficients a are then known as the **transmission parameters**. The inverse relations to these express the output quantities in terms of the input, thus:

$$V_2 = b_{11}V_1 + b_{12}I_1$$
$$I_2 = b_{21}V_1 + b_{22}I_1.$$

The coefficients b are known simply as the **inverse transmission parameters**, and the matrix $[b]$ is clearly the inverse of $[a]$. We should note that, unlike the z and y parameters, the a and b parameters have mixed dimensions: for example, b_{11} and b_{22} are dimensionless ratios while b_{12} represents a resistance and b_{21} a conductance. We may deduce from this that reciprocity, which requires that in a passive four-terminal network $z_{12} = z_{21}$ and $y_{12} = y_{21}$, does not require the same symmetry in the matrices $[a]$ and $[b]$. It can be shown that for these the requirements of reciprocity are that the determinant of the matrix has value -1; that is

$$a_{11}a_{22} + a_{12}a_{21} = -1 \qquad (4.4)$$

and similarly for the b parameters.

To find the transmission parameters of a circuit it is possible to assign each a physical definition as in the previous sections. Consider the circuit of Fig. 4.3 (a), for example. Equations 4.3 show that the coefficient a_{11} is the ratio of input to open-circuit output voltage, which by inspection is

$$a_{11} = (R_1 + R_3)/R_3.$$

The coefficient a_{12}, on the other hand, is the ratio V_1/I_2 when $V_2 = 0$, i.e. with the output short-circuited. The total impedance between the input terminals is then R_1 in series with the parallel combination of R_2 and R_3, and from this I_1 can be written in terms of V_1; by current division between R_2 and R_3, I_2 follows and gives

$$a_{12} = -\frac{R_1R_2 + R_2R_3 + R_3R_1}{R_3}.$$

For a_{21} we need the ratio I_1/V_2 for $I_2 = 0$, and since then V_2 is simply I_1R_3 we have

$$a_{21} = 1/R_3.$$

Finally, a_{22} is the ratio I_1/I_2 with the output short-circuited, and current division gives it immediately as

$$a_{22} = -(R_2 + R_3)/R_3.$$

From these four expressions it is readily confirmed that the reciprocity condition of eqn 4.4 is satisfied.

Since any set of parameters can be expressed in terms of any other, the a parameters could alternatively be found by rearranging a set of equations for which the coefficients are already known. Thus in the case above the z parameters found in Example 4.1(a) could be used to yield the a parameters after rearranging eqns 4.1 in the form of eqns 4.3.

Quite often the transmission eqns 4.3 are written with the notation A, B, C, D instead of $a_{11}, a_{12},...,$ and with the assumed direction of I_2 reversed. The parameters B and D then carry the opposite sign to that of a_{12} and a_{22}. This convention, which lacks the symmetry of the original choices but is in practice more appropriate when the output is connected to a passive load, has particular advantage when applied to networks connected in cascade (Section 4.7) and is common in some areas of electrical engineering. The corresponding inverse transmission parameters are then known as A', B', C', D'.

4.2.4 Hybrid parameters

The previous three sections have dealt with four of the six possible parameter sets which describe a four-terminal network. Another set is defined by the equations

$$\begin{aligned} V_1 &= h_{11}I_1 + h_{12}V_2 \\ I_2 &= h_{21}I_1 + h_{22}V_2. \end{aligned} \tag{4.5}$$

in which the coefficients are called the **hybrid parameters**. Like the transmission parameters, the h parameters are of mixed dimensions; they also relate to a choice of variables which is mixed as between input and output. However, they are important because they are widely used in circuit models of transistors (Section 4.5.4). The inverse set, and the last of the six, known simply as the **inverse hybrid** or g **parameters**, is defined by the equations

$$I_1 = g_{11}V_1 + g_{12}I_2$$
$$V_2 = g_{21}V_1 + g_{22}I_2. \tag{4.6}$$

from which it can be seen that the matrix of g parameters is the inverse of the h matrix. (The use of the symbol g here must not be confused with its common use for conductance: the g parameters, like the hybrid, have mixed dimensions.)

The h parameters, like the z and y parameters, are often named according to their physical significance, thus:

h_{11} is the short-circuit input impedance;
h_{12} is the reverse open-circuit voltage gain;
h_{21} is the forward short-circuit current gain;
h_{22} is the open-circuit output conductance.

Note that in these the two which describe ratios of voltages and of currents are called *gains*, although we did not apply this term to the similar ratios encountered in the previous section. This is merely a matter of custom; it arises mainly from the use of h parameters to represent transistors (Section 4.5), for which output-to-input (i.e. 'forward') ratios can exceed unity. We return to the idea of gain in Section 4.5.6.

It can be shown that for h parameters the reciprocity condition, satisfied in circuits containing only linear passive elements, is $h_{12} = -h_{21}$ (and $g_{12} = -g_{21}$), with the assumed direction for I_2 as in Fig. 4.2. This relation is not satisfied in amplifying circuits generally, because the linearized form of these must include controlled sources.

4.3 T and Π equivalents

We have seen in the foregoing sections that any four-terminal network of linear passive elements can be represented, so far as its external behaviour is concerned, by three parameters only, since two of the four coefficients needed are related by a reciprocity requirement. It follows that such a network can be represented by an equivalent circuit comprising three components only (here we use component to mean an impedance or admittance, each of which may of course be complex and represent more than one physical element). Two arrangements, shown in Fig. 4.4, are possible: the T equivalent as in (a), and the Π equivalent as in (b). These equivalents are so named by convention, although they take the forms otherwise known as star and delta which were discussed in Section 2.5.3 and the transformation given there can be used to convert from one to the other. The components of the chosen equivalent for an actual circuit can be found by equating expressions for, say, the z parameters in each case and solving the resulting three equations.

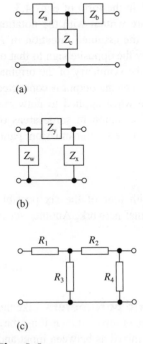

(a)

(b)

(c)

Fig. 4.4
(a) The general T equivalent;
(b) the general Π equivalent;
(c) the circuit for Example 4.3.

Example 4.3   Find the T equivalent circuit for the four-terminal network of Fig. 4.4(c).

We can use the definitions of z parameters given in Section 4.2.1 to write those for the T circuit of Fig. 4.4(a). They are

$$z_{11} = Z_a + Z_c$$
$$z_{12} = z_{21} = Z_c$$
$$z_{22} = Z_b + Z_c$$

from which it follows that for the T circuit generally

$$Z_a = z_{11} - z_{12}$$
$$Z_b = z_{22} - z_{12}$$
$$Z_c = z_{12}.$$

For the circuit of (c), the same definitions yield

$$z_{11} = R_1 + \frac{R_3(R_2 + R_4)}{R_2 + R_3 + R_4};$$
$$z_{12} = z_{21} = \frac{R_3 R_4}{R_2 + R_3 + R_4};$$
$$z_{22} = \frac{R_4(R_2 + R_3)}{R_2 + R_3 + R_4}.$$

Substituting these into the expressions for Z_a etc. then gives

$$Z_a = R_1 + \frac{R_2 R_3}{R_2 + R_3 + R_4};$$
$$Z_b = \frac{R_2 R_4}{R_2 + R_3 + R_4};$$
$$Z_c = \frac{R_3 R_4}{R_2 + R_3 + R_4}.$$

These define the T equivalent of the given circuit. It can readily be confirmed that the values of any set of two-port parameters are the same in both circuits. Although the equivalent has been found here by way of its z parameters, the same results would of course be obtained by following a similar process for any other set.

4.4 Four-terminal circuits with controlled sources

In following sections we shall see that transistors as amplifying devices can be represented for many purposes as four-terminal networks which include controlled sources (Section 1.6.4). One such circuit which is used to represent a bipolar junction transistor is shown in a general form in Fig. 4.5(a); it contains a current source having a value fixed by the

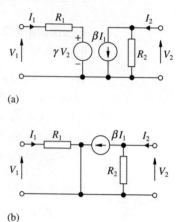

(a)

Fig. 4.5

Four-terminal networks with
controlled sources: (a) a general
version; (b) a simplified equivalent for a junction transistor.

(b)

input current according to a constant of proportionality β, and a voltage
source fixed by the output voltage according to a constant which we
shall here call γ. To find any set of parameters for this circuit we may
apply the same methods as were used above for purely passive circuits.
The impedance parameters, for example, can be found as follows from
the definitions in Section 4.2.1.

The input impedance z_{11} is the ratio V_1/I_1 for $I_2 = 0$; for this
condition the voltage V_2 is evidently given by $-\beta I_1 R_2$. Kirchhoff's
voltage law around the input mesh then gives

$$V_1 = I_1 R_1 + \gamma V_2 = I_1(R_1 - \beta \gamma R_2)$$

and z_{11} is therefore $R_1 - \beta \gamma R_2$. The transfer impedance z_{12} is the ratio
V_1/I_2 for $I_1 = 0$, in which case the current source has zero value and can
therefore be taken to be an open circuit, while V_1 is simply γV_2; a
current I_2 then produces a voltage $V_2 = I_2 R_2$ at the output terminals, V_1
becomes $\gamma I_2 R_2$, and z_{12} is therefore γR_2. On the other hand the transfer
impedance z_{21} is the ratio V_2/I_1 for $I_2 = 0$, i.e. with the output open
circuit, in which case V_2 becomes $-\beta I_1 R_2$ as before, so $z_{21} = -\beta R_2$.
Finally the output impedance z_{22} is defined for $I_1 = 0$, so the current
source again becomes an open circuit and z_{22} is therefore simply R_2.
Collecting these results, we have

$$z_{11} = R_1 - \beta \gamma R_2$$
$$z_{22} = R_2$$
$$z_{12} = \gamma R_2$$
$$z_{21} = -\beta R_2.$$

Similarly, the definitions of the hybrid parameters (Section 4.2.4) would
yield in the present case

$$h_{11} = R_1$$
$$h_{22} = 1/R_2$$
$$h_{12} = \gamma$$
$$h_{21} = \beta.$$

Since $z_{12} \neq z_{21}$ and $h_{12} \neq -h_{21}$ (if we discount the possibility that $\gamma = -\beta$) it is clear from both sets of parameters that the circuit does not satisfy the reciprocity conditions of Section 4.2: although it is linear, the controlled sources mean that it is not bilateral and the reciprocity principle therefore fails (Section 2.5.5). The significance of this can be better seen by taking the case $\gamma = 0$, giving the circuit of Fig. 4.5(b) which is a common version based on the fact that for junction transistors the quantity we have called γ is often negligibly small (Section 4.5.4). In this case we have z_{12} and h_{12} both zero, and this implies that a current or voltage applied to the output terminals would produce no result at all at the input; on the other hand the value of h_{21} shows immediately that there is a short-circuit gain of β from input to output. A circuit like this, with transfer parameters which are zero in one direction, is sometimes said to be **unilateral**.

In general, circuits containing internal sources are not bilateral and do not satisfy the reciprocity conditions; amplifying circuits fall into this category and are discussed in the following sections. Although the induced e.m.f.s. which arise from mutual inductance (Section 3.8) could be regarded as controlled sources, mutual inductance is bilateral in behaviour and obeys the reciprocity principle.

Circuits which contain internal sources of any kind are properly defined as **active** (Section 1.6). The term is most often used, however, for circuits in which these are controlled sources (and for the electronic devices which they represent). The reason for this is that independent sources, wherever they may be connected, need not be regarded as an integral part of any circuit: when connected to a linear passive circuit they do not affect the intrinsic reciprocity of its parameters, and the effect of any one such source may be found, according to the principle of superposition, with the others removed (Section 2.5.5). Controlled sources are different, for each acts by definition so long as its controlling variable has a non-zero value. In the example of Fig. 4.5 considered above, the current source βI_1, for example, must act whenever I_1 flows and it is incorrect to remove it except under conditions for which $I_1 = 0$; similarly, the voltage γV_2 always acts unless V_2 happens to be zero. Thus the references in Chapter 2 to the removal of sources for the purposes of circuit reduction apply *only* to independent sources; controlled sources must in general be retained. With this precaution, and except for the principle of reciprocity, the methods of analysis described in earlier chapters may be safely applied to active circuits.

In general controlled sources do not on their own represent physical entities: almost all are simply convenient attributes by which to describe the characteristic behaviour of certain devices, and in particular of amplifying devices based on the transistor (Section 4.5). We should note too that these devices are in fact highly non-linear, whereas two-ports containing controlled sources and passive elements are essentially linear; such circuits can represent the actual devices only in certain respects, and these are outlined in later sections.

In the previous section it was seen that any two-port having only linear passive elements could be represented by an equivalent circuit containing only three such elements; more generally, any linear two-port can be represented by three passive elements and one controlled source, or by two of each.

4.5 Small-signal analysis of electronic circuits

It is difficult to offer a universal definition of an electronic device. The term is often taken to mean a device which handles information in an electrical form (whether as a continuous signal or digitally, as a series of pulses), or one which is based on the properties of semiconducting materials. Common usage includes exceptions to both of these meanings: for example, a diode is an electronic device which can rectify an a.c. source to produce d.c. for any application, whether or not dealing with information; lasers and microwave devices do not necessarily use semiconductors. In this book we confine our attention to those semiconductor devices which can readily be described in electrical circuit terms, and here we consider in particular the behaviour of the transistor in its so-called analogue applications, as a four-terminal network with the property of amplification.

4.5.1 Transistor characteristics

The transistor is basically a three-terminal device in which one terminal (or electrode) controls the flow of current between the other two. So far as circuit analysis is concerned, the simpler of its two main forms is the **field-effect transistor** or FET, of which there are several variants. In all of them a channel of p- or n-type semiconductor carries a current between two electrodes known as the **source** (S) and **drain** (D), the current being controlled by the voltage applied to an electrode known as the **gate** (G) which itself takes virtually no current in normal operation. (The name 'source' in this context refers to the flow of charge carriers, and has no direct relation to its more usual meaning in circuit theory.) This behaviour can be conveniently seen from the voltage–current characteristics of the device, which are typically as shown in Fig. 4.6 for the version known as a junction FET or JFET with an n-type channel; characteristics for other versions differ mainly in the range of v_{GS}, the voltage between gate and source, needed to vary the current from zero to a maximum. (A summary of the types of transistor and their symbols is given in Appendix C.) We need not here discuss the physical processes which give rise to these characteristics.

The other main type of transistor is the **bipolar junction transistor** or BJT, so called because current flows across two junctions which separate three layers of p- and n-type material, between

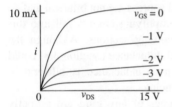

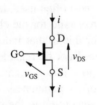

Fig. 4.6
Characteristics for an n-channel JFET.

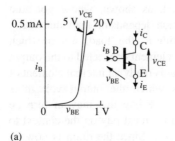

(a)

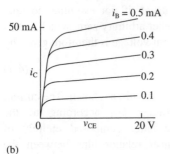

(b)

Fig. 4.7

(a) Input and (b) collector characteristics for a bipolar junction transistor.

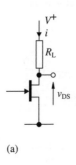

(a)

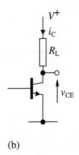

(b)

Fig. 4.8

Transistors with load resistance.

electrodes known in this case as the **emitter** (E) and **collector** (C). The arrangement can be either p-n-p or n-p-n; in each case the controlling electrode is that connected to the middle region, known as the **base** (B). The base of the BJT corresponds to the gate of the FET, but from the circuit point of view the main difference between the two kinds of transistor is that the base, in contrast to the gate, takes a significant (although relatively small) current when a voltage is applied to it under normal operating conditions; hence the BJT needs another set of characteristics relating base voltage and current in order to describe its behaviour completely. The two sets of characteristics for the BJT are usually presented in the form shown for the n-p-n version in Fig. 4.7; the p-n-p form differs only in the relative polarities of the terminals. The curves shown in (a) are called the **base** or **input characteristics**. Because their dependence on v_{CE} is usually weak, so that they lie close together, they are often shown as a single characteristic. This is essentially the characteristic of a diode (Section 11.1), formed here by the base and emitter, and shows that for a normal range of current the voltage v_{BE} is a fraction of a volt, typically 0.6–0.7 V. The curves in (b) are called the **collector** or **output characteristics**. The shape of these characteristics as shown in the figure, and their range of values, are meant merely to be typical. In practice they vary considerably according to the purpose for which the transistor is designed; they also vary from sample to sample because of the extreme sensitivity of the fabrication processes, whether for integrated circuits or for separate transistors.

The variables specified in Figs 4.6 and 4.7 make it clear that the source and the emitter respectively have been chosen as the common terminal, so that the input, or controlling, terminals are evidently assumed to be G and S or B and E, while the output terminals are D and S or C and E. When transistors are actually used in this way they are described as **common source** or **common emitter**. However, these modes of connection are not essential, and other choices can be made for particular purposes; it follows that other sets of characteristics could be presented which might be more convenient. But since any set of characteristics must remain true, the information it contains can be applied to any circuit connection; those shown in the figures are widely used and there is no need to rehearse all the possibilities.

4.5.2 Load line and operating point

The characteristics of Figs 4.6 and 4.7 show that a voltage or current applied to the input or controlling terminal of a transistor determines the current which flows between the other two in response to a d.c. voltage, making the device in effect a controlled current source, but it is not clear as yet what is to be regarded as the output. How, for example, might we obtain an output voltage in proportion to the controlled current? The answer to this is to pass the current through a resistance in series with the transistor, as shown in Fig. 4.8, known as the **load resistance** R_L, which produces a voltage drop in proportion to the current i or i_C. Consider the case of the circuit with a JFET shown in (a): the drain can now be taken to be an output terminal, as shown, whose potential reflects the input in some proportion which depends on the value of R_L. (There is a slight ambiguity here in calling R_L the 'load', a term which has previously been used for an external element connected to the output terminals; now we are applying it to what is really an internal element of the circuit, although it is that across which the output voltage is developed. The output terminals in a case like this may in turn be connected to another, external, load and we must rely on the context to clarify the distinction whenever necessary.) Since the drain is now an output terminal, the drain-to-source voltage v_{DS} may conveniently be regarded as the output voltage: it is usual, if not essential, to take potentials relative to 'ground' or some other arbitrary zero, which in this circuit we take to be the source. The output voltage is therefore given by

$$v_{DS} = V^+ - iR_L \tag{4.7}$$

where V^+ is the d.c. voltage applied to the circuit, as shown. The current i depends on the gate-to-source voltage v_{GS} according to the characteristics of Fig. 4.6. Were the latter composed entirely of equispaced straight lines, showing linear relationships between the variables, we could easily express this by some equation and combine it with eqn 4.7 in order to relate the output v_{DS} to the input v_{GS}. In fact, however, the characteristics evidently represent relationships which are non-linear (for some transistors more so than those shown); in that case a graphical approach is more helpful, in which we combine the linear relation of eqn 4.7 (which is based simply on Ohm's law applied to R_L) with the characteristics. The equation can be rearranged as

$$i = -v_{DS}/R_L + V^+/R_L$$

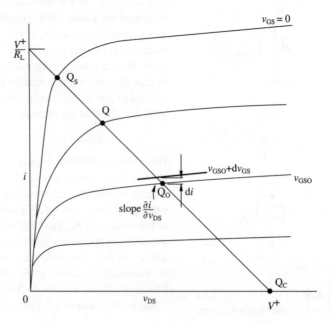

Fig. 4.9
JFET characteristics and load
line.

and therefore represents the line shown in Fig. 4.9, known as the **load line**. The intersection of this line with the characteristic for a particular value of v_{GS}, for example the point Q in the figure, gives the values of i and v_{DS} which result, since these values satisfy both eqn 4.7 and the characteristic itself. (This graphical device is by no means particular to transistors or even to electronic circuits: it is applicable to any case of a non-linear element combined with a linear.) If v_{GS} is varied Q merely moves along the load line.

The point Q can clearly be set by means of v_{GS} to two extreme positions on the load line: in the figure one is shown as Q_C, where the input voltage v_{GS} is such as to reduce i to zero, a state known as **cut-off** in which v_{DS} takes its maximum value, V^+; and the other is Q_S, where i is at its maximum, which occurs when v_{DS} reaches its minimum value (which may be less than a volt) and is known as **saturation**. These extreme states are important in applications where transistors are used as switches or in digital electronics: for in the former state the transistor is in effect an open circuit and can be said to be 'off', while in the latter state it is virtually a short circuit and is 'on'. Here we consider instead the use of the transistor in **analogue** applications where electrical quantities are for most of the time varying smoothly (like, for example, the a.c. quantities which were considered in Chapter 3) rather than suddenly switching between different values, or occurring as pulses. Typical of this kind of application is the amplification of a continuous input waveform of, say, voltage (often representing some other physical quantity such as the sound received by a microphone, or the temperature of a thermocouple) to provide an output which is ideally a faithful reproduction of the input, varying in proportion to it but of greater amplitude. In electronics such varying quantities, usually conveying information, are in general called **signals**.

If the input signal to a transistor is liable to vary in either direction it follows that we should choose a sensible starting position for the point Q in Fig. 4.9 which is well removed from both cut-off Q_C and saturation Q_S, for if the output reaches either of these limits in response to the input it cannot go further and the amplification would fail; a point such as Q_O, midway between the two limits, is clearly the choice which allows the greatest range of amplification in both directions. It is also clear that in order to start from the point Q_O the input signal must be superimposed on that value of the input voltage, shown as v_{GSO} in the figure, which will set the output to Q_O; further, the output signal is itself superimposed on the starting value at Q_O. The point Q_O is known as the **operating** or **quiescent point**, and corresponds to a set of d.c. values for the variables i, v_{GS} and v_{DS}; they can be fixed by the choice of the d.c. supply voltage V^+, the load R_L and the input voltage v_{GSO}, which is known as a **bias** voltage and can be provided, for example, from the same d.c. supply by means of a potential divider.

Similar arguments apply to the BJT, as shown in Fig. 4.8(b), when a load line is combined with the characteristics of Fig. 4.7(b); the only difference is that the input quantity could now in principle be taken either as the voltage v_{BE} or as the current i_B (the two being related by the characteristics of Fig. 4.7(a)). For our present purpose we may choose the latter because its relationship to i_C is more nearly linear and because the range of v_{BE} required is very small and critical. This choice means that we regard the BJT, unlike the FET, as a current-controlled device and that it is i_B rather than v_{BE} which is specified at the operating point, and which can be said to provide the required bias. In practice, however, it is better to specify the operating point by a value for i_C (or i_E, which is nearly the same) and biassing arrangements are usually designed to do that. The reason for this preference is that the factor which relates i_C to i_B, although nearly constant for a given specimen and always large, is highly unpredictable; and it is i_C which provides the output and which must therefore be set to a suitable point within its range. In any case, when we come to deal with voltage amplification we must naturally take the input to be a voltage rather than the base current. The fact that v_{BE} cannot vary much from about 0.7 V or so under normal conditions does not mean that it cannot be varied at all, but simply that the input arrangements should not allow voltages very different from this to be applied directly between base and emitter.

Various arrangements can be used to provide the bias and to set the operating point for transistors generally; these we consider only incidentally, for we are more concerned with the relationship between output and input signals, which are simply changes, with a variation in time as yet unspecified, superimposed on the operating point.

It can be seen from Fig. 4.9, and correspondingly from Fig. 4.7(b), that the greatest effect of the input on the output is obtained when the load line between the limiting points Q_C and Q_S lies in the region of more or less horizontal characteristics; this is known as the **pinch-off region** for the FET and as the **active region** for the BJT. It was pointed out above, in introducing the idea of a load line, that the characteristics cannot be assumed to be linear in general; they are therefore not necessarily evenly spaced in these regions, so it follows that a change in the input does not in general produce a proportional change in the current and hence in the output voltage. Nevertheless we may legitimately regard the changes as linearly related if they are small enough, and we are therefore led to the idea of **small signals** or increments, superimposed on the operating-point values, which can be simply related once the operating point is chosen.

4.5.3   Incremental relations for the FET

For the FET, the current i is determined only by the gate-to-source voltage and, to some extent, the drain-to-source voltage. We may therefore write

$$i = f(v_{GS}, v_{DS}) \tag{4.8}$$

where f is some function which represents the characteristics of Fig. 4.6, whatever their precise form may be. Mathematically, a small increment in i may then be written as

$$di = \frac{\partial i}{\partial v_{GS}} dv_{GS} + \frac{\partial i}{\partial v_{DS}} dv_{DS} \qquad (4.9)$$

which means that it is linearly related to the increments in v_{GS} and v_{DS} by the partial differential coefficients $\partial i / \partial v_{GS}$ and $\partial i / \partial v_{DS}$. The values of these in any given case will depend on the operating point chosen, but once the latter is known they may be found from the characteristics in the following way. Since $\partial i / \partial v_{GS}$ is defined at constant v_{DS}, it must be the ratio of a small intercept di in i along a vertical line through the operating point, to the corresponding increment dv_{GS} in v_{GS}, as shown in Fig. 4.9. This coefficient is known as the **mutual conductance** or transconductance of the transistor and is usually given the symbol g_m. The other, $\partial i / \partial v_{DS}$, being defined at constant v_{GS}, is the ratio of the small change in i to that in v_{DS} for an element of a given characteristic; in other words the slope of the characteristic at the operating point chosen, as shown in the same figure. Although this coefficient also has the dimensions of conductance, it is by convention usually expressed in terms of its inverse, which is called the **drain resistance** and given the symbol r_d. Equation 4.9 can now be written

$$di = g_m dv_{GS} + (1/r_d) dv_{DS}.$$

At this stage, for convenience, we drop the differential notation and simply write i for di, etc. But in doing so we must bear in mind that we are still dealing with relations between small increments; to help in this it is common practice to make all subscripts lower case: thus v_{ds} now signifies a small change in v_{DS}. Finally, then, eqn 4.9 becomes

$$i = g_m v_{gs} + (1/r_d) v_{ds}. \qquad (4.10)$$

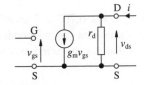

Fig. 4.10
Incremental equivalent circuit for the FET.

This is a linear equation, but at first sight it bears little relation to any of the pairs of equations we have encountered for four-terminal networks in general. Nevertheless it actually describes correctly the behaviour of the circuit shown in Fig. 4.10, which contains only one finite passive element in addition to the controlled source. As a four-terminal circuit it is somewhat artificial because its input is not connected to the rest of the circuit but appears only as the voltage v_{gs} which controls the current source. However, it is our first example of an **incremental** or **small-signal** circuit: it is a kind of equivalent circuit or **model**, whose currents and voltages represent the small changes which may occur in the actual transistor. We should note that, as with all four-terminal networks, we cannot find these currents and voltages without knowing what is connected to the terminals, i.e. to the transistor itself; for example, no mention has been made in this section of a load resistance (Section 4.5.2) which may evidently relate any changes in i and v_{DS}. We return to this question in Section 4.5.5 below.

Example 4.4 A certain n-channel JFET has characteristics which in the pinch-off region can be approximately described by the relation

$$i = 8\left(1 + \frac{v_{GS}}{2}\right)^2$$

for $v_{GS} \leq 0$, when i is in milliamperes and v_{GS} in volts. Find (a) the cut-off value of v_{GS}, (b) the bias value of v_{GS} needed to give half the maximum current at the operating point, (c) the values of mutual conductance and drain resistance at this operating point, and (d) the minimum voltage v_{DS} for a load of 2.2 kΩ and a d.c. supply voltage of 20 V.

(a) The cut-off value of v_{GS} is that value which gives $i = 0$, and is therefore given by $v_{GS} = $ **−2 V**.

(b) The maximum current flows when $v_{GS} = 0$ and is evidently 8 mA. To set the operating point at half this we need

$$\left(1 + \frac{v_{GS}}{2}\right)^2 = \frac{1}{2}$$

which gives $v_{GS} = \sqrt{2} - 2 = $ **−0.586 V** as the necessary bias.

(c) The mutual conductance is by definition

$$g_m = \frac{\partial i}{\partial v_{GS}} = 16\left(1 + \frac{v_{GS}}{2}\right)\Big/2 = 8\left(1 + \frac{v_{GS}}{2}\right)$$

which with the value of v_{GS} from (b) gives $g_m = 8 \times 0.707 = $ **5.66 mS**. The expression given for the characteristics shows that i is independent of v_{DS} in the pinch-off region, which implies that the characteristics are flat and hence that $r_d = $ **0** .

(d) The minimum value of v_{DS} occurs when the maximum current of 8 mA flows in the 2.2 kΩ load, giving $v_{DS} = 20 - (8 \times 2.2) = $ **2.4 V**. (This corresponds to the saturation point Q_S shown, for more general characteristics, in Fig. 4.9.)

Small-signal circuit models are extremely useful in dealing with analogue behaviour, and we shall see that other examples are more immediately recognizable as four-terminal networks. In practice it would be intolerable if we had to evaluate, graphically or otherwise, parameters like g_m and r_d whenever an operating point was chosen, and often we find that certain values (perhaps quoted by a manufacturer) apply reasonably well over a considerable range of the characteristics. Also, while these circuits strictly apply only to mathematically small increments, they can be applied in practice to signals of appreciable size so long as these are restricted to a region around the operating point where the characteristics are more or less linear.

4.5.4 Incremental relations for the BJT

The approach of the previous section can be applied to the BJT, but now there are two sets of characteristics to be considered, as in Fig. 4.7. Keeping to the variables chosen for that figure, and arbitrarily choosing to express the 'input' voltage v_{BE} and the 'output' current i_C in terms of the other two, we may write

$$v_{BE} = f_1(i_B, v_{CE})$$
$$i_C = f_2(i_B, v_{CE}) \tag{4.11}$$

and small increments in these can be expressed, as before, in terms of partial differential coefficients, thus:

$$dv_{BE} = \frac{\partial v_{BE}}{\partial i_B} di_B + \frac{\partial v_{BE}}{\partial v_{CE}} dv_{CE}$$
$$di_C = \frac{\partial i_C}{\partial i_B} di_B + \frac{\partial i_C}{\partial v_{CE}} dv_{CE}. \tag{4.12}$$

This pair of equations may be compared with eqns 4.5: these too express input voltage and output current in terms of the other two quantities; it follows that the coefficients in eqns 4.12 are what were called hybrid parameters. They are therefore given the symbol h, with subscripts which identify the significance of each; as for the FET in the previous section, we now drop the differential notation and write equations eqns 4.12 as

$$v_{be} = h_{ie}i_b + h_{re}v_{ce}$$
$$i_c = h_{fe}i_b + h_{oe}v_{ce}. \tag{4.13}$$

The h parameters here carry the subscript 'e' to indicate that they relate to the common-emitter connection (i.e. the connection to which Fig. 4.7 applies); the other subscript letters refer to the definitions of Section 4.2.4: thus, h_{ie} is an input impedance (or resistance), h_{re} a reverse voltage gain, h_{fe} a forward current gain, and h_{oe} an output admittance (or conductance). All of these now represent differential coefficients at some chosen operating point, and as in the case of the FET they can be evaluated from the characteristics. (Note that the operating point actually appears twice in Fig. 4.7, since to every point such as Q in (b) there corresponds a point such as P in (a).) Thus, h_{ie} is the inverse slope of a base characteristic as shown in Fig. 4.7(a); h_{re} is the ratio of increments in v_{BE} and v_{CE} measured along a horizontal in the same figure; h_{fe} is the ratio of current increments measured along a vertical in (b); and h_{oe} is the slope of a characteristic in (b).

Continuing the argument as for the FET, we may now ask whether there is a circuit which can represent eqns 4.13. Because the h

parameters we get from the transistor characteristics do not satisfy the requirements of reciprocity (Section 4.2.4), such a circuit cannot contain passive elements only. There are, however, several possible choices containing one or two controlled sources (Section 4.4). One which uses the h parameters directly is shown in Fig. 4.11(a). Because the base

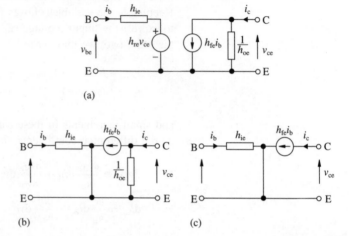

(a)

Fig. 4.11

Incremental equivalent circuits for the BJT: common-emitter h parameters.

(b)   (c)

characteristics of Fig. 4.7(a) depend only weakly on v_{CE} and are often shown as a single curve (Section 4.5.1), it follows that h_{re} is then very small according to its definition above; it is quite commonly neglected, and this gives the circuit of Fig. 4.11(b). Similarly, the collector characteristics of Fig. 4.7(b) are often nearly horizontal in the active region, implying that i_C depends very weakly on v_{CE} and that h_{oe} is negligible. Setting both h_{re} and h_{oe} to zero yields the circuit of Fig. 4.11(c), which is adequate for many purposes.

Other small-signal circuits or models to represent the BJT can be more easily deduced from a different choice of dependent variables than from those which led to eqns 4.13. For example the equations for the currents i_b and i_c lead to a set of admittance parameters (still for the common-emitter connection); by convention these equations may be written in the form

$$i_b = (1/r_\pi)v_{be}$$
$$i_c = g_m v_{be} + (1/r_{ce})v_{ce}. \qquad (4.14)$$

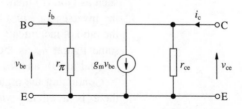

Fig. 4.12

An incremental equivalent circuit for the BJT: common-emitter y parameters.

It can be seen that here again the effect of v_{ce} on i_b has been neglected, making one of the parameters zero (the one corresponding to y_{12} in eqns 4.2), that we have once more a mutual conductance g_m such as was defined for the FET in the previous section, and that as expected the reciprocity condition is not satisfied, since $y_{12} \neq y_{21}$. The circuit which is used to represent eqns 4.14 is shown in Fig. 4.12.

So far we have considered transistors with the common-source or the common-emitter connection for which the characteristics of Figs 4.6 and 4.7 are appropriate. Other connections can of course be used, and lead to other sets of incremental equations and small-signal circuits. Figure 4.13 shows a circuit derived from impedance parameters for the common-base connection of the BJT; it satisfies the equations

$$v_{eb} = (r_e + r_b)i_e + r_b i_c$$
$$v_{cb} = (r_b + \alpha r_c)i_e + (r_b + r_c)i_c. \qquad (4.15)$$

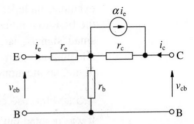

Fig. 4.13
An incremental equivalent circuit for the BJT: common-base z parameters.

It can be shown that here the dimensionless quantity α is in effect a current gain from emitter to collector, corresponding to the base-to-collector current gain h_{fe} for the common-emitter case. But whereas the latter, which is sometimes known as β (compare the circuit in Fig. 4.5), is typically high — in the order of 100 — the value of α is a little less than unity (which shows that the common-base connection does not amplify current). The two are related by the equations

$$\alpha = \frac{\beta}{1 + \beta}; \qquad \beta = \frac{\alpha}{1 - \alpha}$$

which are derived from their definitions together with Kirchhoff's first law as applied to the three terminals of the transistor.

Although there are many possible choices of circuits and sets of parameters, only a few are commonly used. We should remember that a given transistor can be represented by several different sets of characteristics, which in turn yield many incremental parameters, and that all of these remain valid whatever the connection; thus it is perfectly possible to apply parameters defined for the common-emitter connection, say, to a transistor used with its collector as the common terminal, for example. Equally, all parameters are interrelated: it is possible to convert one set not only to a different set for the same connection (e.g. h_e to y_e), as for any four-terminal circuit, but also to a similar set for a different connection (e.g. h_e to h_b). The choice for any particular application need be dictated only by convenience, and in practice might be dictated mainly by what information was most readily available.

The incremental models so far considered have contained only resistance (or conductance) as passive elements. The reason for this is that we have assumed that the voltage–current characteristics from which they derive are d.c. relationships which remain true however the quantities vary in time. This would not be correct if the transistor showed reactive behaviour. For many practical purposes electronic semiconductor devices have negligible inductance, but transistors have capacitance between their terminals which typically becomes significant at frequencies above 1 MHz or so. More accurate circuit models for transistors therefore include one or more capacitances. High-frequency transistors are designed to reduce these and can operate at frequencies around 100 MHz and over.

Incremental circuits are not the only models which can be used to describe the behaviour of transistors: the Ebers–Moll model, for example, includes a diode to represent the non-linear characteristic of the base-to-emitter junction. Nor are transistors necessarily restricted to small signals; large-signal (and therefore non-linear) operation is not uncommon where appreciable output power is needed, and of course in digital applications where the transistor is essentially a switch.

4.5.5 The use of incremental models

It was pointed out earlier that, as for any four-terminal network, currents and voltages in small-signal models cannot be evaluated without knowledge of what is connected to the input and output. The terminals of a transistor model correspond to the actual terminals of the device, so we must connect to them the incremental representations of the rest of the actual circuit. Because passive elements are entirely linear, their incremental behaviour is no different from their overall behaviour; for if we write

$$v = Ri$$

and R is constant, then

$$\mathrm{d}v = R\mathrm{d}i$$

and so on. The d.c. sources which fix the operating point of a transistor are, however, a different matter: by definition an ideal voltage source cannot change its value, so the increment in its terminal voltage is necessarily zero; likewise the increment of current from an ideal current source is zero. It follows that in any incremental circuit *an ideal d.c. voltage source must be represented by a short circuit, and an ideal d.c. current source by an open circuit.*

Consider, for example, the simple transistor circuit of Fig. 4.8(a) which shows a FET with a load resistance R_L. To construct its incremental version we take the incremental circuit of Fig. 4.10 and add to it the resistance R_L connecting the drain, not to the d.c. voltage V^+ as in the actual circuit, but directly back to the source so that the d.c. source

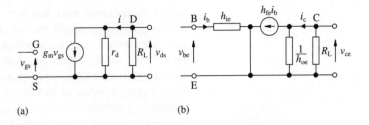

Fig. 4.14
Incremental equivalents of the transistors and loads in Figure 4.8.

(a)   (b)

is replaced by a short circuit; the result is the circuit shown in Fig. 4.14(a), which is the incremental or small-signal equivalent of Fig. 4.8(a). In it the voltage v_{ds} is the output signal and appears across R_L, while the voltage v_{gs} represents the input signal.

For the BJT circuit of Fig. 4.8(b) we must first decide on the model to be used for the transistor itself, say that of Fig. 4.11(b); then we retain R_L connected to the collector but again short-circuit the voltage V^+. The result is the circuit shown in Fig. 4.14(b), which is the incremental, or small-signal, equivalent for the actual circuit of Fig. 4.8(b). Here too the output voltage, in this case v_{ce}, appears across R_L. The input is again taken to be a voltage, namely the incremental voltage v_{be} between base and emitter: this is usual when the aim of the circuit is voltage amplification, although in other contexts the base current may be regarded as the input.

The circuits shown in Fig. 4.14 are *linear*, although they represent the operation of highly non-linear devices, and they are easily analysed in the usual way. The output signals, which we have defined above as the variation of the voltage between the output (drain or collector) terminal and ground, happen in these simple circuits to be respectively v_{ds} and v_{ce} because both source and emitter are at ground potential. In general this may not be so, but the incremental equivalent circuit is always constructed on the same principles; and however the actual output may be defined, it can readily be identified on that circuit. Note from Fig. 4.14 how the load R_L might now be regarded as an external element connected to the output terminals of the transistor model; when, as is not uncommon, an output in Fig. 4.8 is connected to another load which is physically external, then that will simply appear in parallel with R_L in Fig. 4.14.

4.5.6 Gain

In Section 4.2.4 the term 'gain' was first encountered in defining certain hybrid parameters. Its general application to circuits originated in amplification: where an output voltage, say, is an enlarged version of an input voltage (whatever its waveform) and the ratio of output to input voltages therefore exceeds unity, the circuit can be said to have voltage gain. More generally the term may be applied to any ratio of like quantities at two different points: for example, in Section 4.5.4 above h_{re} was defined as a reverse gain, i.e. a ratio of input to output voltages, and was seen to be much less than unity. (It will be as well to note here what will be more evident later, that even in circuits which obey reciprocity a reverse gain is *not* simply the reciprocal of the corresponding forward gain, for the two are usually defined for different terminal conditions.) From now, unless otherwise made clear, we use 'gain' to mean a forward gain, that is the ratio of an output value to the corresponding input value, whether or not it exceeds unity. Although we are here considering active circuits which can amplify, the same terminology can be applied to passive circuits which cannot. Sometimes however a 'gain' less than unity is referred to as an **attenuation** or **loss**; this is more usual when ratios are measured on logarithmic scales (Section 5.2). Because gain relates like quantities at different terminal pairs, whereas transfer immittance (Section 4.2.2) similarly relates unlike quantities, both are examples of **transfer functions**, a rather general term which we shall meet in the following chapter.

Although the present discussion is restricted to small-signal gain, as given by incremental circuit models, the principles remain valid for any circuit which is linear, that is, for which the two quantities vary in proportion. The effect of non-linearity is to make gain dependent on the amplitude of the input signal.

In general the voltage or current gain of a four-terminal network, unlike its parameters, is not a constant, for it depends on the connections to the terminals. Those parameters which were described as gains in previous sections were defined for particular circumstances: h_{fe}, for example, is the (incremental) *short-circuit* current gain for the common-emitter transistor; that is the current which would flow in a short circuit at the output terminals, for unit input current. Figure 4.14(b) shows immediately that when the output is connected to a load R_L, rather than a short circuit, some of the current $h_{fe} i_b$ would flow in h_{oe}, so reducing the output current in R_L and therefore the actual current gain A_i; indeed if the output were open circuit the current gain would necessarily be zero. If, on the other hand, we assume that h_{oe} is zero as in Fig. 4.11(c), then the current $h_{fe} i_b$ must flow in the output and the gain A_i is therefore always h_{fe}, even, in theory, when the output is open circuit (this last being of course impossible in practice, since it would imply infinite output voltage and a collector unconnected to the d.c. supply). If the output of the actual amplifier is connected to an external load R then that appears in parallel with R_L in the incremental circuit and takes a share of its current, with a reduced gain. The question of current gain does not arise for a FET: since the gate current is commonly assumed to be zero, A_i is in theory infinite for a simple circuit such as that of Fig. 4.14(a).

Voltage gain likewise depends on the output conditions. None of the parameters so far applied to transistors represents a forward voltage gain directly (although g_{21} in eqns 4.6 is such a ratio) but it is not difficult to calculate voltage gain in any of the circuits we have met. In the JFET amplifier of Fig. 4.14(a), for example, the output voltage v_{ds} is obviously given by $-g_m v_{gs} R'_L$ where R'_L is the parallel combination of R_L and r_d. It follows that the voltage gain is given by

$$A_v = \frac{v_{ds}}{v_{gs}} = -\frac{g_m R_L r_d}{R_L + r_d} \tag{4.16}$$

in which the negative sign indicates the phase inversion shown by many simple amplifiers, arising from the fact that a positive increment in gate potential causes the drain potential to fall (because the increased current gives a larger voltage drop in R_L). The equation shows that the voltage gain is zero for a short-circuit output and increases with R_L (without limit, in theory, if r_d is assumed to be large enough to ignore; but there are of course various limits to the value of R_L in practice, not least the need for a suitable operating point). Any external resistance connected to the output appears in parallel with R_L and hence reduces the effective load resistance and the voltage gain: we return to this point below.

For the BJT amplifier of Fig. 4.14(b) it is clear that whereas the current gain is h_{fe} when the output is short-circuited, the output voltage and therefore the voltage gain are then zero. But for a load resistance R_L as shown, we may write the output voltage as

$$v_{ce} = -h_{fe} i_b \frac{R_L / h_{oe}}{R_L + 1/h_{oe}}$$

by noting that it is developed by the current $h_{fe}\, i_b$ flowing in the parallel combination of R_L and the conductance h_{oe}. Since the input mesh shows immediately that i_b is given by v_{be}/h_{ie}, it follows that the voltage gain is

$$A_v = \frac{v_{ce}}{v_{be}} = -\frac{h_{fe} R_L}{h_{ie}(1 + h_{oe} R_L)}. \tag{4.17}$$

Here again the negative sign indicates phase inversion.

Although the load R_L in eqns 4.16 and 4.17 is not part of the transistor model in either case, it may be considered a fixed part of the amplifier, and in the absence of an external load connected to the output terminals the gain given by each equation is, so far as the whole amplifier is concerned, an *open-circuit* gain. If an external load R is connected, as shown in Fig. 4.15(a) for the BJT amplifier, then in the incremental equivalent it appears in parallel with R_L, as in (b), giving an effective load

$$R_L'' = R R_L / (R + R_L).$$

The voltage gain is then most easily found by substituting R_L'' for R_L in the appropriate equation. Unless otherwise made clear, the term voltage gain for any four-terminal network is taken to mean its open-circuit value, that is in the absence of an external load such as R; current gain likewise usually means the *short-circuit* value. We return in the following section to the effect of an external load on the gain of a four-terminal network, amplifying or otherwise.

An amplifying device also has a **power gain**, which may be defined as the ratio of the output power due to the signal (i.e. the power

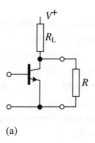

(a)

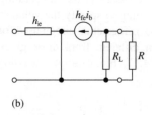

(b)

Fig. 4.15

An amplifier with external load: (a) actual circuit; (b) incremental equivalent.

consumed in the load resistance, whether that is taken to be R_L or some external resistance R, excluding that due to the d.c. current and voltage at the operating point) to the corresponding input power. For a circuit containing no reactance the power gain is simply given by $|A_v A_i|$. Because as we saw above a short-circuited load gives zero voltage gain while an open circuit gives zero current gain, it follows that there is always a value of the load which maximizes power gain. It is also possible to define power gain as a so-called **insertion gain**, which is the ratio of the output power to the power which the input source would deliver to the same load were the two directly connected in the absence of the amplifier; this does not in general give the same value as the previous definition.

The power gain of any four-terminal network containing a controlled source cannot, of course, violate the principle of energy conservation. The reason for the apparent anomaly in the case of a transistor amplifier is that the incremental model is only a partial representation of the actual circuit: it does not include the d.c. supply which in reality is the source of all the power in the circuit except that of the input signal.

It must be emphasized that in this section we have considered so far only the most basic amplifier configuration, without even any arrangement for the bias which is always necessary to fix an operating point (in Fig. 4.8, for example, there is nothing to specify the voltages v_{GS} and v_{BE} in the absence of an input signal). The actual circuit used for an amplifier may produce values of gain different from those given by expressions such as eqns 4.16 and 4.17. For example, biassing

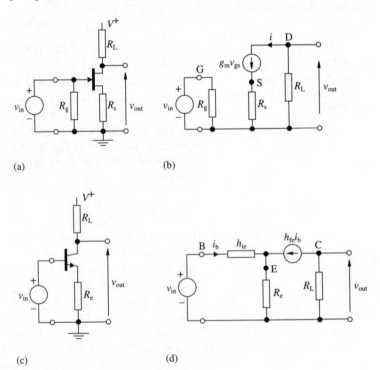

Fig. 4.16
Amplifying circuits: (a) a JFET with bias and (b) its incremental equivalent; (c) a BJT with stabilizing resistor and (d) its incremental equivalent.

arrangements quite often include a resistance connected between source, or emitter, and ground. Consider the circuit with an n-channel JFET shown in Fig. 4.16(a). If the current is to have a value i at the operating point, the resistance R_s can be chosen so that the voltage drop iR_s has precisely the magnitude of the bias voltage v_{GS} needed to allow the current i to flow; the gate, which in the absence of an input signal is kept at ground potential by the (quite large) resistance R_g, is then negative to the source by the correct amount to set the operating point. Now let us find the small-signal voltage gain of this circuit. The incremental equivalent is shown in (b) with an input signal v_{in} connected between the gate and ground and the resistance r_d ignored for simplicity (as it quite commonly is). In this case the voltage v_{gs} is no longer the same as the input, but is the difference between v_{in} and the voltage drop from source to ground; that is

$$v_{gs} = v_{in} - iR_s = v_{in} - g_m v_{gs} R_s.$$

The output voltage is, as before, the drop across R_L although that is no longer the same as v_{ds}; it is given by

$$v_{out} = -iR_L = -g_m v_{gs} R_L.$$

From these two equations the voltage gain is easily shown to be

$$A_v = \frac{v_{out}}{v_{in}} = -\frac{g_m R_L}{1 + g_m R_s} \tag{4.18}$$

which has a magnitude less than that of eqn 4.16 by the factor $(1 + g_m R_s)$.

If the purpose of R_s is only to provide bias without appreciable reduction in gain, a **bypass capacitor** can be connected in parallel with R_s having a sufficiently large value as to be virtually a short circuit at the frequency of the signal to be amplified; R_s does not then appear in the small-signal circuit, although it still provides the required d.c. bias, and the gain is again given by eqn 4.16. This is not invariably done, for there are reasons other than biassing why such a resistance may be used. They can be illustrated by considering the BJT amplifier of Fig. 4.8(b) with a resistance R_e connected between the emitter and ground, as shown in Fig. 4.16(c); in this case R_e does not serve to bias the transistor (at least not on its own), since the emitter must be negative to the base at the operating point, but it is otherwise analogous to R_s. To investigate the effect of R_e we may use the simplest incremental model for the BJT, that of Fig. 4.11(c) which neglects both h_{re} and h_{oe}, to give the circuit shown in Fig. 4.16(d) as the small-signal equivalent of (c). In this the current in R_e is evidently $i_b(h_{fe} + 1)$ and by taking Kirchhoff's voltage law around the input mesh it follows that

$$i_b = \frac{v_{in}}{h_{ie} + (h_{fe} + 1)R_e}.$$

The output voltage is given by $-h_{fe} i_b R_L$, as before (although that is no longer v_{ce}), and from these expressions we find that the voltage gain is now

$$A_v = -\frac{h_{fe}R_L}{h_{ie} + (h_{fe} + 1)R_e}. \tag{4.19}$$

Here the effect of R_e, like that of R_s for the JFET, is evidently to reduce the gain. But inspection of eqn 4.19 reveals more: since h_{fe} is typically in the order of 100, if R_e is at least comparable to h_{ie} then the denominator approximates to $h_{fe}R_e$ in which case the gain becomes

$$A_v = -R_L/R_e. \tag{4.20}$$

This result is remarkable not so much for its simplicity as for the fact that the gain is now completely independent of the transistor parameters. One reason for using R_e is precisely that; for resistance values in practice are relatively predictable and stable, while transistor parameters, and especially h_{fe}, are notoriously liable to large variations from specimen to specimen and are quite strongly temperature dependent. The use of R_e serves to provide a stable and predictable gain, at the cost of reducing its magnitude (it is in fact an example of negative feedback, which is discussed in Chapter 8 and of which these effects are characteristic). The resistance R_e can also form part of a biassing arrangement, since it can be chosen in conjunction with a suitable bias voltage on the base such that when the desired emitter current flows in it the voltage v_{BE} has the appropriate value. Since that is always quite close to 0.7 V for a typical transistor in its active region, the emitter current at the operating point is fixed essentially by R_e and the base potential. If the consequent reduction in gain is undesirable, R_e can be bypassed by a capacitance as described above for the FET amplifier.

Example 4.5 For a certain type of BJT the parameter h_{ie} has the value 500 Ω and h_{fe} has values between 50 and 200, the other two h parameters being negligible. Such a transistor is to be used with a load resistance of 2.2 kΩ to form a simple amplifier. Find the expected range of its open-circuit voltage gain (a) when $R_e = 0$ and (b) when $R_e = 1$ kΩ. Estimate also (c) the values I_C and V_{CE} at the operating point in (b) when the base potential is biassed to 4 V, the d.c. supply being 20 V.

(a) From eqn 4.19 with $R_e = 0$ the open-circuit voltage gain is $-h_{fe}R_L/h_{ie}$ which has the extreme values $-50(2.2/0.5)$ and $-200(2.2/0.5)$; hence the range is

220 $<| A_v |<$ 880.

(b) With $R_e = 1$ kΩ the denominator in eqn 4.19 has the extreme values 51.5 kΩ and 201.5 kΩ; the extreme values of gain are therefore

$-50(2.2/51.5)$ and $-200(2.2/201.5)$ which give the range

$$2.14 <| A_v |< 2.18.$$

This demonstrates the dramatic effect of R_e in reducing both the magnitude of the gain and its dependence on h_{fe}.

(c) The operating point may be deduced with reference to Fig. 4.16(c) and the values given. Since the base potential is 4 V, that of the emitter is close to 3.3 V and the emitter current is therefore $3.3/R_e = 3.3$ mA. Hence, to a close approximation, the collector current is also

$$I_C = 3.3 \text{ mA}.$$

The value of V_{CE} is the supply voltage less the drops in R_L and R_e, so

$$V_{CE} = 20 - 3.3(2.2 + 1) \approx 9.4 \text{ V}.$$

Throughout this section resistances are the only passive elements considered to be part of the amplifier circuit and therefore to appear in the expressions for gain. In fact capacitance is used a great deal also (the bypass capacitance mentioned above is an example) and in any case occurs inevitably in small amounts. Depending on the frequency of the signals to be considered, some of the capacitances in an amplifier circuit may be taken as short circuits and others as open circuits, but in general they must be considered as reactances; voltages and currents in the incremental or small-signal equivalent should then be treated (in the steady state) as phasors, as in any other a.c. circuit. The resistances R_L, R_S, and R_e may then become impedances Z_L, Z_S, and Z_e; also, the form of the various expressions for gain may change according to the position of capacitance in the circuit. Inductance is, of course, similarly dealt with when necessary, although in electronic circuits generally it is less common than capacitance in appreciable amounts.

4.6 Input and output impedance

4.6.1 General

Input and output impedances, like gain, were encountered among the various parameters of four-terminal networks (Section 4.2), but were there defined for open- or short-circuited input or output. These, although often considered, are not the normal working conditions of such circuits, which would usually include a source of some kind at the input and an element or device providing a finite passive load at the output. We shall here assume a source impedance Z_S (whether of a voltage or of a current source) and an external load impedance Z_L. The input impedance Z_{in} is defined in general as V_1/I_1, the impedance of the entire circuit, load included, at the input terminals, as shown in

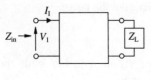

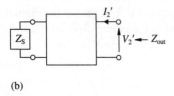

(a)

(b)

Fig. 4.17
Input and output impedances.

Fig. 4.17(a). It can be found from the impedance parameters if these are known. Thus in the present case the second of eqns 4.1 gives

$$-Z_L I_2 = z_{21} I_1 + z_{22} I_2$$

with reference to Fig. 4.2, or

$$I_2 = -\frac{z_{21}}{Z_L + z_{22}} I_1$$

and the first of eqns 4.1 then becomes

$$V_1 = \left(z_{11} - \frac{z_{12} z_{21}}{Z_L + z_{22}} \right) I_1$$

from which the input impedance follows as

$$Z_{in} = \frac{V_1}{I_1} = z_{11} - \frac{z_{12} z_{21}}{Z_L + z_{22}}. \tag{4.21}$$

This expression confirms, as it should, that z_{11} is the value of Z_{in} for an open-circuited output, $Z_L \to \infty$.

If the impedance parameters are not known, it is usually a simple enough matter to find Z_{in} by the ordinary combination rules and/or Kirchhoff's laws. In the case of the small-signal equivalents of active circuits, described in the foregoing sections, the controlled sources which they include must not be removed unless they happen to have value zero (Section 4.4) and Z_{in} cannot in general be found by combination rules alone.

The output impedance Z_{out} may be defined in general as the effective impedance of the circuit between the output terminals when the input source is removed (in the sense that it is to be replaced by its internal impedance Z_S) and the load is disconnected, as in Fig. 4.17(b); in other words it is the Thévenin impedance of the circuit and its input source, with respect to the output terminals (Section 2.5.4). It is *not* the actual ratio V_2/I_2 (which is simply the negative of the load impedance), but it can be regarded as the ratio of a voltage V_2', supposed to be applied to the output terminals, to the current I_2' which it would cause to flow, as shown, with the input source removed.

For any linear circuit whose impedance parameters are known Z_{out} can quickly be found, by analogy with eqn 4.21 for the input impedance, as

$$Z_{out} = \frac{V_2'}{I_2'} = z_{22} - \frac{z_{21} z_{12}}{Z_S + z_{11}} \tag{4.22}$$

which as would be expected gives z_{22} as the value of Z_{out} when $Z_S \to \infty$, that is when the input is open circuit. In general, however, Z_{out} is found by the means discussed in Section 2.5.4: that is, either by finding the impedance V_2'/I_2' from first principles or by finding the ratio of open-circuit voltage to short-circuit current at the output terminals. In the former case only *independent* sources are removed: as in finding Z_{in}, the controlled sources of active circuits must be retained (unless they then happen to reduce to zero) and the impedance deduced by Kirchhoff's laws; the usual combination rules will not do in such cases.

Example 4.6 (a) Find the input impedance of the circuit shown in Fig. 4.13 when the output is short-circuited.

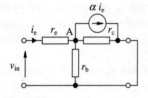

Fig. 4.18
The circuit for Example 4.6(a).

The circuit is now that shown in Fig. 4.18; r_b and r_c are in parallel and by Kirchhoff's first law at node A take a combined current of $(1 - \alpha)i_e$. The equation of the input mesh is therefore

$$v_{in} = i_e \left\{ r_e + (1 - \alpha) \frac{r_b r_c}{r_b + r_c} \right\}$$

and the input impedance is

$$Z_{in} = \frac{v_{in}}{i_e} = r_e + (1 - \alpha) \frac{r_b r_c}{r_b + r_c}.$$

(b) Find the input impedance and the output impedance (with respect to an external load) of the small-signal BJT amplifier circuit shown in Fig. 4.14(b).

Kirchhoff's second law around the input mesh gives

$$v_{be} = h_{ie} i_b,$$

from which the input impedance is simply

$$Z_{in} = h_{ie}.$$

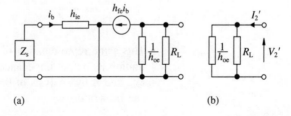

Fig. 4.19
The circuit for the output impedance in Example 4.6(b).

(a)   (b)

To find the output impedance, the input source is removed by replacing it with its impedance Z_S as in Fig. 4.19(a). Kirchhoff's law for the input mesh now shows that $i_b = 0$, so in this case the controlled current source is also removed, and replaced by an open circuit to give the circuit as in (b). From this the output impedance is evidently given by

$$Z_{out} = V'_2/I'_2 = 1/(h_{oe} + 1/R_L) = R_L/(1 + h_{oe}R_L).$$

This expression reduces simply to the load resistance R_L if h_{oe} is assumed to be negligible.

Alternatively, Z_{out} could be found as the ratio of open-circuit output voltage to short-circuit output current in Fig. 4.14(b).

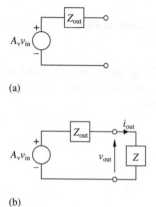

(a)

(b)

Fig. 4.20
The Thévenin equivalent of a voltage amplifier (a) with open-circuit output and (b) with an external load.

It will be noticed that in the above example Z_{in} does not depend on an external load impedance, nor does Z_{out} depend on the source impedance Z_S. The reason for this, in essence, is that the circuit is unilateral (Section 4.4).

The small-signal output impedance of the simple JFET amplifier represented by Fig. 4.14(a) is easily shown, when the effect of r_d is neglected, to approximate to the load resistance R_L as for the BJT case in Example 4.6(b). Its input impedance, on the other hand, is in theory infinite on our assumption that the gate takes no current. In practice the input impedance of a FET circuit, while large, may be governed not so much by the very small 'leakage' current to the gate (which is typically in the order of nanoamperes or less) as by components added to the circuit for biassing or otherwise.

The input and output impedances of four-terminal networks are among their most important properties. They afford a means of expressing the effect of load on gain which is especially useful for amplifying circuits. If for example an amplifier has an open-circuit voltage gain A_v then its Thévenin equivalent for an input voltage v_{in} must consist of the e.m.f. $A_v v_{in}$ in series with Z_{out} as shown in Fig. 4.20(a). If the output is connected to a load of impedance Z, as in (b), the output voltage is evidently given by

$$v_{out} = A_v v_{in} Z / (Z + Z_{out})$$

and the voltage gain with the load connected is therefore

$$\frac{v_{out}}{v_{in}} = A_v \frac{Z}{Z + Z_{out}}. \tag{4.23}$$

This same representation of the amplifier can sometimes be helpful in finding its output impedance, thus: if the output current to the load Z is i_{out}, and if by analysis of the actual amplifier circuit the output voltage can be written as

$$v_{out} = A_v v_{in} - c i_{out}$$

then inspection of Fig. 4.20 shows that the coefficient c must be the output impedance.

It is also important to bear in mind that if the input to any circuit is provided by a source of voltage v_S the actual value of v_{in} will in general be less than that, because of the combined effect of the source

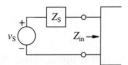

Fig. 4.21
The general voltage source input.

impedance and the input impedance. Thus in Fig. 4.21 the input voltage is, by potential division,

$$v_{in} = v_S Z_{in}/(Z_S + Z_{in}). \tag{4.24}$$

These principles are, of course, perfectly general although we have discussed them for the most part in the language used for electronic circuits in particular. Thus for example the voltage drop due to load current at the terminals of a generator or at any point in a power transmission system can easily be expressed in terms of an impedance which, whatever name it may be given in such contexts, is simply the output impedance of the generator or system.

4.6.2 Optimum amplification

It follows from eqns 4.23 and 4.24 that the voltage gain of an amplifier between source and external load can be written as

$$\frac{v_{out}}{v_S} = A_v \left(\frac{Z_{in}}{Z_S + Z_{in}} \right) \left(\frac{Z_L}{Z_{out} + Z_L} \right) \tag{4.25}$$

which shows that an effective voltage amplifier should in general have a high input impedance and a low output impedance, since its gain will then approach the open-circuit value A_v under most conditions. It may be recalled at this point that the simple FET amplifier has a very large input impedance, so to that extent is a good voltage amplifier. On the other hand, the BJT amplifier was seen in Example 4.6 to have input impedance h_{ie}, which is often in the order of 1 kΩ or so, while the output impedance was that of the load resistance, which may be a few kilohms; so in the form considered it is inferior to the FET as a voltage amplifier. We shall see in Section 4.6.3 that there are ways of improving it in this respect.

If we wish to amplify current rather than voltage the requirements are precisely the reverse: the output impedance should now be large and the input impedance low (in other words, the admittance requirements for current amplification become, because of duality, what the impedance requirements were for voltage). This can be seen from Fig. 4.22, where (a) shows the Norton equivalent circuit of an amplifier whose short-circuit current gain is A_i. It is evident that the greatest current into the load Z_L is obtained when Z_{out} is large. Similarly when a current source is connected to the input as shown in (b), the maximum input current is obviously obtained when Z_{in} is low. The overall current gain for the amplifier can be written

$$\frac{i_{out}}{i_S} = A_i \left(\frac{Z_S}{Z_{in} + Z_S} \right) \left(\frac{Z_{out}}{Z_L + Z_{out}} \right)$$

or, in terms of admittances,

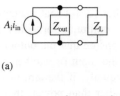

(a)

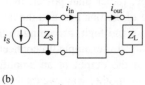

(b)

Fig. 4.22
Current amplification.

$$\frac{i_{out}}{i_S} = A_i\left(\frac{Y_{in}}{Y_S + Y_{in}}\right)\left(\frac{Y_L}{Y_{out} + Y_L}\right) \qquad (4.26)$$

which corresponds exactly to 4.25 above. Since it was noted above that the basic BJT amplifier typically had an output impedance greater than its input impedance, it follows that it can make a reasonably good current amplifier, depending on the impedances of source and load.

Since the optimum conditions for voltage gain are opposite to those for current gain, we may expect the best power gain to occur between the two extremes. The optimum is easily defined if we consider the matching conditions discussed in Sections 2.9 and 3.11.4: the maximum power input to an amplifier occurs when $Z_{in} = Z_S^*$; likewise the output power is maximized when $Z_{out} = Z_L^*$. The output power is therefore greatest when both input and output are matched. If, as is often nearly true, all these impedances are resistive the matching conditions may be written

$$Z_{in} = Z_S = R_S; \; Z_{out} = Z_L = R_L.$$

It then follows from eqn 4.23 that the voltage gain with respect to the amplifier input, that is v_{out}/v_{in}, is given by $A_v/2$; similarly it can readily be shown that the current gain i_{out}/i_{in} is $A_i/2$. Hence the power gain with respect to the input terminals is given by $A_v A_i/4$, and this is the factor by which the amplifier can increase the maximum power which is available from the input source. (For an amplifier with negligible input current, such as the simple FET versions considered earlier, the current gain A_i is very high, but the large Z_{in} means that the power taken by the input from a source is correspondingly small.)

The matching of four-terminal networks in general is considered further in Section 4.7.

4.6.3 Impedance transformation: source and emitter followers

The input and output impedances of an amplifier (or for that matter of any four-terminal network) may be far from optimum for either its source or its load. It is not always possible to ensure that the output of one part of a circuit or system is matched to the input of another part connected to it which is effectively its load; equally, if the aim of the designer is to maximize an output voltage, rather than power, then it may not happen that the output impedance of the one part and the input impedance of the other are respectively low and high, according to the criteria discussed above. It may be necessary, for example, to connect a device represented by a low impedance to the output of an amplifier with a relatively high output impedance; put another way, we may wish to draw a large current from the output without losing too much voltage. In Example 4.6 it appeared that the output impedance of a simple amplifier was in the order of its load resistance, which would quite commonly be a few kilohms; so if the output signal had to provide a current of only a milliampere (r.m.s., let us suppose) the signal voltage

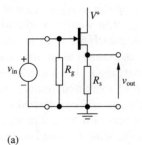

(a)

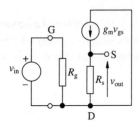

(b)

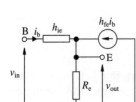

(c)

(d)

Fig. 4.23
Impedance conversion: (a) the source follower and (b) its incremental equivalent; (c) the emitter follower and (d) its incremental equivalent.

at the output could fall by several volts r.m.s. which would be an appreciable fraction of its open-circuit value. Fortunately in such cases it is often possible to interpose between an output and its load a device which is more or less matched to both or has impedances which satisfy the requirements of both. There are several kinds of devices which achieve this, and their use is not confined to amplifying circuits; but here we consider two, the **source follower** and the **emitter follower** which are similar in principle, are based respectively on the FET and the BJT, and are often used with amplifiers.

Consider the JFET circuit shown in Fig. 4.23(a). It may be compared with that of Fig. 4.16(a), but now the drain is connected directly to V^+ so that $R_L = 0$, and the output is taken not from the drain but from the source. The small-signal equivalent with r_d ignored is shown in (b). To find first its voltage gain, we have the relations

$$v_{out} = g_m v_{gs} R_s$$

and

$$v_{in} = v_{gs} + v_{out}.$$

These combine to give

$$A_v = \frac{v_{out}}{v_{in}} = \frac{g_m R_S}{1 + g_m R_S}$$

which shows immediately that the gain is always less than unity. Clearly, then, the circuit is of no use as a voltage amplifier. Its merit lies in its input and output impedances. The first of these is obviously very large, on the usual assumption that the FET takes no gate current. For the second, we may use the ratio of open-circuit output voltage, which is $A_v v_{in}$, to short-circuit output current which is simply $g_m v_{in}$ since with the source short-circuited to ground v_{gs} is equal to v_{in}. We then have

$$Z_{out} = \frac{A_v}{g_m} = \frac{R_S}{1 + g_m R_S}. \tag{4.27}$$

Now consider a typical value of g_m, say 5 mS (compare Example 4.4), and a resistance R_s of say 2.2 kΩ, giving the factor $(1 + g_m R_s)$ the value 12, so that A_v is 11/12 or 0.917 and Z_{out} is (2.2/12) kΩ or 183 Ω. For $R_s = 10$ kΩ these figures become respectively 0.98 and 196 Ω. As R_s increases the expressions above show that A_v approaches the limiting value of unity and Z_{out} the value $1/g_m$ which in this case is 200 Ω.

We thus have a circuit with a voltage gain not much less than unity, which means that is does little harm to the gain of a voltage amplifier connected to its input; at the same time it has a very low output impedance by the standards of amplifier circuits (we saw in Example 4.6 that the output impedance of a simple amplifier might typically be a few kilohms). The source follower can therefore be connected to the output of a voltage amplifier, from which its high input impedance takes little current and so causes no loss of voltage, and its output in turn connected to a load which, so long as it has a moderately high impedance, will cause no great loss of voltage in the small Z_{out} of the source follower.

In other words the circuit allows a voltage amplifier to supply current to a load with a much smaller loss of voltage than would be otherwise possible. In this role it is often known as a **buffer**, since it comes between an amplifier and its load; it is also an example of an impedance transformer, an idea which we shall encounter again. The name 'source follower' arises from the fact that the gain of about unity ensures that whatever change of potential is applied to the gate will be followed more or less faithfully by the source. The circuit may also be said to have a **common-drain** connection, since the drain, being at ground potential in the small-signal equivalent circuit, is common to both input and output.

The BJT equivalent of the source follower is the **emitter follower** or **common-collector** circuit, shown at its simplest in Fig. 4.23(c) with its approximate small-signal equivalent in (d). Here it is not difficult to show that the voltage gain is

$$A_v = \frac{(h_{fe} + 1) R_e}{h_{ie} + (h_{fe} + 1)R_e}$$

which again is less than but near to unity for typical values. To find Z_{out}, we have the open-circuit output voltage as $A_v v_{in}$ and, by inspection of (d), the short-circuit output current is $(h_{fe} + 1)(v_{in}/h_{ie})$; hence we have

$$Z_{out} = \frac{A_v h_{ie}}{h_{fe} + 1} = \frac{R_e h_{ie}}{h_{ie} + (h_{fe} + 1)R_e} \tag{4.28}$$

which for typical values is approximately h_{ie}/h_{fe}, and is therefore small. The input impedance Z_{in} is given by v_{in}/i_b and the equation of the input mesh is

$$v_{in} = h_{ie} i_b + R_e(h_{fe} + 1) i_b$$

which gives

$$Z_{in} = h_{ie} + R_e(h_{fe} + 1).$$

For typical values of R_e, say a few kilohms, this is a large resistance, certainly very much greater than h_{ie}, which was seen earlier (Example 4.6) to be the input impedance of the simple BJT amplifier. The emitter follower, then, provides the same benefits as the source follower: high input impedance and low output impedance, which enable a voltage amplifier to sustain its gain when loaded, at the expense of only a slight loss in open-circuit gain.

4.7 Interconnection

For many purposes four-terminal networks are the building blocks from which larger circuits are made. Just as single elements or two-terminal combinations of elements can be connected in series or in parallel, so too can the input and output terminal pairs of a four-terminal network be connected to those of other such networks in various ways; several inputs, for example, may be connected in series or in parallel and so too the outputs. The behaviour of the combination depends on that of the individual networks together with the connection constraints (equal voltages for parallel, equal currents for series), and sometimes a set of network parameters for the whole is a simple combination of the individual sets. One of the most important cases is the relatively simple arrangement by which the output of one network is directly connected to the input of another as shown in Fig. 4.24. It is known as **cascade** connection, and is the means by which signals can be amplified, or otherwise processed, in successive **stages** of one circuit or in successive circuits; the discussion of the source and emitter followers in the previous section assumed such an arrangement. The output voltage and current of one stage are the input quantities for the next, and the load on one stage is the input impedance of the next (which may in turn depend on later stages).

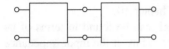

Fig. 4.24
Cascade connection.

The gain of several four-terminal networks in cascade is the product of the individual gains only if each of these is evaluated for the correct load presented by the following input. It is *not* correct in general to use the open-circuit voltage gains or the short-circuit current gains, unless, for the reasons shown in Section 4.6.2, each input impedance is large enough (for voltage) or small enough (for current) compared with the preceding output impedance. In calculating a correct overall gain for cascaded networks, the transmission parameters in the ABCD form (Section 4.2.3) are especially useful; for in that case the output voltage and current of one network (or stage) are precisely the input values of the next, so the transmission matrix for the whole circuit is simply the product of the ABCD matrices of the constituents (and likewise for the inverses). The *a* or *b* parameters can also be used, but with them the output current of one stage is the negative of the input to the next and an extra matrix must be inserted in the product to allow for that.

Strictly, the input impedance of any one stage may depend on the properties of every subsequent stage up to and including the final load Z_L; similarly the output impedance of any one may depend on the properties of every preceding stage, including the source impedance Z_S. A considerable simplification can be achieved, under certain conditions, for a succession of identical stages which together form a ladder network (as discussed for the d.c. case in Section 2.7 and more generally in Chapter 10). If each stage is so designed that its input impedance is Z_L when it is loaded with an impedance Z_L, then if follows that when the final load in Z_L the load on *every* stage is equivalent to Z_L and that sections may be added or removed with no effect on the input impedance of the whole. (It also follows that the same Z_L would then be the input impedance of an infinite ladder of such sections.) In the same way, if the initial source has impedance Z_S and the output impedance of any stage is also Z_S when its input is that source, then Z_S is the output impedance of the whole, however many sections there may be.

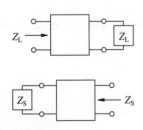

Fig. 4.25
Iterative impedances.

These two conditions are illustrated in Fig. 4.25. It is possible to design a network which satisfies them for any given Z_L and Z_S; equally, there are values of Z_L and Z_S which satisfy them for a given network and these are known as the **iterative impedances** of the network. Like the parameters defined in Section 4.2, they are characteristic of the network and do not depend on connections to it. It follows from the foregoing that a finite ladder network whose output is connected to an impedance Z_L having the appropriate iterative value must behave exactly as if it were infinite so far as the currents and voltages in its sections are concerned; it is said to be **terminated** in its iterative impedance. A network so terminated is also said to be **matched**, for reasons which will be encountered in Chapter 10.

The iterative impedances of a network can be found in terms of its impedance parameters by using eqn 4.21. Thus, if the input impedance is to be Z_L we require that

$$Z_L = z_{11} - \frac{z_{12}z_{21}}{Z_L + z_{22}}$$

from which the iterative impedance at the input is given by

$$Z_L = \frac{1}{2} \left\{ z_{11} - z_{22} \pm \sqrt{(z_{11} - z_{22})^2 - 4(z_{12}z_{21} - z_{11}z_{22})} \right\} \quad (4.29)$$

while by symmetry that at the output, Z_S, is obtained from the same expression with subscripts interchanged.

Example 4.7 Find the iterative impedances of the network shown in Fig. 4.3(a).

As in Example 4.1, we have for this network

$$z_{11} = R_1 + R_3$$
$$z_{22} = R_2 + R_3$$
$$z_{12} = z_{21} = R_3.$$

Substituting these into eqn 4.29 gives

$$Z_L = \frac{1}{2} \left\{ R_1 - R_2 \pm \sqrt{(R_1 + R_2)(R_1 + R_2 + 4R_3)} \right\}$$

in which the negative root may be rejected as giving a negative real value; thus

$$Z_L = \frac{1}{2} R_1(a + 1) + \frac{1}{2} R_2(a - 1)$$

where

$$a = \sqrt{\frac{R_1 + R_2 + 4R_3}{R_1 + R_2}}.$$

To find Z_S, the output iterative impedance, it is necessary in this case only to exchange R_1 and R_2, to give

$$Z_S = \frac{1}{2} R_1(a-1) + \frac{1}{2} R_2(a+1).$$

If it is required to design a passive network for given iterative impedances Z_S and Z_L, then since two equations like 4.29 are available a T or Π network (Section 4.3) can be found by choosing one branch and solving for the remaining two.

In the particular case of a circuit with equal iterative impedances their value, designated Z_0, is known as the **characteristic impedance** of the network and is comparable to the characteristic impedance of a transmission line (Section 10.3.3). This equality is evidently obtained when the network is completely symmetrical or has a symmetrical equivalent, so that $z_{11} = z_{22}$; in Example 4.7 above it occurs for $R_1 = R_2$. Equation 4.29 shows that the characteristic impedance is then given by

$$Z_0^2 = z^2 - z_{12}z_{21} = \Delta_z \tag{4.30}$$

in which $z = z_{11} = z_{22}$; it can be seen from this and eqns 4.1 that Δ_z is the determinant of the matrix of impedance parameters. To design a symmetrical T or Π network for a given Z_0 one value must be chosen and the other then follows either from eqn 4.30 or from the basic definition of Z_0 as an iterative impedance.

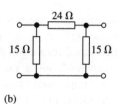

(a)

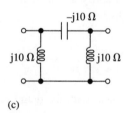

(b)

(c)

Fig. 4.26
Circuits for Example 4.8.

Example 4.8   Find a Π network having a characteristic impedance of $(10 + j0)$ Ω.

We seek a symmetrical network as shown in Fig. 4.26(a). If Z_1 is chosen to be $(15 + j0)$ Ω, say, then from the definition of Z_0 we may deduce Z_2 from the requirement that with a load impedance of 10 Ω at the output the input impedance is also 10 Ω. Noting that 10 Ω in parallel with 15 Ω is equivalent to 6 Ω, we thus require that

$$10 = \frac{(Z_2 + 6)15}{Z_2 + 21}$$

or

$$Z_2 = (24 + j0) \ \Omega,$$

which gives the circuit shown in (b) as one which has the specified Z_0.

We should note from the above example that not all choices of Z_1 would allow a solution. Thus the value required for Z_2 becomes infinite for $Z_1 = (10 + j0)$ Ω and there is no solution for real $Z_1 < 10$ Ω; this is easily appreciated by observing from the diagram that Z_1 must always be larger than the required input impedance. On the other hand, however, it is not necessary to restrict the choice to real values. Suppose that in the above case Z_1 is taken to be j10 Ω while Z_0 is still required to be $(10 + j0)$. Since these two in parallel have the value $j100/(10 + j10)$, or $j10/(1 + j)$, we now have

$$10 = \frac{\left(Z_2 + \dfrac{j10}{1+j}\right)j10}{j10 + Z_2 + \dfrac{j10}{1+j}}$$

which gives

$$Z_2 = -j10 \ \Omega$$

and shows that the circuit shown in (c) is another which satisfies the original requirement. It should of course be no surprise that combinations of imaginary elements can produce a real result; but this example of a network of entirely reactive elements with a purely resistive characteristic impedance represents an important class of circuits, including transmission lines, which is considered in Chapter 10.

There is a general relation which can often simplify the calculation of characteristic impedance: it can be shown that

$$Z_0 = \sqrt{\frac{z_{11}}{y_{11}}} \tag{4.31}$$

and this in turn is sometimes written, in agreement with the definitions of z_{11} and y_{11} (Sections 4.2.1 and 4.2.2), in the form

$$Z_0^2 = Z_{oc}Z_{sc}. \tag{4.32}$$

Here Z_{oc} and Z_{sc} are the open- and short-circuit input impedances; they have the advantage in practice of being easily measured for any four-terminal network even when its component parts are not identifiable.

The idea of iterative and characteristic impedances has been introduced here by considering a chain of identical two-ports from which one or more could be removed without affecting the input or output impedance of the whole. An alternative requirement may be to maximize the power transfer between source and load by way of a number of two-ports not necessarily identical. This means matching each pair of output terminals to the input terminals of the next stage, thus requiring that the output impedance of each stage is equal to the input impedance of the next (strictly its conjugate, as shown in Section 3.11.4, but here we assume resistive impedances for simplicity).

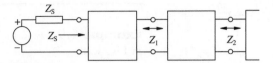

Fig. 4.27
Image impedances.

This condition is shown in Fig. 4.27. Evidently the input impedance of the first stage should match the output impedance Z_S of the source; if the output impedance of this stage is then Z_1 the input impedance of the next should also be Z_1, and so on, until the output impedance of the last stage, which should be equal to the load impedance Z_L. Thus, associated with each two-port is a pair of impedances so defined that one of them is the input impedance when the other is connected to the output, and the second is the output impedance with the first connected to the input. They are known as the **image impedances** and, like iterative impedances, are fixed properties of any four-terminal network. For maximum power transfer it is the image impedances which have to be matched at every stage. A single four-terminal network can be used to match a given source to a given load by choosing its image impedances accordingly. A special case of this is the transformer (Section 4.9).

Equations for the image impedances are easily set up, in terms of the impedance parameters, from the expression for input impedance, eqn 4.21. If we write Z_1 and Z_2 for the image impedances at input and output respectively, then

$$Z_1 = z_{11} - z_{12}z_{21}/(Z_2 + z_{22})$$

and

$$Z_2 = z_{22} - z_{12}z_{21}/(Z_1 + z_{11})$$

which may be solved to give

$$Z_1^2 = \frac{z_{11}}{z_{22}}(z_{11}z_{22} - z_{12}z_{21}) = \frac{z_{11}}{z_{22}}\Delta_z \tag{4.33}$$

and

$$Z_2^2 = \frac{z_{22}}{z_{11}}(z_{11}z_{22} - z_{12}z_{21}) = \frac{z_{22}}{z_{11}}\Delta_z. \tag{4.34}$$

For a symmetrical network $z_{11} = z_{22}$, so the image impedances are equal and given by

$$Z_1^2 = Z_2^2 = \Delta_z;$$

they are also, as comparison with eqn 4.30 confirms, equal to the characteristic impedance Z_0. For such a network, therefore, the iterative and image impedances have one and the same value, namely the characteristic impedance.

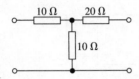

Fig. 4.28
The circuit for Example 4.9.

Example 4.9   Find the image impedances of the T network shown in Fig. 4.28.

According to the definitions in Section 4.1 the impedance parameters of this network are, by inspection,

$$z_{11} = 20 \ \Omega, \ z_{22} = 30 \ \Omega, \ \text{and} \ z_{12} = z_{21} = 10 \ \Omega.$$

Hence

$$\Delta_z = 20 \times 30 - 10^2 = 500$$

and eqns 4.33 and 4.34 then give the image impedances

$$Z_1 = \sqrt{(20/30)500} = \mathbf{18.3 \ \Omega}$$

and

$$Z_2 = \sqrt{(30/20)500} = \mathbf{27.4 \ \Omega}.$$

It may easily be confirmed that connecting 27.4 Ω to the output gives an input impedance of 18.3 Ω while similarly 18.3 Ω at the input yields an output impedance of 27.4 Ω.

As well as the various parameters and properties which have already been defined for four-terminal networks, there are yet others which can be useful. One is the **propagation constant**, which is the natural logarithm of gain for a network terminated in its iterative impedance and is in general complex to account for both magnitude and phase; its significance will become more apparent when ladder networks are considered in Chapter 10. Another is the **insertion loss**, defined as the reduction in the power delivered to a load from a source due to the insertion of the network between the two. It can arise both from power loss due to dissipation in any resistance which the network may have, and from the effect of mismatching; it corresponds to the insertion gain defined in Section 4.5.6.

Fig. 4.29
Coupled inductances with a common earth connection.

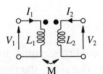

Fig. 4.30
Coupled circuits.

4.8   Coupled circuits

An important class of four-terminal networks is that in which one circuit with input terminals and another with output terminals are connected only by mutual inductance (Section 3.10). (We may include cases where there is a single electrical connection which cannot carry current and serves only to relate the electrical potential of the two, for example to connect both to earth as shown in Fig. 4.29.) Each of the circuits thus coupled may contain many meshes and may also include active elements, but for simplicity consider meantime a passive two-port consisting only of two coupled inductances as shown in Fig. 4.30; this can if necessary be considered as part of a larger circuit.

By referring to Section 3.10 we may write for this network

$$V_1 = j\omega L_1 I_1 + j\omega M I_2$$
$$V_2 = j\omega M I_1 + j\omega L_2 I_2$$

(4.35)

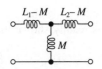

Fig. 4.31
The T equivalent for coupled inductances.

in which M is a positive quantity and the dots are as shown. (A different allocation of the dots, implying negative M, would not affect the principle of what follows.) Comparison with eqns 4.1 shows that the impedance parameters of the network are

$$z_{11} = j\omega L_1, \ z_{22} = j\omega L_2, \ \text{and} \ z_{12} = z_{21} = j\omega M.$$

We may note at this point that these parameters correspond to the equivalent T network shown in Fig. 4.31, in which M now represents a **self-inductance** of that value and there is no mutual inductance. This equivalent can be useful on occasion but is rather artificial, since M may exceed L_1 or L_2, making one of the self-inductances negative; it is in that case an example of an unrealizable equivalent (Section 3.6).

Various properties of the circuit can be deduced from eqns 4.35. Suppose that the voltage V_1 is applied to the input terminals while the output terminals are connected to a load impedance Z_L, so that according to the directions specified we may write

$$I_2 = -V_2/Z_L.$$

By substituting this in the equations the ratio of output to input voltage can be found (by convention the term 'gain' is avoided here); it is given by

$$\frac{V_2}{V_1} = \frac{MZ_L}{L_1Z_L + j\omega(L_1L_2 - M^2)}.$$

(4.36)

Similarly the equations can be rearranged with the same substitution to give the current ratio as

$$\frac{I_2}{I_1} = -\frac{j\omega M}{j\omega L_2 + Z_L}.$$

(4.37)

The input impedance is the ratio of V_1/I_1 and can be shown in the same way to be

$$Z_{in} = \frac{V_1}{I_1} = j\omega L_1 - \frac{(j\omega M)^2}{j\omega L_2 + Z_L}$$

(4.38)

a result which also follows from eqn 4.21. The output impedance is given similarly, or from eqn 4.22, by

$$Z_{\text{out}} = j\omega L_2 - \frac{(j\omega M)^2}{j\omega L_1 + Z_S} \tag{4.39}$$

if the input source has impedance Z_S.

These equations become more informative when we consider the relations between L_1, L_2, and M discussed in Section 3.10.2. In the case of imperfect coupling we have, from eqns 3.44 and 3.45,

$$M^2 = k^2 L_1 L_2 = k_1 k_2 L_1 L_2. \tag{4.40}$$

From the definitions of k_1 and k_2 in that section, together with the definition of flux linkage in Appendix B.3, it is readily shown that

$$k_2 \rho L_2 = M = k_1 L_1 / \rho, \tag{4.41}$$

where ρ is the ratio of the turns in coil 1 to those in coil 2. For perfectly coupled inductances $k_1 = k_2 = 1$ and these relations become

$$M^2 = L_1 L_2 \tag{4.42}$$

and

$$\rho L_2 = M = L_1 / \rho. \tag{4.43}$$

It also follows from eqn 4.41 that even for imperfect coupling the inductances are in the ratio

$$L_1 / L_2 = \rho^2 \tag{4.44}$$

so long as $k_1 = k_2$. (This is to be expected from the fact that the inductance of a coil with a given geometrical arrangement varies as the square of the number of its turns: see for example eqn B.35 in Appendix B.)

Now consider the effect of perfect coupling on eqns 4.36 to 4.39. The voltage ratio in 4.36 becomes, from 4.42 and 4.43,

$$\frac{V_2}{V_1} = \frac{M}{L_1} = \frac{1}{\rho} \tag{4.45}$$

for any Z_L, while the current ratio, eqn 4.37, can be written

$$\frac{I_2}{I_1} = -\frac{j\omega \rho L_2}{j\omega L_2 + Z_L}. \tag{4.46}$$

The input impedance, eqn 4.38, becomes

$$Z_{\text{in}} = \frac{j\omega L_1 Z_L}{j\omega L_2 + Z_L} \tag{4.47}$$

and the output impedance in eqn 4.39 similarly becomes

$$Z_{\text{out}} = \frac{j\omega L_2 Z_S}{j\omega L_1 + Z_S}. \tag{4.48}$$

Of these last four equations, two at least are noteworthy: eqn 4.45 shows that for perfect coupling the voltage ratio is simply given by the turns ratio, whatever the nature of the input source and the value of the output load; and eqn 4.48 shows that if $Z_S = 0$ then $Z_{out} = 0$. Thus if the input is a perfect voltage source so too is the whole circuit, and its value is V_1/ρ so far as the load is concerned.

We may go further. Equation 4.46 shows that if L_2 were infinitely large, so that $\omega L_2 \gg | Z_L |$ for any load impedance Z_L, then we should have

$$\frac{I_2}{I_1} = -\rho \tag{4.49}$$

and the current ratio would be precisely the inverse of the voltage ratio (eqn 4.45), neither depending on load or input. The negative sign here arises from the chosen current directions in Fig. 4.30; they accord with the general convention but in the case of a passive load obviously mean that I_2 is negative and I_1 positive. We remove the sign later, for convenience, but here it serves to remind us of what can be deduced from the discussion in Section 3.10.4, that the fluxes produced by the two currents must physically be in opposition, and in that sense the current themselves can be regarded as in opposition. Not only that: eqn 4.49 shows that (for the assumed conditions of perfect coupling and very large L_2) the ampere-turn products of the two coils *exactly cancel*. This surprising result, which would usually indicate that there could be no resultant flux at all, is entirely understandable if we reflect that the assumption of infinitely large L_2 implies that L_1 and M are also infinite since the ratio ρ remains unrestricted: it then follows from the definition of inductance that the resultant flux which is necessary to induce e.m.f.s of values V_1 and V_2 in the two coils can be produced by a vanishingly small value of ampere turns. The physical significance of this is perhaps clearer if we first suppose an open-circuited output; with very large L_1 the input current is negligibly small but produces the necessary flux. If now output current is allowed to flow, by connecting a load, its ampere turns are cancelled by a rise in the input current according to eqn 4.49, so that the net flux is unchanged.

We have now shown that a network comprising only a pair of very large and perfectly-coupled inductances has the property of transforming the voltage at its input terminals into a different value at the output according to the turns ratio ρ, without phase shift, at the same time transforming the input current in the inverse ratio. Such a two-port is known as an **ideal transformer**. It has one more interesting property stemming from the fact that it converts voltage and current in inverse proportions, and revealed by eqn 4.47. The latter shows that for large L_1 and L_2 the input impedance becomes

$$Z_{in} = L_1 Z_L / L_2 = \rho^2 Z_L. \tag{4.50}$$

Hence, so far as the input source is concerned, an actual load Z_L becomes an effective load $\rho^2 Z_L$. This incidentally confirms that the phase angle between V_1 and I_1 is the same as the angle between V_2 and $-I_2$, which is that of Z_L. In the same way eqn 4.48 shows that in the ideal case the input source of impedance Z_S becomes, so far as the load is concerned, a source of impedance Z_S/ρ^2. The ideal transformer therefore serves to convert not only current and voltage but also impedance, according to relations which can be summarized thus:

$$\frac{V_2}{V_1} = \frac{1}{\rho} = -\frac{I_1}{I_2}$$

$$Z_{\text{in}} = \rho^2 Z_L \qquad\qquad\qquad\qquad\qquad (4.51)$$

$$Z_{\text{out}} = \frac{Z_S}{\rho^2}.$$

Practical transformers, devices which more or less approximate to the ideal, are variously used for each of these conversions; although any one of them may be the aim in a given application, all three inevitably occur together. It may be questioned how all these relations can be sustained in extreme conditions, for example how can the current ratio be correct when the output is open-circuited? The answer of course is that one or other of them fails in those limits where the conditions on which they are derived break down. Actual transformers are considered in the next section.

There is one further point which is worth noting. The ratio ρ can of course be made smaller than unity, in which case the output voltage exceeds the input and it might be thought that the transformer was also a voltage amplifier. In this limited sense it does indeed amplify voltage, but it is a bilateral passive circuit and quite unlike the active electronic devices which were considered in Section 4.5: it is entirely reciprocal, its current gain varies inversely as its voltage gain, and its power gain is unity in the ideal case. (The power gain of an actual transformer is its efficiency and is of course less than unity because of some loss in its inevitable resistance.)

Example 4.10 Two coils have self-inductances 2.2 H and 0.022 H respectively; their mutual inductance is 0.2 H. Find the output-to-input voltage and current ratios when a voltage of frequency 50 Hz is applied to the first coil and a resistance of 10 Ω is connected to the second. Compare these with the ratios which would obtain were the same inductances perfectly coupled.

The voltage ratio is given by eqn 4.36, in which we now have $\omega = 2\pi \times 50 = 314$ rad s^{-1}, and

$$L_1 L_2 - M^2 = 0.0484 - 0.04 = 0.0084 \text{ H}^2.$$

Hence the equation gives

$$V_2/V_1 = MZ_L/(L_1Z_L + j314 \times 0.0084)$$
$$= 2/(22 + j2.64)$$
$$= \mathbf{0.0903|{-6.8°}}.$$

The current ratio is, according to eqn 4.37,

$$I_2/I_1 = -j\omega M/(j\omega L_2 + Z_L)$$
$$= -j314 \times 0.2/(j314 \times 0.022 + 10)$$
$$= -j62.8/(j6.908 + 10)$$
$$= \mathbf{-5.17|55.4°}.$$

(In this the negative sign could of course be incorporated in the phase angle, which would then becomes $(55.4–180)°$, but there is more to be said for the form given, in which the phase is that relative to the ideal case.)

If the same inductance values were perfectly coupled then the mutual inductance would become

$$M = \sqrt{L_1L_2} = 0.22 \text{ H}$$

and the voltage ratio would be, from eqn 4.45,

$$V_2/V_1 = M/L_1 = 0.22/2.2 = \mathbf{0.1}$$

which is the ideal value. Equation 4.37 shows that the current ratio depends directly on M when only the coupling changes, so the value above increases by the factor $0.22/0.2$ to become

$$I_2/I_1 = \mathbf{-5.68|55.4°}.$$

It may be noted from this example that whereas the voltage ratio is quite near to the ideal value, the current ratio is grossly different and is hardly improved by the coupling, because it is affected chiefly by the magnitude of the inductances according to eqn 4.37. It is also important to bear in mind that the actual ratios are not constants of the transformer but depend on its load impedance.

It would be wrong to suppose that in practice all four-terminal networks which contain coupled circuits are aimed at the ideal of perfect coupling and large inductances: Section 5.3.7, for example, describes an application which requires the coupling to be less than perfect. It is also worth noting at this point that, while it may appear from the foregoing that the relations for the ideal transformer are independent of frequency, the arguments used break down for the d.c. case, $\omega = 0$ (because the reactance of an infinite inductance is then indeterminate); it would be surprising if it were otherwise for a device which depends on electromagnetic induction, and it is a familiar fact that transformers will not ordinarily function on d.c.

4.9 Transformers

4.9.1 General

The transformer, the practical realization of the ideal coupled circuit described above, is one of the most important of all electrical devices. It is its ability to transform a.c. voltage and current which allows the economic transmission of electrical power in bulk, because the voltages required for such transmission (ranging up to several hundreds of kilovolts) are quite unsuitable for individual consumers, for whom the 240 V of the domestic supply in the United Kingdom is typical, and they are also greater than the economic range for generation (around 30 kV). Although it is tempting, indeed usual, to regard transformation of voltage as the essential feature in a power system, this is largely because such a system acts as a very good voltage source; in fact it is the initial need to transform current which dictates high-voltage transmission, for the currents needed to transmit sufficient power at the low voltages needed by consumers would be prohibitively high. In short, insulators for high voltage are much cheaper than conductors for high current, and the difference is more than enough to pay for transformers.

Apart from their indispensability in power systems, transformers in a wide range of sizes are used throughout electrical engineering. Voltage transformation is often used between source and load where the voltage supplied by the one differs from that needed by the other. High values of current or voltage can be changed to lower values for measurement purposes by 'instrument' transformers. In the former case the device is usually known as a **current transformer**: its input is connected in series with the current to be measured and its output current, less by a factor n, say, is passed through the measuring device. The impedance of the latter, according to the relations in the previous section, is reduced by a factor n^2 so far as the input is concerned; hence, like any good ammeter, the whole device need have very little effect on the parent circuit. Because they also transform impedance (a property we encountered in electronic terms in Section 4.6.3) transformers can be used to match a source to a load in order to achieve maximum power transfer (Section 3.11.4). There are, in addition, situations where there is no requirement to transform voltage, current, or impedance, but in which a transformer is used simply as an isolating device because its two circuits are electrically separate in the sense of being insulated from each other.

For special purposes transformers can be designed with particular properties, in some cases involving significant departures from ideal behaviour. In most applications, however, and in particular those which involve the transmission of power at more or less constant voltage, the behaviour of a transformer should be as near to the ideal as is economically possible. This requires not only that the inductances should be very large and closely coupled, but also of course that they should be as pure as possible, that is the resistance inevitably associated with inductance in practice should be small. Not only does resistance affect the relations given in the previous section (although often to a small degree), more importantly it also causes dissipation of power in the transformer which must be minimized whenever efficiency is a prime consideration. It is easy to show that an ideal transformer transmits power from input to output without loss, but this is equally true of any two pure inductances, however coupled; it is the presence of resistance, as opposed to reactance, which causes power loss in any four-terminal (or other) network.

4.9.2   The non-ideal transformer

Large values of inductance and a high degree of coupling can be achieved by using cores which have a very high value of permeability μ (Appendix B.3). In these a given loop of current can produce much more magnetic flux, and therefore a larger inductance (see Section 1.2.5), than in other materials. Most materials have a permeability quite near to μ_0, the value in vacuum: the few elements and their alloys which have much higher values are known as **ferromagnetic** materials because iron is the most familiar of them. Various alloys are available, some of which are designed to have exceptional properties for special applications, but in practice those used for most transformers are steels designed to have good magnetic performance at acceptable cost: they are commonly referred to simply as 'iron'. To reduce eddy-current losses (Appendix B.3.5), iron cores are almost invariably made up of thin insulated sheets or **laminations**.

To achieve the highest values of inductance a core should form one or more closed paths for magnetic flux, a so-called **magnetic circuit** (Appendix B.3.4). The coils which form the two (or occasionally more) inductances of a transformer are wound closely around the core; they are known conventionally as **windings**, that intended for the input as the **primary**, that for the output as the **secondary**. Various geometrical arrangements are possible, two of which are shown in Fig. 4.32. In addition there are common arrangements, not included here, particular to transformers for polyphase systems (Chapter 6); but exactly the same principles apply to these and they can when necessary be regarded as a combination of single-phase transformers such as we have tacitly been assuming so far.

A disadvantage, for some purposes, of iron-cored transformers is their non-linearity, caused by the fact that the permeability of a ferromagnetic material is far from constant, depending quite strongly on the level of flux density; for small-scale applications where linearity is important and other considerations less so, **air-cored** transformers are sometimes used. For most applications, however, the effects of non-linearity are not serious in a well-designed transformer and are a small price to pay for the benefits of a magnetic core; iron cores are in any case an economic necessity for large transformers. In what follows here we

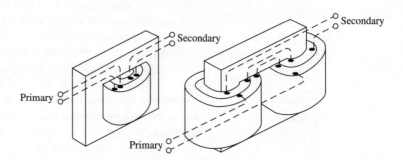

Fig. 4.32
Transformer cores.

assume linear behaviour, and hence constant values of all circuit parameters; in the next section we consider some further points relating to iron-cored transformers in particular.

On the assumption of linearity, the behaviour of a transformer with imperfect coupling and finite inductances can be deduced from eqns 4.35, with terms added to include the inevitable resistance of the windings. However, for engineering purposes, a slightly different approach which leads to relatively simple equivalent circuits (other than the T equivalent of Fig. 4.31) is often preferred; in effect it divides the circuit into elements which separately represent the ideal transformer on the one hand and imperfections on the other, in the following way.

The inductance of each winding is first divided into mutual and leakage components as outlined in Section 3.10.2, so that L_1 comprises a mutual component $k_1 L_1$ and a leakage component $(1 - k_1)L_1$, and likewise for L_2. According to eqn 4.41 the mutual components, which are by definition perfectly coupled, can be written as ρM and M/ρ; hence any two inductances can be represented by the circuit of Fig. 4.33(a), comprising two self-inductances in addition to a perfectly

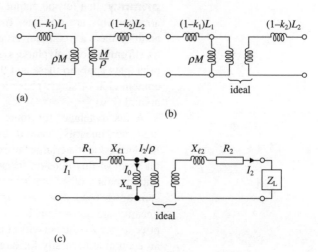

Fig. 4.33
Development of an equivalent circuit for the non-ideal transformer.

coupled pair. The next stage is to extract from the latter an ideal transformer, having infinite inductances. This is most easily done by noting that with an open-circuited output the input impedance of an ideal transformer is infinite, whereas the input impedance of two perfectly-coupled but finite inductances with an open-circuit output is simply the self-inductance ρM. The correct impedance is therefore obtained by inserting this inductance in parallel with an ideal transformer, as shown in Fig. 4.33(b). It is not very difficult to show that this equivalent circuit not only satisfies eqn 4.38 for the input impedance with *any* load, but also satisfies all the other equations for coupled inductances. In it the ideal transformer behaves precisely according to eqn 4.51 so far as *its own* voltages and currents are concerned: the relations between the actual terminal quantities are clearly not so simple. It remains only to represent the fact that actual inductors inevitably have resistance; this is easily done by adding R_1 and R_2 to the input and output meshes. (Capacitance too can be added when need be, but in practice the 'stray' capacitance of transformers is negligible for many applications.) It is common to label the circuit with reactances rather than inductances, since most transformers operate at a given frequency; we denote the leakage reactances $\omega(1 - k_1)L_1$ and $\omega(1 - k_2)L_2$ as $X_{\ell 1}$ and $X_{\ell 2}$, and the reactance $\omega \rho M$ as X_m. The result of all this is the circuit shown Fig. 4.33(c), which correctly represents any transformer for which linearity can be assumed.

The reactance $\omega \rho M$ is known as the **magnetizing reactance** and its current I_0 as the **magnetizing current**, i.e. that part of the input current I_1 to which can be attributed the magnetic flux which is mutual. The rest of I_1, given by [†]I_2/ρ, flows into the ideal transformer and can be thought of as cancelling the mutual flux due to the output current I_2. (This is simply another way of viewing the fact that causing I_2 to flow by connecting a load impedance to the output reduces the input impedance of the coupled circuit.) Note that I_0 is not constant here, for the voltage across it depends on the total input current, that is on the load Z_L. This circuit can be cast into a number of other forms; some of these are precise alternatives, others are simplifications based on approximation. The former arise from the impedance transformation included in eqns 4.51; for just as the load impedance Z_L presents a transformed value $\rho^2 Z_L$ to the primary circuit, so too is any other impedance in the secondary transformed. Thus the impedance $R_2 + jX_{\ell 2}$ presents, to the input of the ideal transformer, the impedance $\rho^2(R_2 + jX_{\ell 2})$; the former can therefore be removed from the secondary and replaced by the latter in the primary, as shown in Fig. 4.34(a). Again, while it is natural to shown the magnetizing reactance X_m on the primary side because it was inserted to give the correct input impedance, it can equally well be shown instead as X_m/ρ^2 (corresponding to an inductance M/ρ across the output of the ideal transformer; and the

---

[†]From here we take I_2 to have its more natural direction as an output current, which simply removes the negative sign from the current ratio.

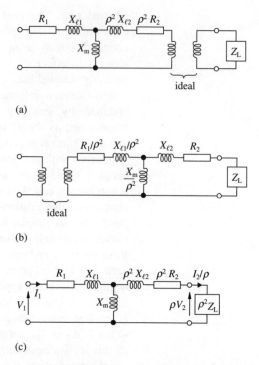

Fig. 4.34
Alternative forms for the circuit
of Figure 4.33 (c).

impedance $R_1 + jX_{\ell 1}$ in the primary can then be transformed by the same multiple and similarly moved to the secondary, as shown in (b). Thus impedances can be moved from either side to the other subject only to the ρ^2 transformation; they must of course retain the same relative positions to each other: the process can be regarded as moving the ideal transformer to different parts of the circuit. An impedance which is thus moved from its own side to the other is said to be **referred** from the first to the second. In the limit all impedances, the load or source impedance included, can be referred to one side and the ideal transformer omitted: for example, referring all to the primary gives the equivalent circuit of Fig. 4.34(c). Loss of the ideal transformer means only that the output quantities are ρV_2 and I_2/ρ instead of V_2 and I_2. The behaviour of the transformer with an actual load Z_L could be correctly predicted from this circuit.

Example 4.11 If the coils in Example 4.10, with the same load resistance, have a turns ratio of 10, find the leakage reactances $X_{\ell 1}$ and $X_{\ell 2}$ and hence confirm from the equivalent circuit of Fig. 4.34(c), with the resistance of the windings neglected, the current ratio at 50 Hz. Compare this with the value of the ratio at 400 Hz and with the ideal value.

Using the turns ratio given here with the inductance values from the earlier example gives, from eqn 4.41,

$$0.22k_2 = 0.2 = 0.22k_1$$

and so

$$k_1 = k_2 = 0.90909$$

from which the definition of leakage reactance gives, at 50 Hz,

$$X_{\ell 1} = \omega(1 - k_1)L_1 = 314 \times 0.0909 \times 2.2 = \mathbf{62.8\ \Omega}$$

and

$$X_{\ell 2} = \omega(1 - k_2)L_2 = 314 \times 0.0909 \times 0.022 = \mathbf{0.628\ \Omega}$$

In the circuit of Fig. 4.34(c) we then have

$$X_m = \omega\rho M = 628\ \Omega; \quad \rho^2 X_{\ell 2} = 62.8\ \Omega; \quad \rho^2 Z_L = 1000\ \Omega.$$

Since the current I_1 divides into I_0 and I_2/ρ in the two right-hand branches the current ratio can be quickly found by current division. Thus:

$$
\begin{aligned}
I_2/\rho I_1 &= jX_m/\{jX_m + \rho^2(jX_{\ell 2} + Z_L)\} \\
&= j628/(j628 + j62.8 + 1000) \\
&= 0.517\underline{|55.4°}
\end{aligned}
$$

and hence the current ratio may be written

$$I_2/I_1 = \mathbf{5.17\underline{|55.4°}}$$

which agrees with the previous calculation, except that the negative sign is now removed.

At 400 Hz all reactances are increased by a factor 8, so that X_m becomes 5027 Ω and $X_{\ell 2}$ becomes 5.027 Ω; we now find

$$
\begin{aligned}
I_2/\rho I_1 &= j5024/(j5024 + j502.4 + 1000) \\
&= 0.895\underline{|10.3°}
\end{aligned}
$$

and so the current ratio is now

$$I_2/I_1 = \mathbf{8.95\underline{|10.3°}}$$

which can be compared with the ideal value given directly by the turns ratio, that is $\mathbf{10\underline{|0°}}$.

The above example illustrates one effect that frequency has on the question of a transformer: because the magnetizing reactance increases, the current ratio at the higher frequency is much nearer to the ideal value. However, we should find on investigation that the voltage ratio

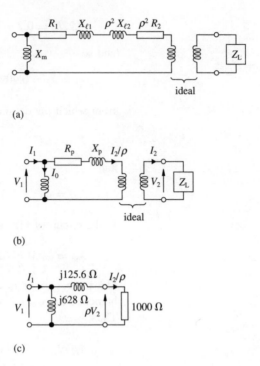

Fig. 4.35
(a), (b) Approximate equivalent
circuits for the iron-cored transformer; (c) circuit for Example
4.12.

was made worse because of the corresponding increase in leakage reactance.

4.9.3 The iron-cored transformer

There is little to choose between the equivalent circuit of 4.33(c) and its alternative forms shown in Fig. 4.34; any of them can be used according to convenience, even for a transformer which is far from ideal. The iron-cored transformer, however, being often quite close to the ideal, allows certain useful approximations. If the leakage reactances and resistances are small, and the magnetizing reactance large, then there will be little error in moving the latter to the input terminals. Figure 4.35(a) shows this done for the version of 4.34(a). It is now a small step to combine the various elements into the simplified circuit shown in 4.35(b) in which R_p represents $R_1 + \rho^2 R_2$, the resistance of the whole transformer referred to the primary, and X_p represents $X_{\ell 1} + \rho^2 X_{\ell 2}$, the leakage reactance of the whole likewise referred. This circuit is very much easier to analyse than earlier versions, because its magnetizing current depends only on the input voltage. It also has the advantage that its open-circuit voltage ratio is precisely ρ, and for this reason it is common to specify the open-circuit voltage ratio (sometimes called the nominal ratio) of a transformer rather than its turns ratio.

Further approximations, such as neglecting I_0 and/or the resistances, are sometimes made. For a transformer near to the ideal there is only a small voltage drop due to resistance and leakage reactance, so the voltages V_1 and ρV_2 are nearly equal even when it is supplying a load;

and since the magnetizing current I_0 is small there is little difference, for normal loads, between the currents I_1 and I_2/ρ. The relationship between V_2 and I_2 (and hence also that between V_1 and I_1) depends, of course, on the load.

Example 4.12 Estimate the current and voltage ratios at 50 Hz of a transformer whose windings and load have the parameters given in Examples 4.10 and 4.11, according to the circuit of Fig. 4.35(b) with resistance neglected.

From Example 4.11 the values of $X_{\ell 1}$ and $\rho^2 X_{\ell 2}$ are each 62.8 Ω so the circuit is now that shown in Fig. 4.35(c) when the 10 Ω load is referred to the primary side. By potential division the voltage ρV_2 is given by

$$\rho V_2/V_1 = 1000/(1000 + \text{j}125.6)$$
$$= 0.992\underline{|-7.2°}$$

and the voltage ratio is therefore

$$V_2/V_1 = \mathbf{0.0992\underline{|-7.2°}}.$$

The current ratio can be found by current division, which gives

$$I_2/\rho I_1 = \text{j}628/(1000 + \text{j}125.6 + \text{j}628)$$
$$= 0.502\underline{|53°}$$

and hence

$$I_2/I_1 = \mathbf{5.02\underline{|53°}} .$$

Comparing the results of this example with those in Example 4.10, we see that the errors here due to approximation are about 10% for the voltage ratio and about 3% for the current ratio. The latter is in each case very different from the ideal value, owing mainly to the relatively low inductance values which make the magnetizing current a substantial part of the input current for the given load; for a large iron-cored transformer in practice this discrepancy would be much smaller for a typical load.

There is an addition which is sometimes made to the circuit in order better to represent an iron-cored transformer. Any conducting material which carries alternating magnetic flux has induced in it eddy currents; a ferromagnetic material suffers in addition the phenomenon of hysteresis (Appendix B.3). Both of these effects cause dissipation of power in the core which shows as a change in the effective impedance of the primary, that is in the impedance presented to the magnetizing current: it is as if the core formed an additional secondary winding with a resistive load.

The combined loss can be represented as a resistance (or conductance) added to the equivalent circuit. Both effects are highly non-linear, depending on both the frequency and the flux in a given core. However, since the flux is fixed by the input voltage the loss is likewise fixed by a given input voltage and frequency; for many transformers these are constant and in consequence the core loss can be represented by a fixed resistance R_m across the input terminals, that is in parallel with the magnetizing reactance, as shown in Fig. 4.36, such that the total loss is V_1^2/R_m. Sometimes R_m and X_m are shown as the components G_m and B_m of a **magnetizing admittance**. The total magnetizing current which flows in this admittance (and which now has a component $V_1 G_m$ in phase with the input voltage) is also known as the 'no-load' current, since it is the input current when the output is open-circuit.

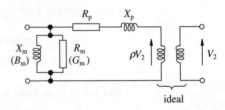

Fig. 4.36
An equivalent circuit with core losses accounted for.

It is a straightforward matter to analyse the behaviour of a transformer, by using any appropriate equivalent circuit, when the load impedance Z_L is specified, as in the above example. However in power systems it is much more convenient to specify a load as a current or a volt-ampere product at a certain power factor (Section 3.9). In such a context the approximate circuits of Fig. 4.35 are appropriate and make the analysis very much easier. If I_2 is given together with the power factor which specifies its phase relative to V_2, it may be required to calculate the value of V_2 for given V_1. The difficulty here can be made clearer from the phasor diagram of Fig. 4.37, which represents the circuit of Fig. 4.36. Here the magnitudes of V_1 and I_2 are known, as is the angle ϕ_2 but not the angle α. From the diagram the magnitude of V_2 can best be found by taking it to be the reference phasor as shown. Alternatively, if only magnitudes are needed, the calculation can be done very simply by making the further approximation that, because the voltage drop $(I_2/\rho)(R_p + jX_p)$ is small, the angle α is also small.

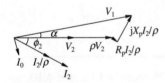

Fig. 4.37
The phasor diagram for the circuit of Figure 4.36.

Example 4.13 A transformer is designed for an input voltage of 11 kV and an open-circuit output voltage of 3.3 kV. Its windings have total resistance and leakage reactance, when referred to the primary, of 2 Ω and 9 Ω respectively. Find the drop in output voltage when a current of 200 A at 0.8 power factor lagging is taken from the output.

It may be assumed from the figures given, which imply a power output in the region of 500 kW, that the transformer is large enough to be sufficiently well represented by the circuit of Fig. 4.36 and the phasor diagram of Fig. 4.37. Making no further approximation as yet, we may resolve the voltage drops into their projections on the real and imaginary axes and write

$$V_1 = V_2' + I_2'(X_p \sin\phi_2 + R_p \cos\phi_2) + j I_2'(X_p \cos\phi_2 - R_p \sin\phi_2)$$

in which V_2' and I_2' represent the magnitudes of ρV_2 and I_2/ρ respectively (i.e. the output quantities referred to the primary). Here ρ is given by the open-circuit voltage ratio, that is $11/3.3$ or $10/3$, so we have

$$I_2' = 200 \times 3/10 = 60 \text{ A},$$

and from the given power factor of 0.8 we have $\cos\phi_2 = 0.8$ and hence also $\sin\phi_2 = 0.6$. Since $V_1 = 11 \times 10^3$ V the voltage equation above becomes

$$11^2 \times 10^6 = \{V_2' + (60 \times 7.0)\}^2 + \{60 \times 6.0\}^2$$
$$= V_2'^2 + 840V_2' + 306000$$

from which

$$V_2' = 10,574 \ V$$

and the output voltage is therefore

$$V_2 = V_2' \times 3/10 = 3172 \ V,$$

implying a drop of **128 V** from the open-circuit value.

Alternatively, if the approximation is made that the angle α is small, then the imaginary term in the above equation can be omitted; this gives

$$V_2' = 10580 \text{ V}$$

and V_2 is now 3174 V, and the drop **126 V**. The actual value of α in this case is $\sin^{-1}(60 \times 6.0/11,000)$, or about $1.9°$.

If on the other hand the load is specified only by the product $V_2 I_2$ and the angle ϕ_2, the calculation of V_2 is more awkward. In such a case it is common to neglect the voltage drop in the first instance by taking V_2 to be V_1/ρ, so that I_2 follows and the calculation for the actual value of V_2 proceeds as before. If need be this can be used to correct I_2 for a further iteration, but if the voltage drop is a small fraction e of the input voltage then the error caused by the initial assumption is only in the order of e^2.

The change in output voltage as a transformer is loaded is sometimes referred to as its **regulation**. Although the output voltage usually falls with increasing current, as in the above example, a capacitive load can cause it to rise (essentially because the reactance of the load then tends to cancel the leakage reactance of the transformer). The maintenance of correct voltages on a power system is so important that large transformers are sometimes fitted with so-called **tap-changers** which can vary the effective number of turns on a winding in order to compensate for variations in load.

The efficiency of large transformers is also very important. It may be calculated, for a given output power, from the power lost in the resistances R_m and R_p (or its counterpart R_s, the total resistance referred to the secondary). Of these two losses, the former, representing the dissipation in the core, depends only on V_1 and is therefore normally nearly constant, but the latter, the loss in winding resistance, varies as the square of the output current and therefore approximately as the square of the output power. As a result the efficiency reaches a maximum at a certain load, and a power transformer is designed to operate at or near this condition.

4.9.4 Measurement of parameters

The values of the four parameters shown in Fig. 4.36 can easily be deduced for an actual transformer by performing an **open-circuit test** and a **short-circuit test**. If one winding (often the low-voltage) is supplied at its normal voltage with the other open-circuit, only the magnetizing current flows; from readings of voltage, current and power at the input values of R_m and X_m can be deduced (and referred to the high-voltage side if need be). If now one winding is short-circuited (often the low-voltage) and the other supplied at reduced voltage to avoid very high currents, most of the input current flows in R_p and X_p (which are much less respectively than R_m and X_m); so the same three readings yield approximate values of R_p and X_p. If need be allowance for the small magnetizing current can be made.

Example 4.14 A transformer with a turns ratio of 14:1 is supplied at 240 V on the low-voltage side with the high-voltage terminals open-circuit; it takes a current of 16 A and 1100 W of power. When the high-voltage side is supplied at 280 V with the other short-circuited, it takes 17 A and 850 W. Estimate the parameters of the equivalent circuit of Fig. 4.36, referred to the high-voltage side.

In the open-circuit test the input power is $V^2/R_m{}'$, so referred to the input

$$R_m{}' = 240^2/1100 = 52.36 \ \Omega.$$

The current in $R_m{}'$ is then $240/52.36 = 4.584$ A; the current in $X_m{}'$ is therefore $\sqrt{(16^2 - 4.584^2)}$ or 15.33 A and it follows that the magnetizing reactance referred to the input is

$$X_m{}' = 240/15.33 = 15.66 \ \Omega.$$

Referred to the high-voltage side these values are simply multiplied by 14^2 to give

$$R_m = 10.3 \text{ k}\Omega; \quad X_m = 3.07 \text{ k}\Omega.$$

In the short-circuit test the input power to the high-voltage side is $I^2 R_p$, so

$$R_p = 850/17^2 = 2.94 \; \Omega.$$

The impedance of R_p and X_p in series is V/I which is $280/17$ or $16.47 \; \Omega$. Hence

$$X_p = \sqrt{(16.47^2 - 2.94^2)} = 16.2 \; \Omega.$$

Problems

4.1 Find the impedance parameters of the potential divider shown in Fig. 2.12(a).

4.2 Find the impedance and admittance parameters for the circuit of Fig. 4.4(c) when each resistance has value $1 \; \Omega$.

4.3 Find the transmission parameters, and their inverse, for the circuit of Fig. 4.4(c) when each resistance has value $1 \; \Omega$.

4.4 Find the transmission parameters for the circuit shown.

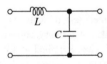

4.5 Find the h parameters for the circuit of Problem 4.4.

4.6 Show that the impedances of the T equivalent for the bridge circuit shown are given by

$$Z_a = Z_b = \frac{R(R_1 + R_4) + 2R_1 R_4}{2R + R_1 + R_4}$$

$$Z_c = \frac{R^2 - R_1 R_4}{2R + R_1 + R_4}$$

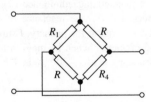

4.7 Find the resistances of the Π equivalent for the

circuit of Fig. 4.4(c) with each resistance $1 \; \Omega$. [Hint: use the answers to Problem 4.2.]

4.8 Find the h parameters of a four-terminal network in terms of its z parameters.

4.9 A four-terminal network has the parameters $z_{11} = z_{22} = 500 \; \Omega$, $z_{12} = 0$, $z_{21} = -5 \text{ k}\Omega$. Find the values of R_1, R_2, γ and β according to the representation in Fig. 4.5(a).

4.10 In a certain BJT the collector current is approximately 50 times the base current for any value of v_{CE} above the saturation value of $0.5 \; \text{V}$. Find the base current needed for saturation when the transistor is connected in series with a load resistance of $2 \text{ k}\Omega$ to a supply voltage of $18 \; \text{V}$.

4.11 A JFET has parameters $g_m = 3 \; \text{mS}$ and $r_d = 50 \text{ k}\Omega$. Find from the circuit of Fig. 4.14(a) its voltage gain v_{ds}/v_{gs} for $R_L = 3.3 \text{ k}\Omega$.

4.12 The BJT of Problem 4.10 has $h_{ie} = 1 \text{ k}\Omega$. Estimate its voltage gain v_{ce}/v_{be} from the circuit of Fig. 4.14(b) with a load $R_L = 5 \text{ k}\Omega$.

4.13 Find the current and voltage gains of the circuit in Fig. 4.14(b) when the transistor has parameters $h_{ie} = 1 \text{ k}\Omega$, $h_{fe} = 50$, and $h_{oe} = 10^{-4} \; \text{S}$, for a load $R_L = 5 \text{ k}\Omega$.

4.14 Find the new voltage gain in Problem 4.13 when the output in Fig. 4.14(b) is connected to external devices of resistance (a) $10 \text{ k}\Omega$ and (b) $1 \text{ k}\Omega$.

4.15 Find the value of the resistance R_s which could provide the bias needed for the operating point specified in Example 4.4, and the consequential voltage gain of the arrangement specified in part (d).

4.16 Show that the voltage gain of the BJT amplifier circuit in Fig. 4.16(d) with a bypass capacitance C connected across R_e is given for frequency ω by the expression

$$A_v = -\frac{h_{fe}R_L(1 + j\omega T)}{h_{ie} + (h_{fe} + 1)R_e + j\omega Th_{ie}}$$

where $T = CR_e$. [Hint: consider eqn 4.19 with R_e replaced by the appropriate impedance.]

4.17 Find the input impedance of the circuit in Fig. 4.16(d).

4.18 Show that the input impedance of the circuit of Fig. 4.3(b) with a load R_L connected across R_2 is

$$Z_{in} = \frac{R_1(R_2 + R_L + j\omega CR_2R_L)}{R_2 + R_L + j\omega C(R_1R_2 + R_1R_L + R_2R_L)}.$$

4.19 Show that the output impedance of the circuit in Fig. 4.16(b) is given by R_L.

4.20 Show that the output impedance of the circuit in Fig. 4.5(a), when V_1 represents a perfect voltage source, is

$$Z_{out} = \frac{R_1R_2}{R_1 - \beta\gamma R_2}.$$

4.21 Find the reduction in the voltage gain of the circuit described in Problem 4.13 when the input signal source has Thévenin resistance 600 Ω.

4.22 What is the maximum power gain of the circuit of Problem 4.13, and into what external load resistance would it be delivered?

4.23 A JFET having $g_m = 3$ mS and negligible drain conductance is to be used as a source follower with output impedance no greater than 250 Ω. What value should be used for the resistance R_s, and what is then the voltage gain?

4.24 Find the input and output impedances, and the voltage gain, of an emitter follower which uses the transistor of Problem 4.13 and a resistance R_e of 1 kΩ.

4.25 Find the iterative and image impedances of the network of Fig. 4.4(c) when $R_1 = R_2 = 1$ kΩ and $R_3 = R_4 = 2$ kΩ. [Hint: use the impedance parameters given in Example 4.3.]

4.26 Find the Thévenin equivalent, with respect to a load on the secondary, of a pair of coupled coils having inductances $L_1 = 1$ H and $L_2 = 0.01$ H, and negligible resistances, when the primary is connected to an ideal voltage source of 200 V at 50 Hz and the coupling coefficient k is (a) 0.8 and (b) 1.

4.27 Find the voltage ratio of the coils of Problem 4.26 for each value of k when the secondary is connected to a load resistance of 10 Ω.

4.28 Find the current ratio of the coils in Problem 4.26 for each value of k with a load as in Problem 4.10.

4.29 Two coupled coils are to be designed with the primary having six times the number of turns of the secondary. If the secondary is connected to a load resistance of 10 Ω, and the coupling can be assumed perfect, what values of inductance are needed for a current ratio within 10% of the ideal magnitude, at 50 Hz?

4.30 What is the maximum phase difference (discounting a negative sign) between the currents in Problem 4.29?

4.31 A transformer with a primary-to-secondary turns ratio of 10 is represented by the circuit in Fig. 4.33(b) when $k_1 = k_2 = 0.9$ and $L_1 = 5$ H. Find the voltage and current ratios when the output is connected to an inductance of 0.1 H. [Hint: refer all impedance to the primary side.]

4.32 Find the new values for the ratios in Problem 4.31 when the reactance ρM in Fig. 4.33(b) is moved to the input terminals.

4.33 A transformer tested with its output terminals short-circuited takes a primary current of 351 A and a power of 15 kW when supplied at a primary voltage of 400 V. Estimate its resistance and leakage reactance referred to the primary, according to the equivalent circuit of Fig. 4.35(b).

4.34 The transformer of Problem 4.33, tested with its output open-circuit, takes a primary current of 9 A and a power of 4 kW when the primary is supplied at 2000 V. Estimate its magnetizing admittance referred to the primary.

4.35 A transformer has a primary-to-secondary turns ratio of 8, a magnetizing admittance of $0.8\lfloor{-80°}$ mS, and resistance and leakage reactance of 1 Ω and 9 Ω respectively, referred to the primary. Estimate its output voltage and efficiency when its input voltage is 2.2 kV and its secondary load is 200 A at 0.8 power factor lagging.

4.36 Find the input volt-amperes and power factor for the transformer of Problem 4.35 under the conditions given.

5 Frequency response

5.1 System functions

In Chapter 3 we saw that the reactances $j\omega L$ and $1/j\omega C$ gave rise to impedances and admittances which are generally complex and depend on the frequency ω (or f). It follows also that circuit properties generally, such as the parameters and gains of four-terminal networks which were discussed in Chapter 4, are often functions of frequency. In practice, while many devices and systems are designed for a fixed frequency, others are intended to cope with a range of frequencies; it is then important to know how their properties vary with frequency. Such variation is known generally as **frequency response**. The term arises from the dependence, in magnitude and relative phase, of a sinusoidal output (the response) on the frequency of an input (the stimulus, or driving function) of given magnitude. But it is not strictly confined to four-terminal networks, nor even to electrical circuits, nor is it a unique property of a given circuit or device. For example, the impedance of a simple two-terminal combination of elements gives the response of the voltage across them to an applied current source which varies only in frequency; its inverse, the admittance, on the other hand, gives the response of the current to a frequency-varying voltage source.

The idea of frequency response has already been met in Section 3.7.3, where it was pointed out that a polar diagram can show the locus of a phasor or other complex quantity as frequency varies, and that the same information can also be shown as separate diagrams of magnitude and phase. Whatever the graphical representation, the steady-state response of a linear system can in general be expressed as a complex algebraic function of the variable ω; for many purposes it is preferable to regard the variable as $j\omega$, since the frequency always appears in this form in the first instance. In order to make the response characteristic of the system (that is, independent of the input quantities) it should always be the ratio of the phasor quantities which are chosen to represent respectively the response and the stimulus. Such a function is known in general as a **system function**, but sometimes simply as a response function or frequency response. It may be an impedance or an admittance, or the ratio of two similar quantities (for example, a voltage or current gain as described in Chapter 4). When the two quantities refer to different points or terminal pairs in a circuit or other system, such as input and output, the system function becomes a **transfer function**; otherwise it is a **driving-point** function which for a circuit must be an immittance (Section 4.2.2). A function representing a gain is bound to be a transfer function, but so too is the ratio of current at one point to voltage at another, or vice versa. Once such a function is known, it is a straightforward matter to plot its real and imaginary parts as a polar diagram having ω as a parameter or, alternatively, to plot its magnitude and phase separately against ω.

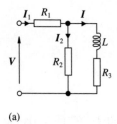

(a)

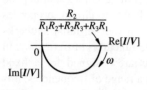

(b)

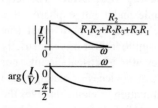

(c)

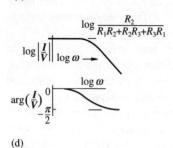

(d)

Fig. 5.1

Example 5.1: (a) circuit; (b) polar diagram; (c) frequency response on linear scales; (d) frequency response on logarithmic scales.

Example 5.1   Find the transfer function I/V for the circuit shown in Fig. 5.1(a).

The function may be found by writing Kirchhoff's second law for the two meshes, or by using current division to find I from V, or by finding V from I. In the last case, the current I gives a voltage $I(R_3 + j\omega L)$ across R_2 which therefore carries current

$$I_2 = I(R_3 + j\omega L)/R_2.$$

This means that the total current through R_1 is given by

$$I_1 = I_2 + I = I(R_2 + R_3 + j\omega L)/R_2$$

from which the total voltage is

$$V = R_1 I_1 + I(R_3 + j\omega L)$$
$$= I\{R_1(R_2 + R_3 + j\omega L) + R_2(R_3 + j\omega L)\}/R_2$$

and the transfer function is therefore

$$\frac{I}{V} = \frac{R_2}{R_1 R_2 + R_2 R_3 + R_3 R_1 + j\omega L(R_1 + R_2)}. \tag{5.1}$$

It is not difficult to verify that the polar diagram of this function is a semi-circle as shown in Fig. 5.1(b). Its similarity to the polar diagram of a simple admittance function which was shown in Fig. 3.10(b) may be noted. The variation of its magnitude and phase with ω is shown in (c). It is a useful exercise to reconcile these diagrams with each other and with the physical behaviour to be expected. Both (b) and (c) show that the magnitude of I/V decreases monotonically with ω because the reactance ωL increases without limit to infinity, and then $I = 0$ for any V; while at d.c., $\omega = 0$, L is a short-circuit and I is fixed by the resistances only. For this latter reason the phase angle of I/V must tend to zero in the limit $\omega = 0$; for very high ω the voltage across R_2 is in phase with V since R_1 and R_2 carry the same current, and the large reactance of L causes the vanishingly small I to lag that voltage by $\pi/2$.

5.2 Response diagrams

5.2.1 Logarithmic scales

In practice it is usual, for reasons which will become evident, to show the curves of Fig. 5.1(c) against a logarithmic scale of ω, and to use also a logarithmic scale for magnitude, in which case they appear somewhat as shown in (d), the scale for phase angle remaining linear. The use of logarithmic scales for frequency and magnitude is so common that specific notations are used for them, which are worth a small digression here. On a logarithmic scale, equal intervals represent a certain multiple: for example, an increase of unity in the common logarithm of a varying quantity represents multiplication by ten. In the case of frequency such an interval is called a **decade** (the interval representing a factor two is called an **octave**, for reasons which arise in the theory of musical scales). In the case of magnitude, an interval of unity in the common logarithm, again representing a factor 10, is historically called the **bel**[†], so that the difference between two values is given in bels by the common logarithm of their ratio. For some practical purposes this measure was found to be too coarse, and the **decibel** (db or dB) was widely adopted. It represents a logarithmic interval of 0.1 and hence multiplication (or division) by the factor $10^{0.1}$ or 1.2589. We might then expect the measure of a ratio in decibels to be ten times its common logarithm.

At this point, for historical reasons, a complication arises. One of the early applications of a logarithmic scale was in the theory of sound, in which intensity, in other words power flow, is a fundamental quantity. In electrical terms power varies as the square of current or voltage in any linear device, hence we may say that a factor $10^{0.1}$ in power is obtained from a factor $10^{0.05}$ in current or voltage, that is to say from an increase (or decrease) of 0.05 in the common logarithm of either. The convention therefore arose, and has persisted, that for voltages or currents the ratio of two values is measured in decibels by 20 times its common logarithm, on the grounds that a decibel then represents, as was originally intended, a factor $10^{0.1}$ in power. The ratio of two powers is of course still given in decibels by ten times its logarithm.

The zero of a scale of bels or decibels obviously occurs when the ratio in question is unity; thus a gain of unity can be expressed as 0 dB. Instead of always working with ratios, it is convenient for some quantities to take unity as a reference value, in which case any other value x is easily expressed as $20 \log_{10} x$ decibels. The value unity has of course no particular significance for a quantity with dimensions, such as admittance in the example above, and in such cases it is often preferred

[†] The word is taken from the name of Alexander Graham Bell, inventor of the telephone.

to choose a suitable reference value to form a ratio, or else to use actual values (but for graphical purposes logarithmically spaced) without the decibel notation. The latter applies to logarithmic frequency scales, which are usually marked with actual frequency values, as opposed to the decibel scale often used for gain, for example. Frequency then appears as a non-linear scale, whereas the decibel scale, being already a logarithmic measure, is uniformly spaced.

Logarithmic measures can also be based on the natural or Naperian logarithm, in which case the unit is called the **neper**[†]; that is to say, the measure of a ratio in nepers is simply its natural logarithm. In this case the definition is usually applied to voltage or current directly, so that the natural logarithm of a power ratio is halved in order to obtain the corresponding voltage or current ratio in nepers. Since a ratio representing n nepers would therefore represent $20 \log_{10}(e^n)$ or $20n \log_{10} e$ or $8.686n$ decibels, it follows that one neper is 8.686 dB, and that one decibel is 0.115 nepers.

The advantages of logarithmic scales include the compression of many orders of magnitude into a manageable compass, the ease of multiplying two or more functions, and the fact that gains greater than unity become positive while attenuation or 'loss' is negative.

5.2.2 Bode diagrams

The appearance of the curves of Fig. 5.1(d) seems to indicate that both magnitude and phase approach asymptotes at both low and high frequencies. To investigate these consider again the function which the curves represent, given by eqn 5.1. It can be written as

$$\frac{I}{V} = \frac{A}{1 + j\omega T} \tag{5.2}$$

in which A and T are constants given by

$$A = \frac{R_2}{R_1 R_2 + R_2 R_3 + R_3 R_1}$$

and

$$T = \frac{L(R_1 + R_2)}{R_1 R_2 + R_2 R_3 + R_3 R_1}.$$

Equation 5.2 shows that for small ω, such that $\omega \ll 1/T$, the admittance I/V approaches the constant value A which, being real, has phase angle zero; while for large ω, such that $\omega \gg 1/T$, the function approaches $A/j\omega T$ which has magnitude $A/\omega T$ and phase angle $-\pi/2$. These limits immediately explain the asymptotes of the phase curve. For magnitude the low-frequency asymptote representing the constant A is again evident, but the limit $A/\omega T$ for high frequencies needs consideration. Its value clearly declines by a factor 10 for every decade of frequency; on a

[†] From the name of John Napier, inventor of logarithms.

decibel scale it must therefore decline by 20 dB per decade. On a diagram with logarithmic scales this means that the curve is asymptotic to a straight line of slope –20 dB per decade. We see here another advantage of logarithmic scales: the *shape* of this diagram is independent of the constants A and T. Any one diagram having this shape is specified completely by two quantities: one is the **level** of the low-frequency asymptote, which in this case is the constant A, and the other is the frequency at which the two asymptotes meet. This is not hard to deduce: equating the limiting expressions for the two lines gives

$$A = A/\omega T$$

which is satisfied by the frequency

$$\omega_0 = 1/T \tag{5.3}$$

and this is known as the **breakpoint** or **corner frequency**. The same terms are applied to any frequency at which asymptotes of differing slopes meet.

The asymptotic version of Fig. 5.1(d) is shown in Fig. 5.2. The line shown joining the two horizontal asymptotes of the phase diagram is not itself an asymptote, since the curve must cross it, but is an approximation to the curve; it can be variously defined, but a line which meets the asymptotes at the frequencies $\omega_0/10$ and $10\omega_0$ is usually deemed to be a good fit. We may note that the actual value of the function at the breakpoint frequency ω_0 is, since $\omega_0 T = 1$ by eqn 5.3,

$$I/V = A/(1+\mathrm{j})$$

which has magnitude $A/\sqrt{2}$ and phase $-\pi/4$. Hence at the frequency ω_0 the phase curve is half-way between its two asymptotes, while the magnitude is less than its asymptotic value by the factor $1/\sqrt{2}$ or, to a good approximation, –3 dB, and this is the maximum error between asymptotes and curve in the present example.

Asymptotic diagrams of this kind for the magnitude and phase of complex system functions are called **Bode**[†] **diagrams**. They are very easy to sketch (even, as we shall see, for more complicated functions than the above example), are very informative, and are often close enough to the true curves for many purposes.

5.2.3 Standard forms

Equation 5.2 takes a form which appears very frequently in system functions. It has already been pointed out that Fig. 5.1(b), the polar diagram of the admittance function which that equation describes, has the same form as Fig. 3.10(b) which represents the admittance of R and L in series, namely

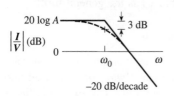
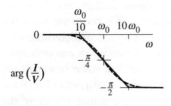

Fig. 5.2
Example 5.1: asymptotic frequency response.

[†] The name is that of the American engineer H. W. Bode.

$$Y = \frac{1}{R + j\omega L} = \frac{1/R}{1 + j\omega(L/R)}.$$

Equation 5.2 represents this function too if

$$A = 1/R$$

and

$$T = L/R.$$

It turns out that for linear systems any system function is rational, which means that it can be written as the ratio of two polynomials, in the variable $j\omega$. A polynomial in the variable x has the general form

$$f(x) = a_n x^n + a_{n-1} x^{n-1} + \dots + a_1 x + a_0$$

where n is an integer and the coefficients a are constant (and in our case real), and whatever the values of these it can always be factorized into some or all of the following: quadratic factors like $1 + bx + cx^2$, linear factors like $1 + dx$, a factor x^r and a real constant; in these b, c and d are also real constant coefficients, and r is an integer or zero. A quadratic factor may itself, of course, have two linear factors like $1 + dx$, depending on the values of the coefficients. We postpone until the following section consideration of those which do not, and consider now functions of $j\omega$ in which both numerator and denominator can only have factors like $(1 + j\omega T)$ together with a constant and some power of $j\omega$. Note that $(1 + j\omega T)$ is dimensionless and that the coefficient of $j\omega$ must therefore have the dimension of time, hence the common choice of the symbol T in such factors. The constant may have whatever physical dimensions are appropriate; in the admittance functions above, the constant A is evidently measured in siemens.

Casting a function into this standard form leads to the very simple construction of its Bode diagrams. Because in the multiplication of complex quantities magnitudes multiply but angles add, the multiplication and division of the factors of a system function requires only the addition and subtraction of their logarithmic magnitudes and of their phase angles; in other words, the Bode diagram of the whole, for both magnitude and phase, can be obtained by superimposing the simple diagrams for its factors. The diagrams representing $(1 + j\omega T)$ in the denominator are easily deduced from the example above by removing the constant, to give the result shown in Fig. 5.3(a); the only difference is that the level of the low-frequency asymptote is now 0 dB. A similar factor in the numerator can be shown by the same arguments as before to give the forms shown in (b); low-frequency asymptotes and breakpoint are as before, but now the high-frequency asymptotes have positive slope and positive phase angle respectively.

To draw the diagrams for a factor $(j\omega)^r$, we note first that its magnitude is unity at $\omega = 1$ and that a decade change in frequency produces a magnitude change of 10^r, which is equivalent to $20r$ dB.

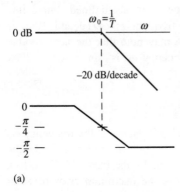

(a)

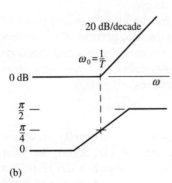

(b)

Fig. 5.3
Bode diagrams for linear factors:
(a) $(1 + j\omega T)^{-1}$; (b) $1 + j\omega T$.

Such a factor is therefore represented by a straight line, without breakpoint, which crosses the 0 dB axis at $\omega = 1$ and has slope $20r$ dB per decade, positive or negative according to the sign of r. The phase of $(j\omega)^r$ is clearly $r\pi/2$, because j represents $\pi/2$, and the phase diagram is therefore a horizontal line showing this constant value, positive or negative. Thus, for example, $j\omega$ on the numerator has slope 20 dB per decade and phase $\pi/2$, and $(j\omega)^3$ on the denominator has slope –60 dB per decade and phase $-3\pi/2$. It remains only to note that a constant factor, like A in the example above, is represented in magnitude by a horizontal line at the appropriate level and has zero phase at all frequencies; its only effect is to raise or lower the resultant magnitude diagram for the whole function.

Example 5.2   Sketch Bode diagrams for the function

$$F(j\omega) = \frac{100j\omega}{1 + j\omega10^{-3}}.$$

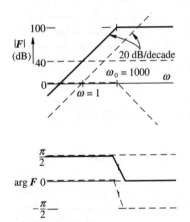

Fig. 5.4
Bode diagrams for Example 5.2.

The denominator is represented by the diagrams of Fig. 5.3(a) with $1/T = 10^3$ s^{-1}. To these must be added, to account for $j\omega$ in the numerator, a positive slope of 20 dB per decade intersecting the axis at $\omega = 1$ rad s^{-1} and a constant phase of $\pi/2$; the constant 100 increases the overall magnitude by $20\log_{10}100$ or 40 dB. The final diagrams are therefore those of Fig. 5.4, in which the component parts are shown in broken lines. Superimposing these is a simple matter if it is noted that breakpoints are unchanged and that two equal and opposite slopes cancel to yield a horizontal asymptote.

Example 5.3   Sketch Bode diagrams for the function

$$G(j\omega) = \frac{1 + j\omega T_1}{(1 + j\omega T_2)^2}.$$

Here we meet a repeated factor, for which the Bode diagrams must be the superposition of two like Fig. 5.3(a). The result of this is very simply obtained: the breakpoint frequency is unchanged; for magnitude the only effect is to double the slope, and the phase angle is doubled everywhere. (For the cube of such a factor, the slope and phase would both be trebled, and so on.) The complete diagrams are found from the superposition of numerator and denominator contributions as shown in Fig. 5.5. The final form depends, as indicated, on the relative values of T_1 and T_2.

It will be evident from the examples given that for every factor of the form $(1 + j\omega T)$ in a system function there is a breakpoint in its Bode diagrams at the angular frequency $1/T$ and, conversely, every breakpoint arises from such a factor. It we write ω_0 for the breakpoint $1/T$ then an

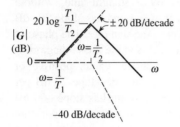

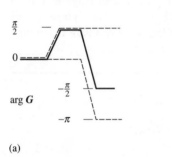

(a)

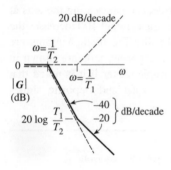

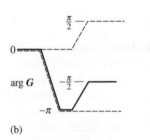

(b)

Fig. 5.5

Bode diagrams for Example 5.3.
(a) $T_1 > T_2$; (b) $T_2 > T_1$.

alternative form for these factors is $(1 + j\omega/\omega_0)$. The name **time constant** is often given to T; the wider significance of this term will be discussed in Chapter 7. It is useful to remember that for passive electrical circuits the time constants which arise in linear factors are in one of the two forms CR and L/R; that of eqn 5.2, for example, is of the latter form.

It is not essential to construct Bode diagrams by superimposing the simple forms for the component factors. An alternative is to apply directly the approach first used, in Section 5.2.2, and consider the limiting forms of the whole function. To do so, we should note that the original asymptotes were obtained by assigning to the factor $(1 + j\omega T)$ one of two limiting values, either 1 or $j\omega T$, according as ω is less than or greater than $1/T$. For every range of frequency it is therefore possible to assign one or other limiting value to each such factor, and the resulting expression immediately fixes the magnitude and phase asymptotes for that range. Using this approach for Example 5.2 would enable us to write the low-frequency limit immediately as $100j\omega$, which gives a magnitude of slope 20 dB per decade and a phase $\pi/2$; similarly the high-frequency limit is $100j\omega/j\omega10^{-3}$, or 10^5, which gives a horizontal asymptote at the level 100 dB and zero phase. These give directly the diagrams obtained by superposition in Fig. 5.4.

Example 5.4   Sketch the Bode diagrams of the function

$$H(j\omega) = \frac{10j\omega(1 + 0.05j\omega)}{(1 + 0.1j\omega)(1 + 0.01j\omega)}$$

The factors show at once that there are breakpoints at 10, 20 and 100 rad s^{-1}. At frequencies below 10, all three of the corresponding factors may be approximated to unity; the form of H is therefore asymptotic to the value $10j\omega$ and its magnitude to 10ω, represented by a line of slope 20 dB per decade having the value 20 dB at $\omega = 1$, while its phase is asymptotic to $\pi/2$, as in Fig. 5.6. Between the frequencies 10 and 20 the factor $(1 + 0.1j\omega)$ takes the value $0.1j\omega$ and the others are unchanged, so H approximates to $10/0.1$, that is the constant value 100 or 40 dB, with angle zero, as shown. Next, at the frequency 20 the factor $(1 + 0.05j\omega)$ changes to $0.05j\omega$, and H becomes $100 \times 0.05j\omega$, which again gives a line of slope 20 dB per decade with phase $\pi/2$. Finally, above 100 rad s^{-1}, the factor $(1 + 0.01j\omega)$ becomes $0.01j\omega$ and H is therefore $100 \times 0.05/0.01$, that is 500 or 54 dB, with zero phase angle once more. The resulting diagrams are as shown in Fig. 5.6. Note that there is no need to confirm that the various asymptotes meet at the breakpoints; knowing that they must do so reduces the calculation needed to locate the asymptotes once the first one is fixed.

It may be observed from the above examples that there is a relation between the diagrams for magnitude and phase: a horizontal asymptote

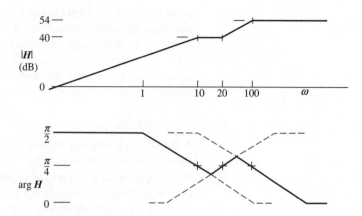

Fig. 5.6
Bode diagrams for Example 5.4.

in the former corresponds to zero phase, a slope of $\pm$ 20 dB per decade to a phase of $\pm \pi/2$, a slope of ± 40 dB per decade to a phase of $\pm \pi$, and so on. This useful relationship, which is based on a formal result known as **Bode's theorem**, allows approximate phase diagrams to be quickly deduced from those of magnitude. It applies to many functions which arise from linear circuits, but it can fail because some circuits give rise to factors having the form $(1 - \mathrm{j}\omega T)$, where T is as before a positive quantity. The high-frequency limit of this has magnitude ωT but its phase is that of $-\mathrm{j}$, i.e. $-\pi/2$, in contrast to $\pi/2$ for the more usual $(1 + \mathrm{j}\omega T)$. In general, systems for which the simple magnitude-phase relation does not hold are said to have **non-minimum phase**.

In the last section it was pointed out that Bode diagrams having the basic form of Fig. 5.2 had a maximum magnitude error of 3 dB and a phase half-way between its asymptotes at the breakpoint. It is, however, worth noting that these statements do not generally hold for diagrams with more than one breakpoint, and that the phase in particular may be far from the value indicated by its asymptotes, especially when the frequency interval between breakpoints is small: this can be seen from Fig. 5.6. If there is any requirement for accuracy the safest, if more laborious, procedure is to calculate the actual values of magnitude and angle at crucial frequencies; but it is also helpful to bear in mind that a polar diagram (Section 3.7.3) gives the same information as a pair of Bode diagrams, in exact form.

5.3 Resonance

5.3.1 Introduction

In the previous section it was mentioned that some system functions contain quadratic factors in $\mathrm{j}\omega$, having the form

$$1 + \mathrm{j}\omega b + (\mathrm{j}\omega)^2 c$$

where b and c are constant coefficients which involve the system parameters. An expression of this form can be factorized in turn, and if its factors have the form $(1 + j\omega T)$ with T real then they behave exactly like those discussed above. If, on the other hand, the parameters are such that $b^2 < 4c$ then its factors do not have that form. This situation can arise in electrical circuits which contain both capacitance and inductance; it produces the important phenomenon of **resonance**, which we consider below.

5.3.2 The parallel resonant circuit

A circuit containing any combination of inductance and capacitance can show resonance, but we consider first the simple case of an inductance L and a capacitance C connected in parallel. Because resistance is always present in practice, a resistance R is included, again in parallel, as shown in Fig. 5.7(a). The admittance of the circuit is

$$Y_p = 1/R + 1/j\omega L + j\omega C$$
$$= 1/R + j(\omega C - 1/\omega L) \tag{5.4}$$

and we see immediately that the admittance becomes real, and has its minimum magnitude $1/R$, when $\omega C = 1/\omega L$. This condition is clearly satisfied when the angular frequency ω has the value

$$\omega_0 = 1/\sqrt{LC} \tag{5.5}$$

which is known here as the **resonant frequency**. The existence of a function which can become real and pass through a turning point in magnitude as the frequency varies is characteristic of resonance; but these two effects do not always coincide precisely, as they do here, so formally we define the resonant frequency of a two-terminal circuit as that frequency for which its admittance and impedance are real and for which, therefore, its terminal voltage and current are in phase.

It can readily be seen from eqn 5.4 how the resonant condition arises: the susceptances (and reactances) of L and C, being of opposite sign and varying in opposing directions with ω, exactly cancel each other at the frequency ω_0. In the present circuit the voltage V is common to all three elements; hence when the reactances cancel at ω_0 the currents taken by L and C must also be equal and opposite, i.e. out of phase by π, and the total current amounts only to that taken by R, as shown on the phasor diagram of Fig. 5.7(b). The currents in L and C actually flow, of course, but not in the external circuit; in fact they make up a circulating current around the loop formed by the two elements. The physical significance of resonance, however, goes further than this. Since the energy stored in L and in C varies as the square of current and of voltage respectively, as shown in Fig. 3.16(a) of Section 3.11.1, and the current in L lags the voltage V by $\pi/2$, the variation of stored energy in the components of our resonant circuit must be as shown in Fig. 5.7(c). If the voltage is written as $V_m \cos \omega t$ then the current in the inductance is

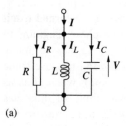

(a)

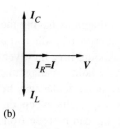

(b)

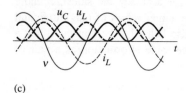

(c)

Fig. 5.7
Parallel resonance: (a) circuit; (b) phasor diagram; (c) variation of stored energy.

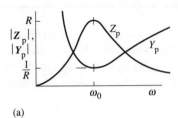

(a)

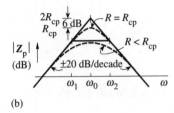

(b)

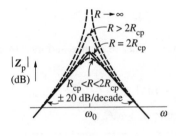

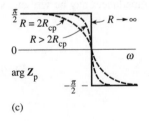

(c)

Fig. 5.8

Parallel resonance: (a) resonance curves; (b) variation of Z for R less than critical; (c) variation of Z for R greater than critical.

$(V_m/\omega L)\sin\omega t$; the stored energy is therefore

$$u_C = \frac{1}{2}CV_m^2\cos^2\omega t$$

in the capacitance and

$$u_L = \frac{L}{2}\frac{V_m^2\sin^2\omega t}{\omega^2 L^2}$$

in the inductance. Now at resonance ω^2 is given by $1/LC$, by eqn 5.5, so it follows that

$$u_L = \frac{1}{2}CV_m^2\sin^2\omega t,$$

which confirms that the maximum energy stored in L is the same as that stored in C; further, the total stored at any instant is

$$u_L + u_C = \frac{1}{2}CV_m^2 = CV^2$$

and is clearly a constant. For the circuit now considered what we therefore find at resonance in the steady state is a constant amount of energy oscillating to and fro between L and C. In Chapter 3 it was pointed out that in an a.c. circuit energy can oscillate between a reactive circuit and a source; but at resonance it now appears that the source at any instant provides only energy for consumption (by the resistance), and that stored energy oscillates only between L and C. The many other examples of resonance of physical systems always arise from this same process: the oscillating exchange of energy between two different forms of storage. (The total stored energy is not precisely constant in every case.)

Because the magnitude Y_p of the admittance of the parallel RLC circuit reaches a minimum $1/R$ at resonance (eqn 5.4), the circuit will take a minimum current V/R from an a.c. voltage source of given magnitude V when the frequency of the source is ω_0. Equally, since the impedance Z_p of the circuit has maximum magnitude R at resonance, the voltage across it reaches a maximum of IR when it is supplied with an a.c. current of constant magnitude I and varying frequency. The frequency response of the circuit may be presented as the variation of either Y_p or Z_p, as shown in Fig. 5.8(a). Such curves are generally known as **resonance curves**; usually the curve which shows a maximum (in this case Z_p), rather than a minimum, is chosen to indicate the response. Because the inductance is a short circuit for d.c., $\omega = 0$, and the capacitance likewise when $\omega \to \infty$, it follows that in this case Z_p is zero and Y_p infinite at both limits; in general, however, limiting values may be finite, depending on the circuit and function considered. The frequency response of a resonant circuit can also, of course, be shown on logarithmic axes like the Bode diagrams discussed in Section 5.2, and we return to this below.

Example 5.5   An inductance of 10 mH, a capacitance of 10 nF, and a resistance of 10 kΩ are all connected in parallel. Find the resonant frequency and, at that frequency, the energy stored and the circulating reactive current when the circuit is energized from a source of voltage 10 V r.m.s.

By eqn 5.5 the resonant frequency is

$$\omega_0 = 1/\sqrt{LC} = 1/\sqrt{(10^{-2} \times 10^{-8})} = 10^5 \text{ rad s}^{-1},$$

or

$$f_0 = 10^5/2\pi = \mathbf{15.9\ kHz}.$$

The total energy stored at resonance is the maximum energy in either L or C; it is therefore

$$\frac{1}{2}CV_{\mathrm{m}}^2 = CV^2 = 10^{-8} \times 100 = \mathbf{10^{-6}\ J}.$$

The circulating current is likewise that in either reactance and therefore has the r.m.s. magnitude

$$I = V/\omega_0 L = 10/(10^5 \times 10^{-2}) = \mathbf{10^{-2}\ A},$$

which is also the current in the capacitance, $V\omega_0 C$.

Let us consider more precisely the form of the frequency response for the same parallel circuit. From eqn 5.4 its impedance may be written as

$$
\begin{aligned}
Z_{\mathrm{p}} = \frac{1}{Y_{\mathrm{p}}} &= \frac{j\omega LR}{j\omega L + R + (j\omega)^2 LCR} \\
&= \frac{j\omega L}{1 + j\omega \dfrac{L}{R} + (j\omega)^2 LC}
\end{aligned}
\tag{5.6}
$$

and we may recognize in the denominator a quadratic factor having, in the notation used in the last section, the coefficients

$$b = L/R, \qquad c = LC.$$

The denominator can therefore be written in the factorized form $(1 + j\omega T_1)(1 + j\omega T_2)$ with real T_1 and T_2 if

$$(L/R)^2 > 4LC,$$

that is if the resistance is less than a critical value given by

$$R_{\mathrm{cp}} = \frac{1}{2}\sqrt{L/C}. \tag{5.7}$$

For $R < R_{cp}$ the two time constants can be shown to be

$$T_1, T_2 = \frac{L}{2R} \pm \sqrt{\frac{L^2}{4R^2} - LC}. \tag{5.8}$$

In the critical case, when $R = R_{cp}$, the factors are identical and

$$T_1, T_2 = 2CR = \sqrt{LC} = 1/\omega_0$$

according to our definition of ω_0, eqn 5.5.

For values of R up to and including R_{cp} we may therefore write the impedance of eqn 5.6 in the form

$$Z_p = \frac{j\omega L}{(1 + j\omega T_1)(1 + j\omega T_2)} \tag{5.9}$$

and it follows that the frequency response can be shown in the form of the Bode diagrams of Fig. 5.8(b). The relative position of the breakpoint frequencies $\omega_1 = 1/T_1$ and $\omega_2 = 1/T_2$ depends only on R, and the two coalesce at the resonant frequency ω_0 when R has the critical value: the error in magnitude between asymptotes and curve is then easily shown to be 6 dB, since the former intersect at the magnitude $\sqrt{L/C}$, or $2R_{cp}$, which is twice the actual magnitude. The low- and high-frequency asymptotes are determined by L and C respectively, since $T_1 T_2 = LC$ and eqn 5.9 therefore gives $j\omega L$ and $1/j\omega C$ as the respective limiting values of Z; this is of course obvious enough from physical reasoning.

Example 5.6   Find the critical value of resistance for the circuit specified in Example 5.5, and the two time constants which would result if the resistance had half this value.

From eqn 5.7 the critical resistance is

$$R_{cp} = \frac{1}{2}\sqrt{L/C} = 0.5 \times \sqrt{(10^{-2}/10^{-8})} = \mathbf{500\ \Omega}.$$

The time constants for $R = 250\ \Omega$ can be found from eqn 5.8 above, or by factorizing the denominator of eqn 5.6 with the appropriate values of the parameters. The factors we need are those of

$$1 + j\omega(10^{-2}/250) + (j\omega)^2 \times 10^{-2} \times 10^{-8}$$

or

$$1 + j\omega(4 \times 10^{-5}) + (j\omega)^2 \times 10^{-10},$$

which are

$$\{1 + j\omega 10^{-5}(2 + \sqrt{3})\}\{1 + j\omega 10^{-5}(2 - \sqrt{3})\}.$$

Hence the time constants are

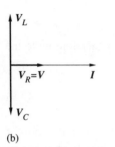

(a)

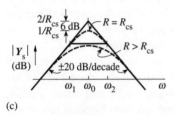

(b)

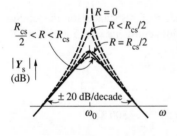

(c)

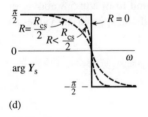

(d)

Fig. 5.9
Series resonance: (a) circuit;
(b) phasor diagram; (c) variation
of Y for R greater than critical; (d)
variation of Y for R less than
critical.

$$T_1 = (2 + \sqrt{3})10^{-5}; T_2 = (2 - \sqrt{3})10^{-5}$$

or

$$T_1 = 37.3\ \mu s; T_2 = 2.68\ \mu s$$

and it is easily confirmed that eqn 5.8 also gives these values.

The question to address now is: what happens to the Bode representation of Z_p when the resistance R exceeds the critical value? The first thing to note is that the low- and high-frequency asymptotes are unchanged, because they do not depend on R: since they still meet at ω_0, that must remain the breakpoint (for now T_1 and T_2 do not exist as real values). It follows that values of R over the critical value R_{cp} can only affect the actual curve and not the asymptotes. Consider, for example, the situation at the breakpoint, that is at the resonant frequency ω_0. As we saw, the asymptotes indicate that Z_p has the value $\sqrt{L/C}$ or $2R_{cp}$ there, whereas the actual value of Z_p at resonance is R. As R increases from the value R_{cp} the peak of the magnitude curve therefore rises, meets the breakpoint when $R = 2R_{cp}$ and then rises without limit, as shown in Fig. 5.8(c). As R rises to infinity, implying a purely reactive circuit with zero parallel conductance, so also does the resonance peak. The phase diagram always shows zero phase at resonance, since Z_p is then real, but as R increases the change between $\pi/2$, the low-frequency limit, and the high-frequency limit of $-\pi/2$ becomes sharper until at infinite R it appears as a sudden switch between the two at ω_0, a consequence which is hardly surprising since a purely reactive circuit can only have phase angles $\pm \pi/2$.

5.3.3 The series resonant circuit

Because a circuit comprising R, L, and C in series is the dual of the three similar elements in parallel, we may expect it to show a comparable frequency response. The impedance of the series circuit shown in Fig. 5.9(a) is

$$\begin{aligned} Z_s &= R + j\omega L + 1/j\omega C \\ &= R + j(\omega L - 1/\omega C) \end{aligned}$$

which falls to the real value R, which is also its minimum magnitude, at the same resonance frequency ω_0 as we found for the parallel circuit. At this frequency energy is again oscillating between L and C, but now it is their voltages which are equal and opposite and cancel each other, as shown in the phasor diagram of Fig. 5.9(b). The admittance, which reaches a peak value $1/R$ at resonance, can be written

$$Y_s = \frac{j\omega C}{1 + j\omega CR + (j\omega)^2 LC} \tag{5.10}$$

This may be compared with eqn 5.6, and it represents a resonance curve analogous to that for the impedance of the parallel circuit (Fig. 5.8(a)). The condition for the denominator in eqn 5.10 to have factors like $(1 + j\omega T)$ with real T is now

$$C^2 R^2 > 4LC$$

which is satisfied if R is *greater* than a critical value

$$R_{cs} = 2\sqrt{L/C}. \tag{5.11}$$

It will be noticed not only that this exceeds the corresponding resistance R_{cp} by the factor four, but also that in contrast to the latter it is the value *above* which the response will have two time constants and therefore two breakpoints. From this point the argument and the results correspond, *mutatis mutandis*, to those for the parallel circuit. For R greater than R_{cs} the Bode diagrams for Y_s are as in Fig. 5.9(c), corresponding to those of Fig. 5.8(b) for Z_p. At the critical resistance, $R = R_{cs}$, the breakpoints coalesce at the frequency ω_0 and the asymptotes intersect at the magnitude $\sqrt{C/L}$ or $2/R_{cs}$; the phase angle is zero at that point and the error in magnitude is 6 dB, as before. For lower values of R the actual peak of magnitude, which is always $1/R$, rises to meet the asymptotes when $R = R_{cs}/2$ and then increases to infinity as R tends to zero, as shown in (d). It is worth noting that the asymptotes are exchanged, as compared with the parallel case, for at low frequencies Y_s tends to $j\omega C$, while the high-frequency asymptotes are those for $1/j\omega L$. The variation of phase is thus as before, from $\pi/2$ to $-\pi/2$, and the transition becomes infinitely sharp for $R = 0$.

It is easy to see why the infinite peak at resonance, of impedance for parallel and of admittance for series, should be obtained for infinite R in the former case and for zero R in the latter: both represent the *removal* of resistance, for in a series circuit the effect of resistance vanishes when it becomes zero, but parallel resistance has no effect when it is infinite because that means zero conductance. The physical significance in both is that the circuit is purely reactive and dissipates no power ($I^2 R$ is zero in the one case, V^2/R in the other), so that at resonance the oscillating energy can build up with no losses: in the terminology which we shall use later, the **damping** is zero.

5.3.4 Another resonant configuration

The three elements of a simple resonant circuit may be neither all in parallel nor all in series. Circuits with L and C in parallel are often used, but in practice the resistance is associated mainly with L and acts in series with it, giving the circuit of Fig. 5.10(a). This circuit can still show parallel resonance, having an impedance maximum at a certain frequency if r is not too large, but the expressions which govern its behaviour do not have quite the simplicity of those found so far. The

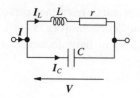

(a)

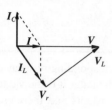

(b)

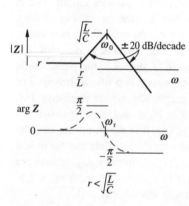

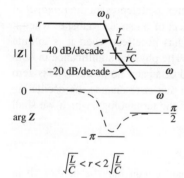

(c)

Fig. 5.10

Parallel resonance with series
resistance: (a) circuit; (b) phasor
diagram; (c) Bode diagrams.

impedance is readily shown to be

$$Z = \frac{r(1 + j\omega L/r)}{1 + j\omega rC + (j\omega)^2 LC} \tag{5.12}$$

which, although is contains the same quadratic as eqn 5.10 for the series circuit, is *not* real at the frequency $\omega_0 = 1/\sqrt{LC}$; nor is its magnitude a maximum at that frequency. The resonant frequency, which was defined in Section 5.3.2 to be that for which Z is real, we now denote ω_r, and it is most easily found by equating the angles of numerator and denominator in eqn 5.12; it is given by

$$\omega_r^2 = \omega_0^2 (1 - r^2 C/L). \tag{5.13}$$

The value of Z is then $r/\{1 - \omega_r^2/\omega_0^2\}$ or L/Cr. The phasor diagram for this resonant condition is shown in Fig. 5.10(b). Evidently when r is much less than $\sqrt{L/C}$ the resonant frequency is very close to ω_0 and the resonant impedance very large. On the other hand, it is clear from eqn 5.13 that Z cannot be real at any non-zero frequency if r exceeds $\sqrt{L/C}$, so the circuit does not then show resonance as we have defined it. The frequency for which $|Z|$ reaches a maximum differs from both ω_r and ω_0, and can be shown to lie between the two (it is therefore, like ω_r, very close to ω_0 when r is small). The maximum turns out to occur at zero frequency if

$$r^2 \geq (1 + \sqrt{2})L/C,$$

so the peak in response which is a feature of resonance also disappears if r is too high. Much of this can be better appreciated from the Bode diagrams, shown in Fig. 5.10(c) for two possible cases. They lack the symmetry of Figs 5.8(b) and 5.9(c) for the parallel and series circuits, but evidently if r is small they approach the forms of Fig. 5.8(b) — as they must, for the circuit then reduces to the 'lossless' parallel form with R infinite — with peak magnitude and zero phase at ω_0. If r increases to the value $\sqrt{L/C}$, the two breakpoints shown coincide at the frequency ω_0; further increase takes the breakpoint at $\omega = r/L$ beyond ω_0, and when r reaches the value $2\sqrt{L/C}$ the breakpoint at ω_0 splits into two, exactly as for the series circuit (because of the same quadratic in the denominator, which for larger values of r has two factors with real time constants).

Example 5.7 A coil has inductance 25 mH and resistance 20 Ω. It is connected in parallel with a capacitance of 100 nF. Find the frequency f_0, the resonant frequency f_r and the value of the impedance Z at resonance.

We have

$$\omega_0 = 1/\sqrt{LC} = 1/\sqrt{(0.025 \times 10^{-7})} = 2 \times 10^4 \text{ rad s}^{-1}$$

or

$$f_0 = 2 \times 10^4/2\pi = \textbf{3.183 kHz}.$$

By eqn 5.13 the resonant frequency is now given by

$$\omega_r^2 = 4 \times 10^8(1 - 400 \times 10^{-7}/0.025) = 3.994 \times 10^8$$

from which

$$\omega_r = 1.998 \times 10^4 \text{ rad s}^{-1}$$

and therefore

$$f_r = \textbf{3.181 kHz}.$$

At this frequency the impedance is real and given by L/Cr, so

$$Z = 0.025/(10^{-7} \times 20) = \textbf{12.5}\underline{|0°}\textbf{ k}\Omega$$

5.3.5 Quality factor

It emerged in the previous sections that a characteristic property of resonance which always appears, provided that the resistance is high enough or, as the case may be, low enough (in general, when the losses are low enough), is a peak in impedance or admittance at some particular frequency. It is this which makes resonance in some circumstances a phenomenon to be avoided, in others one of great practical use. In mechanical systems especially it is often destructive, as in unwanted vibration, but in electrical engineering it has many applications because it makes a circuit **frequency selective**. For example, if a parallel resonant circuit is supplied from a current source of varying frequency, the voltage across it can be large for a certain range of frequency and otherwise small: in essence the circuit selects that range. This is a common requirement, in broadcast reception and in electronic oscillators, for example; and the higher and narrower the peak in the response, the better the selectivity. Because the frequency of the peak can be varied by adjusting L or C, rather like tuning a musical instrument, a circuit which is resonant by design is often called a **tuned** circuit.

It is useful to have a measure of this selectivity, which according to our previous discussion depends on the resistance in the circuit. In parallel resonance, a peak in voltage corresponds to large currents I_L and I_C, equal and opposite, in L and C and each may be compared in magnitude with the total current I, as in Fig. 5.7(b). Similarly, in series resonance the opposing voltages V_L and V_C may be compared with the total voltage V as in Fig. 5.9(b). The magnitude ratio in each case can easily be expressed, as follows:

Parallel	*Series*

$$\frac{I_L}{I} = \frac{V/\omega_0 L}{V/R} = \frac{R}{\omega_0 L} \qquad\qquad \frac{V_L}{V} = \frac{\omega_0 L I}{RI} = \frac{\omega_0 L}{R}$$

or   or

$$\frac{I_C}{I} = \frac{V\omega_0 C}{V/R} = \omega_0 CR \qquad\qquad \frac{V_C}{V} = \frac{I/\omega_0 C}{RI} = \frac{1}{\omega_0 CR}.$$

Each of these ratios is known as the **quality factor** Q of the circuit concerned, or as its **magnification**. High Q is associated with a high peak in the appropriate response function, and is achieved when the circuit losses are small. Because unavoidable loss is in practice mainly due to the resistance associated with inductance, it is usual to use those of the above expressions which involve L rather than C, thus:

$$\textit{Parallel} \qquad Q = R/\omega_0 L \qquad\qquad\qquad \text{(5.14)}$$

$$\textit{Series} \qquad Q = \omega_0 L/R. \qquad\qquad\qquad \text{(5.15)}$$

For the same reason it is common to refer simply to the Q-factor of a coil having inductance L and series resistance r as $\omega L/r$, at any frequency ω, the value of the latter being taken as whatever resonant frequency may be dictated by the choice of C. There is no reason in principle why the Q-factor of a capacitor should not be referred to in the same way; since any resistive imperfection in a capacitor is most conveniently expressed as a parallel resistance, we should choose in this case to express Q as ωCR where now the relevant value of ω would be dictated by the choice of L. Although it is possible, and legitimate, to define quality factor in this way for any frequency, without reference to resonance, we shall from now use Q to signify the value at a particular resonance frequency unless otherwise indicated.

If a coil with series resistance r is used with C in parallel, so that the circuit is that of Fig. 5.10(a), then the simple relations above no longer apply precisely; but, as we shall see, they are nevertheless useful. The resonant state of that circuit was shown in the phasor diagram of Fig. 5.10(b). The magnitudes I_L and I_C are no longer equal, but with reference to the diagram we may express them in terms of the total current I by way of the angle ϕ of the coil's impedance, given by

$$\tan\phi = \omega L/r;$$

thus

$$\frac{I_L}{I} = \sec\phi = \frac{\sqrt{r^2 + \omega^2 L^2}}{r}$$

and

$$\frac{I_C}{I} = \tan\phi = \frac{\omega L}{r}.$$

These show that the definition $\omega L/r$ for the Q-factor of the coil will correctly give the current magnification I_C/I for this circuit at resonance, and that it will also give the magnification I_L/I closely if $r^2 \ll \omega^2 L^2$, that is if $Q \gg 1$. We may also note that, since $\omega_0 L/r$ can be written $\sqrt{L/C}/r$, eqn 5.13 can be written

$$\omega_r^2 = \omega_0^2(1 - 1/Q^2)$$

which shows that ω_r, the actual resonance frequency, is very near to ω_0 for high Q. This result for small r was pointed out in the previous section, but in terms of Q it brings us to a common approximation which applies to all resonant circuits: if Q is high enough the definitions for the simplest cases remain valid for most purposes; thus, the error in setting $\omega_r = \omega_0$ above would be only 1% for $Q = 10$, a figure easily achieved. (But there are natural limits to the values of Q attainable in ordinary electrical circuits: see below.) We may note here, by combining the definitions of eqns 5.14 and 5.15 with the values of critical resistance given by eqns 5.7 and 5.11, that in every case considered the critical condition corresponds to the value 0.5 for Q.

A useful result in terms of Q which applies equally to the simple parallel and series circuits relates the impedance Z_p or admittance Y_s at any frequency ω to that at resonance, i.e. to the resistance in each case. It can be shown from eqns 5.6 and 5.10 that

$$\frac{Z_p}{R_p} = R_s Y_s = \frac{1}{1 + jQ(\omega/\omega_0 - \omega_0/\omega)}. \qquad (5.16)$$

This expression, which might be called the **normalized** impedance or admittance as the case may be, gives resonance curves of the same form as those for Z_p and Y_s but with the advantage of dimensionless magnitude and a peak value of unity; they are therefore applicable to any circuit of either form. Equation 5.16 also shows a useful property of these curves: the imaginary part of the denominator has the same magnitude for any frequency $k\omega_0$ as for the frequency ω_0/k, where k is a constant, but the opposite sign; it follows that at any such pair of frequencies the impedance or admittance has the same magnitude and opposite phase (positive below resonance and negative above). Otherwise put, ω_0 is the geometrical mean of any two frequencies for which the function has the same magnitude, and it also follows that the resonance curves representing eqn 5.16 are symmetrical on a logarithmic scale of frequency.

Example 5.8 Evaluate the Q-factor of the parallel circuit specified in Example 5.5, and hence find the impedance of the circuit at (a) twice and (b) half the resonant frequency.

From that example we may write immediately

$$Q = R/\omega_0 L = 10^4/(10^5 \times 10^{-2}) = \mathbf{10}.$$

Using this in eqn 5.16 gives for (a)

$$\mathbf{Z_p} = 10^4/\{1 + \text{j}10(2 - 1/2)\} = (10^4/\sqrt{226})\underline{|\tan^{-1}15}$$
$$= 665\underline{|-86°}\ \Omega.$$

At half the frequency ω_0 the impedance is the same except for the opposite angle, so for (b)

$$\mathbf{Z_p} = 665\underline{|86°}\ \Omega.$$

5.3.6 General definitions of Q-factor

Useful as are the above definitions for Q, they are not entirely satisfactory, because as ratios of currents or voltages they can be slightly ambiguous, and they need different versions for different circuits; also none of them can be applied to non-electrical systems. All of these disadvantages can be overcome by two more general definitions which we now derive.

A high peak response also implies a *sharp* peak, which falls rapidly at higher and lower frequencies. The 'sharpness' or relative height of a peak is most easily defined in terms of the frequency range between two points defining a certain fraction of the peak amplitude. The two are conventionally taken to be those at which the amplitude is 3 dB less than the maximum, commonly called the '3 dB' or 'half-power' points, the latter term arising from the fact that −3 dB means a factor $1/\sqrt{2}$ in current or voltage and hence a factor $1/2$ in power (Section 5.2.1). An interval of frequency between two such points (or, if the peak should be at zero, between zero and one such point) is called a **bandwidth**, a term which is used widely and by no means confined to resonant circuits; we denote it here by $\Delta\omega$, the interval in angular frequency. For practical purposes it is often given as Δf, i.e. $\Delta\omega/2\pi$, the interval in cyclic frequency measured in hertz. Figure 5.11 illustrates the bandwidth of a typical curve which gives (with linear scales) the magnitude of the general expression in eqn 5.16, representing either the impedance of the parallel RLC circuit, as in Fig. 5.8(c), or the admittance of the series circuit as in Fig. 5.9(d). That

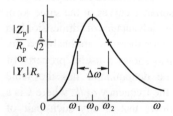

Fig. 5.11

The bandwidth of a resonant circuit.

expression, which has a maximum value of unity, clearly has magnitude $1/\sqrt{2}$ when the imaginary part of the denominator is ± 1, that is when

$$Q(\omega/\omega_0 - \omega_0/\omega) = \pm 1. \tag{5.17}$$

Of the two frequencies which satisfy this condition, the higher, labelled ω_2 in the figure, must relate to the positive sign; this choice leads to the equation

$$\omega_2^2 Q - \omega_2\omega_0 - \omega_0^2 Q = 0,$$

the solution of which is

$$\omega_2 = \frac{\omega_0(1 + \sqrt{1 + 4Q^2})}{2Q}$$

when the negative result is discarded. Similarly the negative sign in eqn 5.17, which must relate to the lower frequency ω_1, leads to

$$\omega_1^2 Q + \omega_1\omega_0 - \omega_0^2 Q = 0,$$

of which the positive solution is

$$\omega_1 = \frac{\omega_0(-1 + \sqrt{1 + 4Q^2})}{2Q}.$$

The bandwidth is therefore

$$\Delta\omega = \omega_2 - \omega_1 = \omega_0/Q$$

and so

$$Q = \omega_0/\Delta\omega. \tag{5.18}$$

This result, which also applies quite closely to the circuit of Section 5.3.4 provided Q is high enough, is an important one. Not only does it specify the value of Q needed to achieve a certain selectivity, it also allows the effective Q of a circuit to be readily evaluated from a response curve. In fact eqn 5.18 is a useful definition for any resonant system, for the idea of Q-factor, introduced above as a magnification of voltage or current, is not confined to electrical circuits. It can be generalized by yet another definition, in the following way.

The connection between Q and power losses in a circuit has been pointed out, as has the association of resonance with stored energy. In Section 5.3.2 we saw that the total stored energy in a parallel RLC circuit is constant at resonance and may be written

$$u_C + u_L = CV_m^2/2 = CV^2$$

where V is as usual a r.m.s. value. The rate of dissipation, on the other hand, is V^2/R, which means that the energy dissipated *per cycle* at the resonant frequency ω_0 is $2\pi V^2/R\omega_0$. Hence we may write

$$\frac{\text{Energy stored}}{\text{dissipation per cycle}} = \frac{CV^2R\omega_0}{2\pi V^2} = \frac{\omega_0 CR}{2\pi} = \frac{Q}{2\pi}.$$

Because the stored energy at resonance is not exactly constant but has a fluctuating component in some cases (for example, the circuit of Section 5.3.4), stricter definition stipulates its maximum value, thus:

$$Q = \frac{2\pi \times \text{maximum energy stored}}{\text{energy loss per cycle}}. \qquad (5.19)$$

This affords a very general definition of Q, applicable in many circumstances. It is especially useful in systems where linear parameters such as R, L, and C are difficult to identify.

Example 5.9 A coil has inductance 10 mH and resistance 20 Ω. It is connected in series with a capacitance of 1 μF. Find the bandwidth of the current taken by the circuit from a variable-frequency voltage source, and confirm that eqn 5.19 holds for this circuit.

The resonant frequency is given by

$$\omega_0 = 1/\sqrt{LC} = 1/\sqrt{10^{-8}} = 10^4 \text{ rad s}^{-1}$$

and the Q-factor of the circuit is then

$$Q = \omega_0 L/R = 10^4 \times 10^{-2}/20 = 5.$$

The width of the admittance peak is therefore, from eqn 5.18, given by

$$\Delta\omega = \omega_0/Q = 10^4/5 = 2000 \text{ rad s}^{-1}$$

and the bandwidth in hertz is then

$$\Delta f = 2000/2\pi = \textbf{318 Hz}.$$

The stored energy at resonance in this case is constant, as for the parallel circuit. We may first confirm this by noting that for peak current I_m the peak voltage V_{Cm} across the capacitance is $I_m/\omega C$ and its peak value of stored energy is therefore $CV_{Cm}^2/2$ or $I_m^2/2\omega^2 C$; at the resonant frequency this becomes $LI_m^2/2$, or, in term of r.m.s. current, LI^2, which is precisely the peak value of stored energy in the inductance. Because the two components of stored energy have the same peak value but are again in antiphase, their sum is constant and equal to the peak value of each. At resonance the energy loss per cycle is $I^2 R/f_0$. Equation 5.19 then gives

$$Q = 2\pi f_0 L I^2 / I^2 R = \omega_0 L / R$$

as expected, and as evaluated above.

There are many applications in which resonant systems must be designed with the highest possible Q, and this means keeping all losses to a minimum. Thus, the (series) resistance of a coil, often the main source of dissipation, must be minimized, but so also must the (parallel) conductance of capacitors and all other series resistance and parallel conductance, of connections for example, associated with the circuit. Equation 5.15 indicates that the Q obtainable from a given coil in an otherwise lossless circuit increases with the frequency ω_0. It does not in practice do so indefinitely because as frequency increases L and R do not remain constant: R increases by the skin effect (Appendix B.1) while L is effectively reduced by the increasing effect of the normally negligible capacitance of a coil. There is therefore a practical limit to the attainable values of Q in a purely passive circuit, in the order of a few hundred. Larger values of Q can be achieved electronically by means of an amplifier with positive feedback, which effectively produces negative resistance to cancel losses (Section 8.3.2). Some non-electrical systems can attain very high natural values: the mechanical behaviour of the quartz crystals used in timepieces corresponds to values of Q around 10^4 or more.

In systems which are designed to take advantage of resonance a high value of Q is usually desirable, but there are many engineering situations where the effects of resonance are unwanted and low values of Q are then advantageous. Natural values of Q can be lowered by introducing dissipation or **damping**, for example by inserting friction in a mechanical system or, in an electrical circuit, resistance; the term is applied equally to unavoidable losses and to those deliberately introduced. Alternatively, and sometimes preferably, other parameters can be adjusted to set the frequency ω_0 outside the range which the system will encounter, so that the resonance peak is simply moved out of harm's way.

5.3.7 Standard forms

In the previous section it was seen that quality factor can be defined rather generally. The functions which arise in resonant systems also have general features. For one thing, as we saw in Section 5.3.1, they all contain a quadratic factor. By examining this factor in eqns 5.6, 5.10, and 5.12, which relate to the three circuits considered, it emerges that each takes the form

$$1 + j\omega/\omega_0 Q + (j\omega/\omega_0)^2 \tag{5.20}$$

in which Q is as defined in each case and ω_0 or $1/\sqrt{LC}$ is the resonant frequency, or near to it if Q is fairly high (say 10 or more). If we define

ω_0 as the **undamped** resonant frequency then the expression 5.20 is perfectly general for any resonant system. It is often written in a slightly different form by introducing a new quantity ζ defined by

$$\zeta = 1/2Q$$

and known as the **damping factor** or damping ratio. The advantage of using ζ is that it has the value unity for what we may now call **critical damping**. In the previous section it was noted that the critical resistance values for the circuits so far considered corresponded to $Q = 0.5$, which means $\zeta = 1$; and in all resonant systems critical values of dissipation parameters correspond to these same values of Q and ζ. In Chapter 7 it will emerge that they can have a still wider significance. For ζ less than unity the damping is below critical and for $\zeta = 0$ there is no damping, which means no losses and an infinite resonance peak. When ζ exceeds unity, the quadratic expression 5.20 has factors with real time constants (as discussed earlier) and does not show the usual effects of resonance. In terms of ζ, 5.20 becomes

$$1 + 2\zeta j\omega/\omega_0 + (j\omega/\omega_0)^2 \tag{5.21}$$

which is easily seen to have two equal factors $(1 + j\omega/\omega_0)$ when $\zeta = 1$. Both the expressions 5.20 and 5.21 are useful general forms. They can of course be written in slightly different ways, such as

$$(\omega_0^2 + 2\zeta j\omega\omega_0 - \omega^2)/\omega_0^2.$$

It is often helpful to remember that these expressions have magnitude 2ζ and angle $\pi/2$ at the frequency ω_0.

Response curves such as those of Figs 5.8(b) and (c) and 5.9(d) can be labelled with values of Q or ζ rather than resistance. From the definitions of Q in eqns 5.14 and 5.15 it can be shown that

$$\zeta = \frac{1}{2R}\sqrt{\frac{L}{C}} \tag{5.22}$$

for the parallel circuit and

$$\zeta = \frac{1}{2}R\sqrt{\frac{C}{L}} \tag{5.23}$$

for the series. The complete expression for the normalized impedance and admittance of the parallel and series circuits respectively, given in terms of Q by eqn 5.16, can now be written as

$$\frac{Z_p}{R_p} = Y_s R_s = \frac{2\mathrm{j}\,\zeta\,\omega/\omega_0}{1 + 2\mathrm{j}\,\zeta\,\omega/\omega_0 + (\mathrm{j}\omega/\omega_0)^2}. \tag{5.24}$$

5.3.8 Other response functions

The particular quadratic factors which have now been represented in standard form arose from the simplest functions which conveniently illustrate resonant behaviour in circuits, namely the impedance of circuits with L and C in parallel branches and the admittance of the series circuit. But a given circuit or system does not have a unique response function, and several of its functions may show a resonance peak. For example, in the parallel resonant circuit discussed in Section 5.3.2, rather than taking the impedance, i.e. the voltage response to a current input, we might wish to consider instead how the ratio of the current in the inductance to the total current varies with frequency. This ratio I_L/I can be shown, by current division or otherwise, to be given by

$$I_L/I = 1/\{1 + \mathrm{j}\omega L/R + (\mathrm{j}\omega)^2 LC\} \tag{5.25}$$

which contains exactly the same quadratic as eqn 5.6 for the impedance but is clearly not the same function. (It is obvious that the two functions must differ, not only because the current ratio is dimensionless but also because its low-frequency limit has magnitude unity rather than zero.) The Bode diagrams for I_L/I, together with a few curves for various values of ζ, are shown in Fig. 5.12(a), and they may be compared with those for the impedance in Fig. 5.8. The current ratio, while it shows the same general resonant behaviour, has different asymptotes and lacks the symmetry of the other. One consequence of this is that the peak magnitude is no longer at the frequency ω_0; it can be shown (by differentiating the modulus of the quadratic in eqn 5.25) to occur at the frequency

$$\omega_m = \omega_0\sqrt{1 - L/2R^2 C}. \tag{5.26}$$

This peak occurs at zero frequency if $R = \sqrt{L/2C}$, which means, according to eqn 5.22, if $\zeta = 1/\sqrt{2}$. Had we chosen to express the quadratic in standard form we should have found

$$\boxed{\omega_m^2 = \omega_0^2(1 - 2\zeta^2) \tag{5.27}}$$

which agrees with eqn 5.26 when ζ is taken from eqn 5.22 for the parallel circuit. We may note from eqn 5.27 that if ζ is small ω_m is quite near to ω_0; thus, $\zeta = 0.05$ (which means $Q = 10$) gives $\omega_m = 0.997\omega_0$, confirming once again that simplifying assumptions can usually be made when the damping is small.

It can easily be shown for this same circuit that the ratio of the capacitance current to the total current is

$$I_C/I = (\mathrm{j}\omega)^2 LC/\{1 + \mathrm{j}\omega L/R + (\mathrm{j}\omega)^2 LC\}.$$

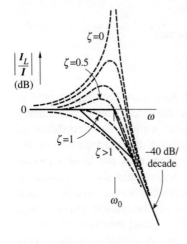

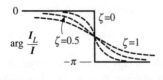

(a)

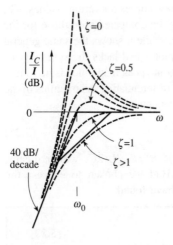

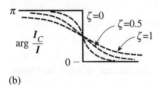

(b)

Fig. 5.12
Parallel resonance: current ratios.

This too, of course, differs from the impedance function, eqn 5.6, but it contains again exactly the same quadratic. Its Bode diagrams are shown in Fig. 5.12(b) together with typical curves; the peak magnitude now occurs at a frequency ω'_m given by

$$\omega'^2_m = \omega_0^2/(1 - 2\zeta^2)$$

which may be compared with eqn 5.27. There is an obviously symmetrical relationship between the two current ratios, and we should find similar results for the voltage ratios in a series circuit: in that case the ratios V_L/V and V_C/V would have the same quadratic denominator as the admittance, eqn 5.10, and their frequency responses would be like those shown for the current ratios in Fig. 5.12(b) and (a) respectively. It may also be noted here that, whereas we have so far considered resonant circuits only as two-terminal networks, any of them could also constitute a four-terminal network: for example, the input might be across the three series elements and the output the voltage across any one of them. In such cases the above current and voltage ratios would count as transfer functions (so, for that matter, would the normalized immittance, eqn 5.24, since it gives the proportion of total current in the parallel resistance and of voltage across the series resistance). So we deduce that resonant behaviour applies just as much to transfer functions as to driving-point functions when the circuit conditions allow it.

For each of the two basic circuits we have now found three different response functions; they can be written in standard form thus:

$$F_1 = \frac{1}{1 + 2\zeta j\omega/\omega_0 + (j\omega/\omega_0)^2}$$

$$F_2 = \frac{2\zeta j\omega/\omega_0}{1 + 2\zeta j\omega/\omega_0 + (j\omega/\omega_0)^2}$$

$$F_3 = \frac{(j\omega/\omega_0)^2}{1 + 2\zeta j\omega/\omega_0 + (j\omega/\omega_0)^2}.$$

Each of these contains the same quadratic factor, with the same ω_0, Q, and ζ for a given circuit. We should find this also to be true for any circuit containing an inductance and a capacitance, and therefore capable of resonance. For example the impedance of the series-parallel circuit considered in Section 5.3.4, given by eqn 5.12, can be written in the form

$$\frac{Z}{r} = \frac{1 + j\omega T}{1 + 2\zeta j\omega/\omega_0 + (j\omega/\omega_0)^2}$$

of which the quadratic would also appear in any other frequency response function for the same circuit. The parameters ω_0, Q, and ζ are therefore characteristic of the circuit or system rather than the particular function considered. The asymptotes of Bode diagrams for a resonant system always show characteristic behaviour for the same reasons: for $\zeta \leq 1$, the magnitude asymptotes change in slope at the breakpoint frequency ω_0 by a total of 40 dB per decade, while those for phase change by π.

Example 5.10 (a) For the series-parallel circuit of Fig. 5.10 find the ratio of the current I_C in the capacitance to the total current I in terms of the undamped resonant frequency ω_0 and a damping factor ζ.
(b) A response function for a certain system contains the factor $(1 + 10^{-4}j\omega - 10^{-6}\omega^2)$. Find the values of ζ and ω_0 for the system.

(a) By current division we may write immediately

$$\frac{I_C}{I} = \frac{r + j\omega L}{1/j\omega C + r + j\omega L}$$

$$= \frac{j\omega C(r + j\omega L)}{1 + j\omega Cr + (j\omega)^2 LC}$$

and by comparing the coefficients of numerator and denominator with those in the standard form of 5.21, this can be written

$$\frac{I_C}{I} = \frac{2\zeta j\omega/\omega_0 + (j\omega/\omega_0)^2}{1 + 2\zeta j\omega/\omega_0 + (j\omega/\omega_0)^2}.$$

(b) By comparing the given factor with the standard form of 5.21, it can be seen from the final term ω_0^2 is given by 10^6 and so

$$\omega_0 = \textbf{1000 rad s}^{-1}.$$

The coefficient of $j\omega$ in the second term must correspond to $2\zeta/\omega_0$ so it follows that

$$2\zeta = \omega_0 \times 10^{-4} = 0.1$$

and

$$\zeta = \textbf{0.05} \, .$$

5.3.9 Multiple resonance

All of the resonant behaviour discussed in the preceding sections has related to circuits having one inductance and one capacitance. Circuits which have two or more of each are likely to show more complicated frequency responses with two or more resonant frequencies. The analysis of such circuits is generally laborious and not especially illuminating, but there are two interesting cases to be mentioned.

For selectivity in communication, a resonant circuit can be incorporated in an amplifying stage, essentially in place of a load resistance (Section 4.5.2). Its gain then becomes a function similar to the impedance or admittance of the two-terminal resonant circuits so far considered. Thus the voltage gain, say, might be similar in form to the impedance of eqn 5.6 and could be written, in the form given in eqn 5.16, as

$$A_v = A_0/\{1 + jQ(\omega/\omega_0 - \omega_0/\omega)\}$$

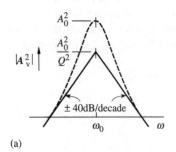

(a)

in which A_0 is the peak magnitude of the gain, occurring at the resonant frequency ω_0. Now if two such amplifier stages are connected in cascade, and if the second does not appreciably load the first (Section 4.7), then the overall voltage gain is

$$A_v^2 = A_0^2/\{1 + jQ(\omega/\omega_0 - \omega_0/\omega)\}^2. \tag{5.28}$$

It can be readily shown that the Bode diagram for the magnitude of this function has low- and high-frequency asymptotes of slope ± 40 dB per decade, as shown in Fig. 5.13(a). At the breakpoint ω_0 the asymptotes show the value A_0^2/Q^2, as opposed to the actual peak value of A_0^2. It can also be shown that, by the definition of bandwidth in Section 5.3.5, eqn 5.28 gives the bandwidth for the two stages as

$$\Delta\omega = (\omega_0/Q)(\sqrt{2} - 1)^{1/2} \tag{5.29}$$

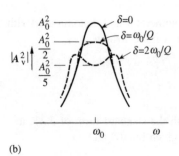

(b)

Fig. 5.13

Resonant circuits in cascade: (a) identical stages; (b) stagger-tuned stages.

which is less than the bandwidth ω_0/Q of each stage. According to the definition of Q in eqn 5.18 it follows from eqn 5.29 that the effective Q-factor of the whole amplifier is given by

$$Q' = Q/(\sqrt{2} - 1)^{1/2}$$

which is greater than that of each stage.

In practice it is often highly desirable for such a frequency-selective circuit to show a larger overall bandwidth than is implied by eqn 5.29 while retaining the steep fall of gain on each side of the peak which results from a high value of Q. This can be done by **stagger-tuning** the two stages so that they have slightly different values of resonant frequency, giving a resultant curve with a more or less flat-topped peak and a maximum somewhat lower than A_0^2. Too great a difference merely gives separate peaks in the overall response with a dip between them. If the difference is chosen to be ω_0/Q (that is, the bandwidth of each original stage) centred on ω_0, then the overall bandwidth turns out to be $\sqrt{2}\omega_0/Q$, more than twice that given by eqn 5.29, and there is a single peak of maximum value $A_0^2/2$. Figure 5.13(b) shows the effect of varying the difference δ between the resonant frequencies of two stages.

A similar kind of response can be obtained from two identical resonant circuits coupled by a transformer, as shown for example in

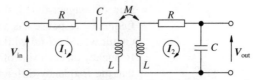

Fig. 5.14
Coupled resonant circuits.

Fig. 5.14. At resonance we may expect the output voltage across the capacitance C in the second mesh to show magnification with respect to the e.m.f. in that mesh and hence with respect to the input voltage. If we let

$$Z = R + j\omega L + 1/j\omega C$$

then the mesh equations can be written

$$V_{in} = ZI_1 + j\omega MI_2$$
$$0 = j\omega MI_1 + ZI_2$$

and the output voltage is

$$V_{out} = I_2/j\omega C.$$

These combine to give the transfer function

$$\frac{V_{out}}{V_{in}} = \frac{M}{C(Z^2 + \omega^2 M^2)} = \frac{M}{C(Z + j\omega M)(Z - j\omega M)}$$

in which the expression for Z can be substituted to give

$$\frac{V_{out}}{V_{in}} = \frac{(j\omega)^2 MC}{\{1 + j\omega CR + (j\omega)^2 C(L+M)\}\{1 + j\omega CR + (j\omega)^2 C(L-M)\}}. \quad (5.30)$$

Here we have two quadratic factors which by comparison with the standard forms of Section 5.3.6 can be seen to imply resonant frequencies

$$\omega_0' = \frac{1}{\sqrt{C(L+M)}}$$

and

$$\omega_0'' = \frac{1}{\sqrt{C(L-M)}}.$$

If we write ω_0 as usual for $1/\sqrt{LC}$, the resonant frequency of each mesh on its own, and put kL for M where k is the coupling coefficient, then these become

$$\omega_0', \omega_0'' = \omega_0/\sqrt{1 \pm k}. \quad (5.31)$$

If the coupling is very weak so that $k \ll 1$ and the circuit has two resonant frequencies quite close together, then the effect is similar to the pair of stagger-tuned stages considered above. That is to say, the frequency response has a rather wider bandwidth than a single resonant circuit would afford while retaining the same rapid change of gain with frequency on each side of the peak. The advantage of this arrangement is that the frequency difference and hence the bandwidth can be adjusted by varying k. As before too large a difference results in separation of the peaks and a loss of gain between them.

5.4 Filters

5.4.1 General

Filters are widely used in electrical engineering, especially for purposes which are broadly described as signal processing and will be encountered in Chapter 9. Here we discuss only analogue filters, as opposed to digital filters which are used for the same purposes but are different in nature. The analogue filter is a four-terminal network whose frequency response gives a gain which is close to unity (or perhaps some other constant, as may be required) over a certain range of frequency, and otherwise close to zero. It is therefore selective in frequency and in this respect has something in common with resonant circuits; the latter, however, typically provide high magnification over a fairly narrow range of frequency whereas filters in general operate rather by providing strong attenuation at some frequencies (often called the **stop band**), leaving a **pass band** of other frequencies, having a bandwidth as wide or as narrow as required, over which signals are unimpaired. Filters can be divided into the following categories according to the frequency range of the pass band. A *low pass* filter passes frequencies from zero up to some finite specified value, a *high pass* those from a finite value upwards; a *band pass* filter passes only the range between two finite values, and a *band stop* passes only frequencies outside such a range. For simplicity we shall assume in what follows that the gain in question is a voltage gain.

An **ideal filter** has a gain of unity over the pass band and zero otherwise, and (for reasons which are discussed in Section 9.3.3) a phase shift proportional to frequency. The ideal responses for the four categories mentioned are shown in Fig. 5.15(a). No actual circuit, of course, has an ideal response; the closer the approximation required the more complicated the circuit needed. The art of filter design is to choose a transfer function which represents a response as close to the ideal as is required, and then to find the circuit which implements it. The function chosen should be **realizable**, that is it must represent what is physically possible (Section 3.6); essentially, this means that it must be a rational function (Section 5.2.3) in which the numerator polynomial is of lower order than the denominator.

By considering the Bode amplitude diagrams of some simple functions we may see how crude approximations may be obtained. Fig. 5.15(b) shows diagrams for the functions

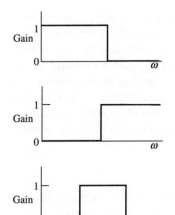

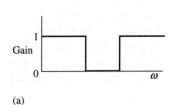

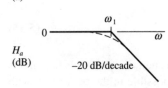

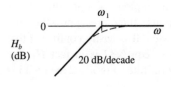

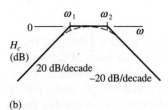

(a)

H_a
(dB)   −20 dB/decade

H_b
(dB)   20 dB/decade

H_c
(dB)   20 dB/decade
                 −20 dB/decade

(b)

Fig. 5.15
Filter characteristics: (a) ideal; (b) first-order approximations.

$$H_a = \frac{1}{1 + j\omega/\omega_1}, \tag{5.32}$$

$$H_b = \frac{j\omega/\omega_1}{1 + j\omega/\omega_1} \tag{5.33}$$

and

$$H_c = \frac{j\omega/\omega_1}{(1 + j\omega/\omega_1)(1 + j\omega/\omega_2)} \tag{5.34}$$

all of which have been expressed for convenience in terms of breakpoint frequencies ω_1 and ω_2 instead of time constants; in filter terminology these are often known as **cut-off** frequencies and denoted ω_c. The diagrams can be seen to represent, after a fashion, the first three of the ideal forms shown in (a), but they diverge from these by falling below unity in the pass band and by retaining a finite value in the stop band. The low-pass function H_a, for example, falls to the magnitude −3 dB or 0.707, instead of unity, as the cut-off frequency is reached and has maximum attenuation of 20 dB per decade beyond that. We might improve the response by increasing the rate of fall in gain above cut-off to 40 dB per decade instead of 20; this could be simply achieved by squaring H_a to give the quadratic function

$$H_a' = \frac{1}{(1 + j\omega/\omega_1)^2}$$

but the effect of this is to worsen the error at the highest frequencies in the pass band, for the magnitude at ω_1 is now −6 dB, or one half. However, a compromise can be achieved because a general quadratic factor, as considered in Section 5.3.6, can produce a variety of response curves depending on the damping ratio ζ: the choice $\zeta = 0.5$, for example, would give a curve passing through unity at the cut-off frequency. This requires the transfer function to be

$$H_a'' = 1/\{1 + j\omega/\omega_1 + (j\omega/\omega_1)^2\}$$

which has the amplitude curve shown in Fig. 5.16. The zero error at ω_1 has now been achieved at the expense of a gain exceeding unity at lower frequencies, which may or may not be acceptable. It is clear from this that, although we might continue to improve the response by taking higher-order polynomials (representing more complex circuits), there would have to be at every stage some criteria by which to judge the optimum choice of coefficients, depending on the weight to be given to the various imperfections of the response — and these include a non-ideal phase response, which we have not yet considered. Similar arguments of course apply to functions representing the other three categories of filters.

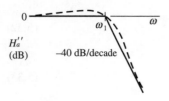

Fig. 5.16

A second-order low-pass filter characteristic

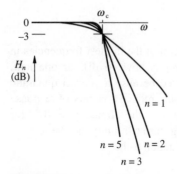

Fig. 5.17

Butterworth characteristics.

Functions with polynomial denominators of higher orders have correspondingly steeper asymptotes above cut-off: if the order of the denominator is greater than that of the numerator by r the high-frequency asymptote falls by $20r$ dB per decade (often called in this context the **roll-off** rate). For given r the variation of the response over the passband can be adjusted by choice of the coefficients. Because there is some arbitrariness in this process, a number of standard choices have gained acceptance, giving weight to different features. A few of these are considered in the following section.

5.4.2 Types of analogue filter

By convention the transfer functions $H(j\omega)$ which represent different choices for filter response are defined for the low-pass version; it turns out that any such function can be converted to a corresponding function for a different version (high pass etc.) by a simple transformation, mentioned below. It is also usual for reasons which need not concern us here to give these functions in magnitude-squared form, that is H^2 or HH^*, which means that each becomes a function of ω^2. The **Butterworth** or **maximally-flat** filter of order n has a transfer function H_n given by

$$H_n^2 = \frac{1}{1 + (\omega/\omega_c)^{2n}} \tag{5.35}$$

in which ω_c is the cut-off frequency. As its name implies, this function is chosen to give H_n as flat a response as possible in the pass band, with zero error at $\omega = 0$: its magnitude at the cut-off frequency is -3 dB for any order n, while the high-frequency asymptote falls at $20n$ dB per decade as shown in Fig. 5.17; the response therefore approaches the ideal magnitude more closely as n increases. It is easily confirmed that the simple low-pass function H_a given in eqn 5.32 is in fact H_1, the transfer function of a first-order Butterworth filter. For $n = 2$, eqn 5.35 is satisfied by the function

$$H_2 = \frac{1}{1 + \sqrt{2}(j\omega/\omega_c) + (j\omega/\omega_c)^2} \tag{5.36}$$

which on comparison with the standard form implies a damping ratio ζ of $1/\sqrt{2}$. Because for any n the polynomial denominator of the function H_n which satisfies eqn 5.35 is defined by a set of coefficients $a_0, a_1, ..., a_{n-1}, a_n$ (of which a_0 and a_n are unity) Butterworth filters can be specified by a table of these sets. The coefficients for the first few orders are given in Table 5.1.

For given n the Butterworth filter gives very small errors over the greater part of the pass band at the expense of considerable errors in the vicinity of cut-off, with a maximum roll-off rate of $20n$ dB per decade. Although the ultimate attenuation at high frequencies for any $H(j\omega)$ is bound to be fixed by the order of the denominator polynomial less that of the numerator, whatever the coefficients, it is possible to achieve faster roll-off in the vicinity of ω_c by allowing a specified maximum error to occur at any frequency. This is achieved in the **Chebyshev filter**, known also as an **equi-ripple** filter because the specified error is reached at several frequencies, giving the appearance of a ripple in the amplitude response diagram. Here the denominator of $H(j\omega)$ is based on the functions known as Chebyshev polynomials, and it may be chosen to allow ripple to appear either in the pass band or in the stop band. In the former case the magnitude-squared function is defined as

$$H_n^2 = \frac{1}{1 + \epsilon^2\, T_n^2(\omega/\omega_c)} \tag{5.37}$$

in which T_n is the Chebyshev polynomial of order n and ϵ is a real number between zero and unity. The value of ϵ governs the amplitude of the pass-band ripple: for any n the gain varies between unity and $1/\sqrt{1 + \epsilon^2}$, taking the latter value at the cut-off ω_c, as shown in Fig. 5.18. The larger ϵ, and therefore the ripple amplitude, the faster the rate of roll-off above ω_c. This rate increases with n, as in the Butterworth case, and so too does the number of ripples in the band: there are n turning points in H_n of which that at $\omega = 0$ is a maximum for n odd and a minimum for n even.

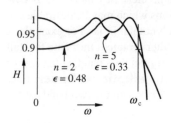

Fig. 5.18
Chebyshev characteristics.

The Chebyshev polynomials T_n are defined by the relation

$$T_n(x) = \cos(n \cos^{-1} x)$$

and the resulting expressions can be tabulated as in Table 5.2.

Neither the Butterworth nor the Chebyshev filters are designed to optimize the phase response, which ideally should show a lag proportional to ω. The form of their transfer functions shows that each has a phase lag varying from zero at $\omega = 0$ to the limiting value $-n\pi/2$ at very high frequencies, but neither is linear; the Chebyshev is increasingly non-linear the higher the choice of ϵ. For some applications phase linearity may be important, and it is optimized in the **Bessel**

Table 5.1 *Butterworth coefficients*

n	a_0	a_1	a_2	a_3	a_4	a_5	a_6
1	1	1					
2	1	$\sqrt{2}$	1				
3	1	2	2	1			
4	1	2.613	3.414	2.613	1		
5	1	3.236	5.236	5.236	3.236	1	
6	1	3.864	7.464	9.142	7.464	3.864	1

filter at the expense of the magnitude response. (The Bessel filter is sometimes said to have a maximally-flat time delay, because a linear phase shift corresponds to a constant time shift.) The name is derived from the fact that the coefficients of the denominator of $H_n(j\omega)$ are now based on Bessel polynomials.

The filters so far considered are all defined by functions with constant numerators (unity in each case) which gives them the maximum ultimate roll-off rate at high frequencies for a denominator of given order n; these are known, for reasons which will become clear in Chapter 7, as **all-pole** functions. However, certain variations which may be desirable can be achieved if the numerator is also a polynomial in $j\omega$. For example, a variation of this kind on the Chebyshev filter produces the same fast fall in magnitude around ω_c together with a very flat pass band, at the cost of introducing ripple in the stop band. An **elliptic** or **Cauer** filter provides in a similar way a very sharp fall between the two bands at the expense of ripple in both.

The foregoing discussion has dealt only with functions for low-pass filters. The corresponding functions needed for the other three categories can be obtained from the low-pass functions, which are conventionally regarded as the prototype, by **frequency transformations**. Thus, a high-pass function can be obtained from a low-pass by substituting $\omega_c^2/j\omega$ for $j\omega$; it can readily be confirmed, for the first-order Butterworth filter, that this substitution in eqn 5.32 yields the high-pass equivalent, eqn 5.33, when the frequency ω_1 is taken as the cut-off frequency ω_c in each case. The substitution of $(j\omega + \omega_0^2/j\omega)$ for $j\omega$ gives a bandpass function with bandwidth ω_c of which the geometric mean frequency is ω_0; similarly the substitution of $\omega_c^2/(j\omega + \omega_0^2/j\omega)$ gives a stopband of width ω_c around the mean frequency ω_0.

Table 5.2   *Chebyshev polynomials*

n	$T_n(x)$
0	1
1	x
2	$2x^2 - 1$
3	$4x^3 - 3x$
4	$8x^4 - 8x^2 + 1$
5	$16x^5 - 20x^3 + 5x$
6	$32x^6 - 48x^4 + 18x^2 - 1$

5.4.3   Realization

The previous section dealt with the functions which would yield desired frequency responses and which by definition are realizable, but not with the circuits which would actually produce them. Finding a filter circuit to realize a given function is a matter of synthesis, and is a large subject. It is possible, and for a low order n not very difficult, to find a passive circuit to give any of the functions we have described. For example the function H_a in eqn 5.32, which was found to represent a Butterworth filter of order one, could be provided by a combination of R and C with $RC = \omega_1$. That of order two, eqn 5.36, contains the kind of quadratic factor which we know from earlier sections to arise from a resonant circuit containing R, C, and L. We must remember, however, that the transfer function of a four-terminal network generally depends on what is connected to its output, so in designing a filter its load and the impedance of its input source should be specified.

Example 5.11   Design a second-order low-pass Butterworth filter for a load resistance R and a cut-off frequency ω_c.

Suppose, for want of any knowledge to the contrary, that the resistance which is necessary to give the function of eqn 5.36 can in this case be the load R itself and that the input source is ideal. The problem now is to choose the disposition of L and C. Since the gain must be unity at d.c., when inductance is a short circuit, we deduce that L cannot be in parallel with R; but the capacitance, which becomes a short circuit at high frequencies, could well be in parallel with R, since the output is required to tend to zero then. By this argument we may consider the circuit of Fig. 5.19. Its voltage transfer function is, by potential division,

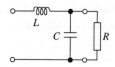

Fig. 5.19
The circuit for Example 5.11.

$$\frac{V_{\text{out}}}{V_{\text{in}}} = \frac{Z}{j\omega L + Z}$$

where Z is the impedance of R and C in parallel, which can be written as $R/(1 + j\omega CR)$. Substitution then gives

$$\frac{V_{\text{out}}}{V_{\text{in}}} = \frac{1}{1 + j\omega L/R + (j\omega)^2 LC}$$

This has exactly the same form as eqn 5.36, so the circuit suggested is correct and we now need only to find values for L and C. Comparison of the two expressions shows that these must be

$$L = \sqrt{2}R/\omega_c$$

and

$$C = 1/\omega_c^2 L = 1/\sqrt{2}R\omega_c.$$

The synthesis of higher-order filters by such a procedure as in this example would of course be a much more demanding exercise. It can be made more systematic by **normalizing** the values so that the cut-off frequency is standardized as $\omega_c = 1$ rad s^{-1} and a chosen impedance as $1\ \Omega$; the values deduced from these can be converted to the actual values required. If both load and source impedances have to be taken as finite the filter is said to be **doubly terminated** and synthesis is not so straightforward.

However, in practice purely passive circuits are not now widely used to realize the functions discussed in the previous section. Active circuits based on operational amplifiers (Chapter 8) are better for the purpose, chiefly because they can easily produce the necessary effects of inductance from circuits whose passive elements are R and C only: see Example 8.8. (It was noted in earlier chapters that inductance is avoided in electronic circuits wherever possible: it tends to be lossy, non-linear, and difficult to incorporate in integrated circuits.) Operational amplifier circuits also have the advantage of low output impedance and high input impedance, which means (as we saw in Chapter 4) that they can be connected in cascade if need be without each stage loading the previous stage and thereby affecting its transfer function. Since any polynomial of order higher than two can be factorized into one or more quadratics, together with a linear factor if n is odd, it follows that the higher-order polynomial functions of the previous section can be realized by second-order active circuits connected in cascade, with an extra first-order circuit if n is odd. The second-order components are, of course, different from each other and from the second-order filter itself. In the Butterworth case, for example, the denominator polynomials for $n = 2$, 3, and 4 are, in factorized form:

$$1/H_2(x) = 1 + \sqrt{2}x + x^2$$
$$1/H_3(x) = (1 + x + x^2)(1 + x)$$
$$1/H_4(x) = (1 + 0.765x + x^2)(1 + 1.848x + x^2)$$

An additional advantage of an active filter is that its gain in the pass band can readily be given any required value in excess of unity.

Ladder networks made up from various sections can act as passive filters, and used to be widely used for the purpose under the general description **wave filters**. Their characteristics are discussed in Section 10.1.

Digital filters are used for the same purposes as analogue filters for signals which have been converted to digital form by encoding samples (Section 9.3.4). Their action is essentially computational, and can be implemented either by means of an integrated circuit designed for the purpose or as a computer program. Digital filtering has a number of advantages over a wholly analogue process, including the possibility of very close approximation to ideal performance.

Problems

5.1 In Fig. 5.1(a), show that the transfer function relating the voltage across R_2 to the input voltage V is given by

$$G(j\omega) = \frac{R_2(R_3 + j\omega L)}{R_1 R_2 + R_2 R_3 + R_3 R_1 + j\omega L(R_1 + R_2)}.$$

[Hint: use the result of eqn 5.1 for the same circuit.]

5.2 In Problem 5.1, let $R_1 = R_2 = R_3 = 1$ kΩ and $L = 1$ mH. Find the breakpoint frequencies and the asymptotic values of G in decibels.

5.3 Find the errors in the asymptotic values of Problem 5.2 at each breakpoint frequency, as a percentage and in decibels.

5.4 Show that the maximum phase shift in a function having the form of $G(j\omega)$ in Problem 5.1 occurs at the geometric mean $\sqrt{\omega_1 \omega_2}$ of the breakpoint frequencies ω_1 and ω_2. Find this maximum for the circuit specified in Problem 5.2.

$$\left[\text{Hint}: \frac{d}{dx}\left(\tan^{-1}\frac{x}{a}\right) = \frac{a}{a^2 + x^2}. \right]$$

5.5 In Fig. 4.16(d) the circuit has values $h_{ie} = 500$ Ω, $h_{fe} = 100$, $R_L = 2.2$ kΩ, and $R_e = 470$ Ω. Use the result of Problem 4.16 to find the breakpoint frequencies and the asymptotic values of its voltage gain in decibels when R_e is bypassed with a capacitance of 1 μF.

5.6 Find the maximum phase shift (discounting the negative sign) in the gain of Problem 5.5 according to the first result of Problem 5.4.

5.7 Find the peak values of energy stored in a capacitance of 1 μF and in an inductance of 1 H when a voltage of 240 V r.m.s. at 50 Hz is applied to each. At what frequency would these energies be equal, and what would then be their value for the same voltage?

5.8 The capacitance and inductance of Problem 5.7 are connected in parallel with each other and with a resistance of 2 kΩ. Find the magnitude of the impedance of the combination (a) at the resonant frequency ω_0, (b) at $\omega_0/2$ and (c) at $2\omega_0$. In each case what would the asymptotic representation give?

5.9 If the three elements of Problem 5.8 are connected in series, find the magnitude of the overall admittance at the same three frequencies, and the corresponding values from the asymptotes.

5.10 If the resistance in the series circuit of Problem 5.9 were reduced to half the critical value, what would be the actual values of admittance at the same three frequencies?

5.11 Find the value of resistance which in the series circuit of Problem 5.9 would give breakpoints at half and twice the resonant frequency.

5.12 With the resistance of Problem 5.11 in the same circuit, what are the actual and asymptotic magnitudes for the admittance at the resonant frequency?

5.13 Find the phase angle of the admittance at half and at twice the resonant frequency, (a) in the circuit of Problem 5.10 and (b) in the circuit of Problem 5.11.

5.14 An inductor has inductance 10 mH and resistance 40 Ω; it is connected in parallel with a capacitance of 50 nF. Find the frequency at which the impedance is real, as a fraction of ω_0, the resonant frequency with resistance neglected, and the value of that impedance.

5.15 Find the extra resistance which, in series with the coil, would remove the peak in the impedance of Problem 5.14.

5.16 Find the quality factors of the parallel circuit in Problem 5.8 and of the series circuit in Problem 5.9.

5.17 Estimate the quality factor, the resonant frequency, and the bandwidth of the circuit in Problem 5.14.

5.18 A certain frequency response is represented by the transfer function

$$G(j\omega) = \frac{0.2j\omega}{1 + 5 \times 10^{-5}j\omega + 4 \times 10^{-8}(j\omega)^2}.$$

Find the values of the undamped resonant frequency, the damping factor, the quality factor, and the response at resonance.

5.19 A certain system has a response function

$$H(j\omega) = \frac{2 \times 10^{10}}{3 \times 10^{10} + 2 \times 10^5 j\omega - \omega^2}.$$

Find the values of the undamped resonant frequency, the damping factor, the frequency and magnitude of the peak response, and the magnitude of the d.c. response.

5.20 A tuned amplifier stage has a Q factor of 10 and a peak gain of 100 at a frequency of 100 kHz. Two such stages are connected in cascade, the second having a negligible loading effect on the first. First the bandwidth and Q factor of the combination, and the magnitude of the gain (a) at resonance and (b) at twice the resonance frequency.

5.21 In the circuit of Fig. 5.14, $R = 180 \ \Omega$, $C = 1$ nF, $L = 1$ mH, and $M = 0.1$ mH. Find the undamped resonance frequencies, and the magnitude of the gain at each of these and at their mean $1/2\pi\sqrt{LC}$.

5.22 Find the magnitude of the gain of a Butterworth filter at 10% above its cut-off frequency, for $n = 1, 2,$ and 3.

5.23 Find the corresponding gains if the filter in Problem 5.22 is a Chebyshev filter with (a) $\epsilon = 0.5$ or (b) $\epsilon = 0.25$.

6 Polyphase circuits

6.1 General

Polyphase electrical systems are those in which both a.c. sources and the impedances connected to them, which we shall call their loads, are arranged in groups, the members of each group being mutually separated in phase for both voltage and current. If the voltages or currents in a group have equal magnitudes and equispaced phases then they are said to be **balanced**: thus a balanced n-phase set of voltage sources consists of n sources of equal magnitude (and fixed frequency) separated in phase from each other by $2\pi/n$ radians. If the load impedances in a group are identical and symmetrically connected, then balanced sources will evidently produce balanced sets of voltages and currents, and such a load is also said to be balanced.

The reasons for using polyphase systems apply in the main to the large-scale generation and use of electrical power, and include economy in the cost of transmitting a.c. power (Appendix D) as well as the possibility of providing a uniform, as opposed to fluctuating, flow of total power to a given device (Section 6.5). The economy of generation increases with n, the number of phases, but at a cost of increasing complexity. The best compromise for general purposes is obtained for $n = 3$ and in practice three-phase systems account for most large-scale supplies of electrical power. In a few applications where more than three phases can be justified (to improve smoothing in large rectifiers, for example) transformers with zig-zag connections (Section 6.7) can be used to give more phases from a three-phase input. Two-phase arrangements are occasionally used, and are considered in Section 6.8. In what follows we shall for the most part consider three phases.

There are two ways of connecting a group of three sources or loads, and they are shown in Fig. 6.1 for the case of three balanced voltage sources. In the **star** connection (sometimes called 'Y' or even 'wye') all are connected to a common point, the star point S, and in the **delta** connection (sometimes 'mesh') they are connected in a closed loop. The terminology has already been met in the star–delta transformation

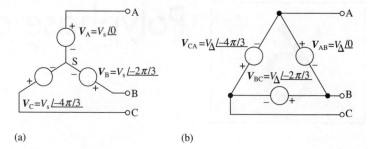

Fig. 6.1
Three-phase voltage sources:
(a) star or Y; (b) delta or mesh.

(a)   (b)

(Section 2.5.3) for impedances. The symmetry of the directions chosen for voltages should be noted; corresponding directions would be chosen for the currents which may flow.

6.2 Three-phase voltage sources

The supplies of power systems approximate to ideal voltage sources: it is a familiar fact that a mains power supply is normally provided at a virtually constant voltage. We therefore consider only such sources here, although there is no reason in principle why there should not also be three-phase current sources.

Both kinds of three-phase connection provide three terminals (in the star connection S may be a fourth) labelled A, B, and C in Fig. 6.1. We shall retain this notation although in practice R, Y, and B are commonly used in reference to the conventional colour coding. The three conductors connected to these terminals are known as **lines**. The voltage between any two, such as V_{AB}, is a **line voltage**; on the other hand the voltage of each 'phase', i.e. of each component source, such as V_A, is a **phase voltage**. This leads to one immediate distinction between the two connections: in delta, phase voltages and line voltages form identical sets, whereas in star they do not. The voltage phasors for the star-connected sources of Fig. 6.1(a) are shown in Fig. 6.2(a). Because V_A, V_B, and V_C are all specified with respect to the common

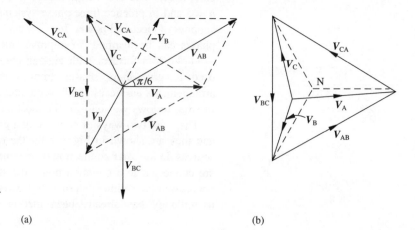

Fig. 6.2
Voltage phasors in the star connection: (a) balanced; (b) with unbalanced phases.

(a)   (b)

point S, it follows that for the star connection the line voltage V_{AB} for example is given by

$$V_{AB} = V_A - V_B = V_s \underline{|0} - V_s \underline{|-2\pi/3} = \sqrt{3}V_s \underline{|\pi/6}$$

as may be confirmed from the diagram. Thus in the balanced star connection the magnitude of the line voltages is greater than that of the phase voltages by the factor $\sqrt{3}$, and there is a phase difference of $\pi/6$ between the two sets. (The line voltage phasors can either be drawn from the origin as usual, or drawn by joining the tips of the phasors V_A, V_B, and V_C as shown; the same set of phasors can thus be drawn either in the form of a star or of a delta, but this should not be confused with the actual circuit connection.) Writing the magnitude of line voltage as V_ℓ and that of phase voltage as V_p in general, then for balanced sets we have in summary:

$$
\begin{aligned}
&Star\colon \; V_\ell = \sqrt{3}V_p; \\
&Delta\colon \; V_\ell = V_p.
\end{aligned}
\tag{6.1}
$$

Either form of connection leads to three external lines between which appear a set of line voltages, and if in practice only these lines are accessible it is obviously not possible to say by which connection they are energized, i.e. what the phase voltages are. It is line voltage which is always directly measurable and unambiguously defined; phase voltage is an ambiguous term in that it can sometimes be neither measured nor deduced. For this reason, in a polyphase system 'voltage' *means the line voltage*, unless otherwise stated, and it is equivalent to phase voltage only in the delta connection. However it is also possible to define another set of voltages which correspond to the phase voltages V_A, V_B, and V_C of the balanced star connection. If the line voltages are balanced there must exist a **neutral point** N with respect to which the three lines are electrically symmetrical: we may consider it to be a point of zero potential, and that is usual though not essential. The neutral point may exist in physical fact, as the star point S in Fig. 6.1(a), or it may be merely notional as in the delta connection of (b). But the **line-to-neutral** voltages can always be specified for a balanced system: since they are simply the phase voltages in the star case, it follows by the same argument as before that their magnitude $V_{\ell n}$ is a fraction $1/\sqrt{3}$ of the line-to-line voltage, that is

$$V_\ell = \sqrt{3}V_{\ell n}. \tag{6.2}$$

It should be noted, first, that 'line voltage' always means, unless otherwise specified, the line-to-line rather than the line-to-neutral voltage; secondly, that the 240 V of the ordinary domestic supply is a line-to-neutral voltage.

If the system should be unbalanced, then there will not in general be a neutral point in the sense of a point of symmetry; however it is always possible if need be to specify the voltages of each line to any convenient common point or potential, such as earth (or zero potential). We should also note that in the star connection the line voltages may be balanced while the phase voltages are not; in this case the neutral point and the star point have different potentials, as shown in Fig. 6.2(b).

Instead of specifying three balanced sources as in Fig. 6.1, it is sufficient to specify one together with the **phase sequence**, that is the sequence in which lines reach a given point of the voltage waveform, for example a positive peak. In Figs 6.1 and 6.2 the phase sequence is clearly ABC (and BCA and CAB).

The sum of any balanced set of polyphase voltages is always zero, and it is this fact which allows three sources to be connected in a closed loop in the delta connection: an unbalanced set may not sum to zero and a current would then circulate in the mesh, limited only by the impedance of the sources. However, by definition, and by Kirchhoff's second law, any set of line voltages must always add to zero; so any imbalance of the sources in the delta connection must be compensated by the voltage drops in their internal impedances when current flows around the mesh.

We should note, finally, that the three sources are not usually physically separate: the three-phase generators which supply the electricity in almost all power systems generate the three voltages in one machine by connecting its conductors in appropriate sections; this is easier and more economical than arranging the correct phase relationship in separate single-phase machines. Three-phase transformers can also be constructed as a unit; they are considered further in Section 6.7.

6.3 Three-phase currents

It was mentioned in Section 6.1 that a balanced load, i.e. one comprising a set of equal impedances, will produce balanced three-phase currents from a balanced set of sources. The load may be in star or delta connection, as in Fig. 6.3, and can evidently be supplied by three lines in each case. Since these lines can be energized by sources of either connection, it follows that the connections of sources and load are independent: they may be the same, or one may be star and the other delta. If both are star-connected then a fourth **neutral line** may connect the two star points.

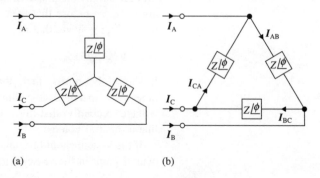

Fig. 6.3
Three-phase currents (a) in star and (b) in delta.

(a)   (b)

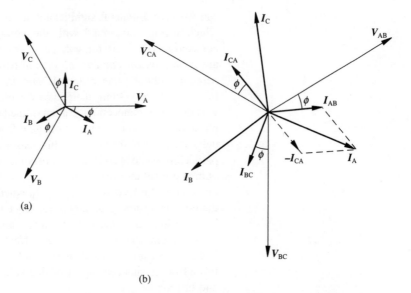

(a)

Fig. 6.4
Balanced current phasors (a) in
star and (b) in delta.

(b)

In the star connection, the **line current** is evidently also the
phase current, for example the current I_A in Fig. 6.3(a). In the delta
connection, Kirchhoff's first law at node A, for example, shows that for
symmetrically-defined directions as in Fig. 6.3(b) the line current I_A
must be given by

$$I_A = I_{AB} - I_{CA},$$

the difference of two phase currents. Figure 6.4 shows the current
phasors resulting when balanced voltages are applied to the loads of
Fig. 6.3. For the delta connection, it is clear from Fig. 6.4(b) that the
magnitude of I_A is related to that of the phase current I_{AB}, say, by

$$I_A = \sqrt{3}I_{AB}.$$

The line currents are shown here as phasors drawn from the origin, but
again, as with voltages in star (Section 6.2), they may be quickly found
simply by joining the tips of the phase currents (the delta-shaped
diagram which results having nothing to do with the actual circuit
connection).

If I_ℓ represents the magnitude of each line current and I_p that of each
phase current, then we may write in summary for balanced conditions:

$$
\begin{aligned}
&Star\colon\ I_\ell = I_p; \\
&Delta\colon I_\ell = \sqrt{3}I_p.
\end{aligned}
\qquad\qquad \textbf{(6.3)}
$$

This may be contrasted with eqns 6.1 for voltage. Again we should note
that while line current is in practice an unambiguous and measurable
quantity, since each line must be accessible, phase current is not so,
since the form of connection of source or load may not be known. Nor

can we give a general significance to the current of magnitude $I_\ell/\sqrt{3}$ which would correspond with the voltage $V_{\ell\mathrm{n}}$ in applying to either connection. Thus, as for voltage, 'current' in any three-phase system always means *line current unless otherwise stated*. It should be observed here that although line voltage V_ℓ and line current I_ℓ are basic quantities in a polyphase system, their ratio represents no one actual impedance whatever the connection; for an impedance can only constitute one phase of a set, and in star its voltage differs from the line value while in delta its current differs. For this reason, when the behaviour of any particular device is considered it becomes much more convenient to deal with phase values rather than line values: for example, the impedance of one phase of a load is simply $V_\mathrm{p}/I_\mathrm{p}$ whether it is in star or delta. For a three-phase system generally, impedance per phase is taken to mean that between line and neutral, as for a star connection.

It is interesting to compare the effects of connecting a given set of three equal impedances Z in star and in delta. In star the line current taken from a balanced supply of line voltage V_ℓ is also the phase current and has magnitude

$$I_{\ell\mathrm{s}} = I_\mathrm{p} = V_{\ell\mathrm{n}}/Z = V_\ell/\sqrt{3}Z,$$

whereas in delta the phase current I_p is V_ℓ/Z and the line current magnitude is therefore

$$I_{\ell\mathrm{d}} = \sqrt{3}I_\mathrm{p} = \sqrt{3}V_\ell/Z = 3I_{\ell\mathrm{s}}. \tag{6.4}$$

This means not only that three given equal impedances connected in delta take three times as much current from a three-phase supply as they take in star, but also that three impedances Z connected in star are equivalent to three impedances $3Z$ connected in delta. This agrees, as it must, with the star–delta transformation of Section 2.5.3.

In an actual system many separate loads are of course connected to the same lines and must therefore be in parallel. There is no difficulty in dealing with parallel loads of the same connection, for evidently their phase impedances can be combined by the usual combination rules. Note that two balanced star-connected loads between the same lines have their phases in parallel even when the star points are not connected, because the two star points must have the same potential (this would also be true of two unbalanced stars so long as their impedances were in the same proportions). The more general question of parallel connections in three-phase systems is considered further in Section 6.7.

The power factor of a balanced three-phase load is simply that of each component impedance, that is $\cos\phi$ where ϕ is the phase angle of each, and is therefore independent of their connection. If only currents and voltages, rather than impedances, are known then the power factor is the cosine of the angle between voltage and current *in any one phase*: in general the angle between a line voltage and a line current is not significant (as might be guessed from the fact that each line voltage can be associated with two different line currents). The question of power in three-phase systems is considered in Section 6.5.

It was noted in the previous section that any set of balanced voltages summed to zero and that any set of line voltages summed to zero whether balanced or not. Likewise any set of balanced currents sums to zero, but whether the same is true of line currents in general depends on the connection. Figure 6.3 shows that line currents in delta, with the directions indicated, must add to zero by Kirchhoff's first law; the same applies to the star connection if the star point S is unconnected. But if S is connected by a neutral line, for example to the star point of a generator (giving the so-called **four-wire** system, to which we return below), then we may assert only that the sum of the three line currents must be the current which flows from S. If the system is balanced, then the line currents must add to zero and no current would flow from S; the star-point connection is then clearly redundant. In practice the neutral line is used to provide single-phase supplies at line-to-neutral voltage; it can also be used to connect star points to earth or to each other. Neutral connections are important in unbalanced conditions, which are considered further in the next section.

Both star and delta connections are used extensively in practice, and in a healthy system they are almost always in balance to a very good approximation. Domestic electricity supplies are usually single-phase but they are nevertheless provided by a three-phase system, in a four-wire arrangement with a neutral line: a typical small-scale consumer forms a load connected between one line and the neutral. Consumers are allocated to the three lines in roughly equal numbers so that in aggregate they form a more or less balanced star-connected load. The resultant neutral current for the system is therefore close to zero although for each consumer the neutral current and the line current must of course be equal. This arrangement has the advantage, from the point of view of safety, of reducing the voltage available to each consumer to a fraction $1/\sqrt{3}$ of the line voltage used in the final stages of transmission: in the U.K. the latter is 415 V, and the domestic supply is therefore at 240 V. (In some countries, including North America, a domestic supply at 230 V or so is further reduced to half that by means of the single-phase three-wire system described in Section 6.8.) Larger consumers receive a complete three-phase supply; indeed many of the larger power-consuming devices, notably a.c. motors rated at more than one kilowatt or so, are typically designed only for three-phase operation partly because of superior performance and partly because a limited number of large single-phase devices, operated at random, would be unlikely to give a good degree of balance. Small single-phase appliances, for a particular three-phase consumer, can be distributed in a four-wire arrangement exactly as for domestic consumers.

Example 6.1 A three-phase generator, having a phase voltage of 6350 V, is star connected and provides a line current of 80 A to a balanced delta-connected load. Find the impedance (in magnitude only) of each phase of the load.

The voltage across each phase of the load is the line voltage of the generator, which is

$$V_\ell = 6350 \times \sqrt{3} = 11000 \; V$$

and the current in each phase of the load is a fraction $1/\sqrt{3}$ of the line current, that is

$$I_p = 80/\sqrt{3} = 46.2 \; A$$

and the load impedance per phase is therefore

$$Z_p = 11000/46.2 = \mathbf{238 \; \Omega}.$$

6.4 Unbalanced three-phase systems

A polyphase system is unbalanced, in general, if either the voltages or currents have different magnitudes and/or irregularly spaced phases. We shall assume that voltage sources are always balanced (which does not necessarily mean that voltages everywhere in the system are balanced), and that any imbalance is therefore due to unequal loads causing unbalanced currents.

In the delta-connected load, the current in each phase is easily found from the appropriate line voltage and impedance. By Kirchhoff's first law each line current is as before the difference of two phase currents. We have already noted in the last section that in delta the three line currents must always sum to zero, and their phasors may conveniently be drawn by joining the tips of the phase-current phasors to form a closed triangle. For example, Fig. 6.5 shows the currents which result when the impedances Z_{AB} and Z_{BC} of a delta-connected load are resistances R while Z_{CA} is an inductive reactance of identical magnitude to R. Although it is only the angle of one impedance which upsets the balance here, the effect is dramatic. It should be noted that the phase currents do *not* add to zero although the line currents must.

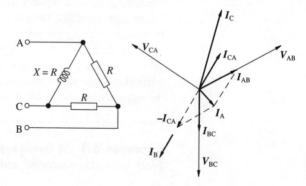

Fig. 6.5
Currents in an unbalanced delta.

For an unbalanced star-connected load, there are two possibilities to consider. If the star point remains unconnected, its a.c. potential must take up such a value as to satisfy the requirement of Kirchhoff's first law, that the three currents add to zero. To find this potential from the line voltages and the three impedances, the equations of the circuit must be solved.

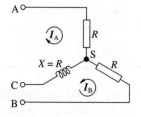

Fig. 6.6
The circuit for example 6.2.

Example 6.2   Find the potential of the star point, in terms of the line voltage, when the three impedances of Fig. 6.5 are reconnected in star with the reactance as Z_C (Fig. 6.6).

By taking V_{AB} as the reference phasor, we may write the line voltages as

$$V_{AB} = V_\ell + j0$$
$$V_{BC} = V_\ell(-1/2 - j\sqrt{3}/2)$$
$$V_{CA} = V_\ell(-1/2 + j\sqrt{3}/2).$$

If I_A and I_B are regarded as mesh currents in the directions shown, Kirchhoff's second law gives

$$I_A R + (I_A + I_B)jR = V_{AC} = -V_{CA} = V_\ell(1/2 - j\sqrt{3}/2)$$
$$I_B R + (I_A + I_B)jR = V_{BC} = V_\ell(-1/2 - j\sqrt{3}/2)$$

which yield the solution

$$I_B = -(V_\ell/10)(5 + 2\sqrt{3} + j\sqrt{3}).$$

The potential of S with respect to B, for example, is easily found as $(-I_B R)$, but it would be more usual to find V_{SN}, its potential with respect to the (in this case imagined) neutral point N. By sketching phasors or otherwise the line-to-neutral voltage V_{BN} can be seen to be

$$V_{BN} = (V_\ell/\sqrt{3})(-\sqrt{3}/2 - j1/2)$$

and the potential difference we want is then

$$V_{SN} = V_{SB} + V_{BN} = -I_B R + V_{BN}$$
$$= (V_\ell/10)(5 + 2\sqrt{3} + j\sqrt{3} - 5 - j5/\sqrt{3}).$$
$$= 0.365 V_\ell \underline{|-18.4°}.$$

In the particular case of one impedance being a short circuit or an open circuit the potential of S is much more easily found, being simply that of a line in the former case. It is possible for a combination of impedances to produce a potential for S which lies outside the triangle defined by the line potentials (that is, by the tips of the line-to-neutral voltage phasors, as in Fig. 6.2 for example).

The other possibility to be considered for an unbalanced star occurs when a neutral line connected to the star point provides a path for current. Nearly always any such connection would be made to a neutral point, i.e. a point normally at zero potential such as the star point of a balanced source; this keeps the star point of the load also at zero potential so that its phase voltages must form a balanced set. With unequal impedances it follows that the currents will then be unbalanced, and in general will not add to zero; there is no constraint on them to do so, for their resultant can flow in the star connection to form the neutral current. The source and the load are said to form a four-wire system (but still, of course, three-phase). Because the voltages are balanced it is a simple matter in such a case to calculate the line currents and hence, by Kirchhoff's first law, the neutral current.

Example 6.3 Find the neutral current taken by the star-connected load of Fig. 6.6, if the star point S is connected to the neutral point of a supply of line voltage V_ℓ.

Because we must first find the currents in terms of the phase voltages, it is easier to choose one of the latter as reference phasor, say V_A, so that by referring to Fig. 6.2 they can be written

$$V_A = V_\ell/\sqrt{3}$$
$$V_B = (V_\ell/\sqrt{3})(-1/2 - j\sqrt{3}/2).$$
$$V_C = (V_\ell/\sqrt{3})(-1/2 + j\sqrt{3}/2).$$

The currents in the three load phases follow immediately from these, and the neutral current is their sum, namely

$$I_N = V_A/R + V_B/R + V_C/jR$$
$$= (V_\ell/\sqrt{3}R)(1 - 1/2 - j\sqrt{3}/2 + j1/2 + \sqrt{3}/2)$$
$$= (V_\ell/R)(0.789 - j0.211).$$

It can be seen that in this example the neutral current, far from being zero, actually has a larger magnitude than each phase current.

6.5 Power in three-phase systems

The general consideration given to a.c. power in Section 3.9 naturally applies to three-phase circuits, but there are some additional relationships which can be useful in the latter. We consider first a balanced system.

If the load impedances have an angle ϕ, then the mean power supplied to each phase of a load (or by each phase of the source) is, for a star connection,

$$P_p = V_p I_p \cos\phi = (V_\ell/\sqrt{3})I_\ell \cos\phi \tag{6.5}$$

from eqns 6.1 and, for a delta connection,

$$P_p = V_p I_p \cos\phi = V_\ell(I_\ell/\sqrt{3})\cos\phi \tag{6.6}$$

from eqns 6.3. The total power P for a balanced system must be $3P_p$, since power is directly additive, and eqns 6.5 and 6.6 then give

$$P = \sqrt{3} V_\ell I_\ell \cos \phi \qquad (6.7)$$

for *both* connections. Although the analysis of three-phase devices is best done in terms of phase values, eqn 6.7 is often convenient because at any point in a three-phase system line quantities are always measurable.

There is one important point to be observed before proceeding. In Section 3.9 we saw that the *instantaneous* power represented by an a.c. voltage and current is a fluctuating quantity; but when expressions like that in eqn 3.54 are written for sets of balanced voltages and currents and then added, it turns out that the fluctuating terms sum to zero. Hence the instantaneous total power due to balanced polyphase voltages and currents does not fluctuate, and eqn 6.7 actually gives the constant value of instantaneous power rather than simply the mean. The smoothness of the power produced by balanced voltages and currents has considerable practical significance, especially for electromechanical conversion in motors and generators of all but the smallest sizes.

If the load is unbalanced, the eqn 6.7 is not applicable. The total power can always be found as the sum of the three phase powers, but it may also be expressed in other ways. If a load is supplied by three lines only, then its terminal conditions can be completely specified by two voltages and two currents (since the three line voltages and the three line currents must sum to zero). It follows that the total power can be expressed as the sum of two components having the form $VI \cos \phi$. By the same argument, the power flow in a four-wire system can be expressed by three such terms (and, by extension, $n - 1$ terms are required to express the total power in any n-wire polyphase system). It can be shown that if one line, say C, is taken as reference, the others being A, B, and N, then

$$P = V_{AC} I_A \cos \phi_A + V_{BC} I_B \cos \phi_B + V_{NC} I_N \cos \phi_C \qquad (6.8)$$

where ϕ_A is the angle between the line voltage V_{AC} and the line current I_A, and so on (*not* the angle of a load impedance). Equation 6.8 enables total power to be deduced from line values only. The three terms are not necessarily all positive, although for passive loads their sum must be positive.

The relations for volt-amperes and VAR in a three-phase system correspond to those for power, $\cos \phi$ being omitted or replaced by $\sin \phi$ as the case may be. All such expressions can alternatively be expressed in terms of complex products as defined in Section 3.9. For an unbalanced load, an overall power factor can be defined as the ratio of total power to total volt-amperes, but this quantity does not represent a significant angle as in the case of a balanced, or single-phase, load.

6.6 Symmetrical components

Example 6.2 served to illustrate the untidiness which can arise in analysing an unbalanced three-phase circuit. The effect of imbalance can

Polyphase circuits

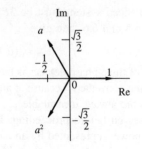

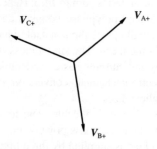

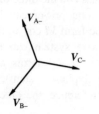

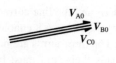

(a)

(b)

(c)

(d)

Fig. 6.7
Symmetrical components: (a) the
'a' operator; (b) positive
sequence; (c) negative sequence;
(d) zero sequence.

often be investigated more easily by the use of **symmetrical components**, which depend upon the fact that any set of three phasors at a given frequency can in general be expressed as the superposition of three sets: one balanced set with the same phase sequence as the original, another with the opposite sequence, and a third set of three equal phasors having identical phase, all of these having the same frequency. The three sets are known respectively as the **positive-**, **negative-** and **zero-sequence** components. Their use is greatly facilitated by defining a new operator 'a' to replace 'j': whereas 'j' represents an anticlockwise rotation of $\pi/2$, 'a' represents a rotation, in the same sense, of $2\pi/3$. In terms of j we may write, referring to Fig. 6.7(a),

$$a = -1/2 + j\sqrt{3}/2$$
$$a^2 = -1/2 - j\sqrt{3}/2$$

(6.9)

and we may note also that

$$1 + a + a^2 = 0.$$

A balanced set of positive-sequence voltages is shown in Fig. 6.7(b); they have the sequence ABC (since they rotate in the usual anticlockwise direction) and can be written

$$V_{A+} = V_+, \; V_{B+} = a^2 V_+, \; V_{C+} = a V_+$$

(6.10)

where A is taken as reference. On the other hand the negative-sequence set in (c) has the sequence ACB and can clearly be written, if we again take A as reference (different in general from the previous A),

$$V_{A-} = V_-, \; V_{B-} = a V_-, \; V_{C-} = a^2 V_-.$$

(6.11)

The third set shown in (d) is the zero-sequence set comprising identical phasors, for which we may write simply

$$V_{A0} = V_{B0} = V_{C0} = V_0.$$

(6.12)

Accepting without proof that unbalanced voltages can be expressed as the superposition of three such sets, we may now write

$$V_A = V_{A+} + V_{A-} + V_{A0} = V_+ + V_- + V_0$$
$$V_B = V_{B+} + V_{B-} + V_{B0} = a^2 V_+ + a V_- + V_0$$
$$V_C = V_{C+} + V_{C-} + V_{C0} = a V_+ + a^2 V_- + V_0$$

(6.13)

These equations show that the unbalanced set is recovered when the column matrix (V_+, V_-, V_0) is multiplied by the square matrix

$$[A] = \begin{bmatrix} 1 & 1 & 1 \\ a^2 & a & 1 \\ a & a^2 & 1 \end{bmatrix}.$$

It follows that in order to deduce the symmetrical components of a given set of unbalanced voltages V_A, V_B, V_C we need only multiply their column matrix by the inverse of $[A]$, that is by

$$[A]^{-1} = \frac{1}{3} \begin{bmatrix} 1 & a & a^2 \\ 1 & a^2 & a \\ 1 & 1 & 1 \end{bmatrix}.$$

The three reference phasors of the symmetrical components are thus given by

$$V_+ = \tfrac{1}{3}(V_A + aV_B + a^2 V_C)$$
$$V_- = \tfrac{1}{3}(V_A + a^2 V_B + aV_C)$$
$$V_0 = \tfrac{1}{3}(V_A + V_B + V_C). \tag{6.14}$$

Precisely similar results, of course, apply to currents. We should also note that the expressions make no distinction between line and phase values: they are applicable to any set of three phasors.

Although these components can be used to investigate the effect of unbalanced impedances generally they are most easily used where the imbalance arises from the extreme cases of asymmetrical short circuits or open circuits: in practice the most severe imbalance in power systems arises not from varying loads but from faults of this kind. Thus a short circuit between lines B and C gives $V_{BC} = 0$, $V_{CA} = -V_{AB}$ and $V_{BN} = V_{CN}$, as in Fig. 6.8(a); and an open circuit in line B, say, gives $I_B = 0, I_C = -I_A$, as shown in (b). Because the energy available in a power system is potentially destructive, the analysis of such abnormal conditions is important.

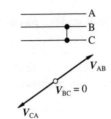

(a)

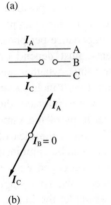

(b)

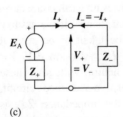

(c)

Fig. 6.8
Faults: (a) line-to-line short circuit; (b) open circuit; (c) equivalent circuit for (a).

Example 6.4 Find the symmetrical components of the three currents of Fig. 6.8(b).

If we take the normal balanced magnitude of each current to be unity, for simplicity, and I_A to be positive real, then we have

$$I_B = 0; I_A = 1; I_C = -1$$

and eqns 6.14 then give

$$I_+ = (1 - a^2)/3 = 1/2 + j/2\sqrt{3}$$
$$I_- = (1 - a)/3 = 1/2 - j/2\sqrt{3}$$
$$I_0 = 0$$

as the reference phasors of the three balanced sets. It can immediately be confirmed that the sum of these three is unity, which, as expected from eqns 6.13, is the value of I_A; it can also be checked that the same equations give I_B and I_C correctly.

There are some particular features of the zero-sequence component which are worth noting. Equations 6.14 show that the zero-sequence phasor is simply the average of the three phasors which form the original out-of-balance set, and it follows immediately that any set of phasors which add to zero has no zero-sequence component. Thus for example, there is no such component in any set of line voltages, nor in the line currents of any three-wire system; hence in any delta connection a zero-sequence current cannot flow in the lines although it may do so in the phases (in which case we may say it circulates around the delta). In a three-wire star connection there can be a zero-sequence phase voltage but no zero-sequence current; while in a four-wire system the neutral current, being the sum of the three line currents, must, by eqns 6.14, be the total zero-sequence current $3I_0$ (or $I_{A0} + I_{B0} + I_{C0}$). The possibilities may be summarized as follows:

3-wire system: zero-sequence possible only in delta phase currents and star phase voltages;

4-wire system: zero-sequence possible in phase (i.e. line) currents, the sum giving the neutral current, and (less probably) in phase voltages.

It is in the nature of power-system components, including rotating machines, that in general they present different impedances to positive-, negative- and zero-sequence voltages and currents. It is, for example, clear from the discussion above that the effective impedance per phase of any part of a system to zero-sequence currents must depend on whether it has a fourth wire between any two star points, whether any such points are earthed, and whether any short-circuit fault includes an earth connection. Also the different sequences do not interact: the currents of a given sequence produce only voltage drops of the same sequence, in the appropriate impedance. From this comes the idea of separate 'circuits' for the different components, which can be combined according to the constraints of the situation. Usually generated e.m.f.s remain balanced and are therefore positive-sequence only, so voltage sources appear only in that circuit. Consider, for example, the line-to-line fault represented in Fig. 6.8(a). Because there is no earth connection at the fault, it can carry no zero-sequence current; hence, if we assume that the current I_A on the unfaulted line can be taken to be zero, it follows from the first of eqns 6.13 that $I_+ = -I_-$. Also since $V_{BN} = V_{CN}$, eqns 6.14 show that $V_+ = V_-$ for line-to-neutral voltages. These two results can only be satisfied if the positive-sequence circuit, containing some generated voltage E_A and the impedance Z_+, is connected to the negative-sequence circuit containing only the impedance Z_-, as shown in Fig. 6.8(c). (All of these are as usual the effective values per phase.) From the complete circuit $I_+, I_-, V_+,$ and V_- can immediately be found, and the various actual voltages and currents at the fault then follow by eqns 6.13, there being no zero-sequence quantities. Other kinds of faults are analysed in comparable ways.

The power in a three-phase system may be expressed in terms of the symmetrical components of voltages and currents. If the components are evaluated for the *phase* voltages and currents of a load, then the total power consumed is

$$P = 3(V_+I_+ \cos \phi_+ + V_-I_- \cos \phi_- + V_0I_0 \cos \phi_0) \qquad (6.16)$$

where ϕ_+, etc. are the angles between the voltage and current phasors of each component in any phase. As might be expected, a voltage and a current of differing sequences do not contribute to the average power. The power can also, of course, be expressed in terms of the voltage and current phasors in complex form, as in eqn 3.57.

6.7 Parallel connections

The fact that a three-phase system is likely to contain many loads in parallel has been mentioned in Section 6.3, and the same is of course true of sources. There is no difficulty in combining passive three-phase loads in parallel, whether balanced or not, and the equivalent load which results is easily found: deltas, or stars with common star points, as shown in Fig. 6.9(a) and (b), can be combined in the usual way, phase by phase, into a single set; stars with unconnected star points, as in (c), can be converted first to equivalent deltas (Section 2.5.3) to give the same form as (a); and by the same token a star in parallel with a delta can best be combined by first converting the star to another delta. In every case the resultant load can be converted from delta to star or vice versa.

Three-phase sources can also be connected in parallel, but in practice, as with other near-ideal sources, it is necessary that certain conditions be satisfied. If single-phase a.c. sources are to be safely connected in parallel, they must have identical voltages in frequency, magnitude, and phase; for otherwise the discrepancy in voltage would cause current to circulate in the sources, limited only by their internal impedances, even when no load is connected. In the three-phase case, it is safe to connect sources in parallel if their line voltages are identical in the same three respects; and this is so whether they are all star, all delta, or mixed. In the first case, the star points may also be connected together if they are at the same potential, as would usually be the case if the sources are balanced.

A rather more demanding situation arises in the case of transformers when both primary and secondary windings are to be connected in parallel. Here it will be helpful to consider the nature of three-phase transformers as compared with the single-phase version described in Section 4.9. Evidently the three sets of primary and secondary coils cannot all share a common core, since one resultant flux could not possibly induce three distinct e.m.f.s. A common construction is that shown in Fig. 6.10, in which the core has three limbs, joined top and bottom to form closed paths for magnetic flux. Three suitably connected

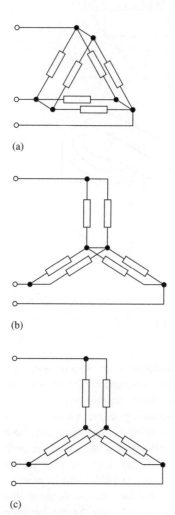

(a)

(b)

(c)

Fig. 6.9
Parallel circuits: (a) deltas; (b) stars with common star point; (c) stars with star points unconnected.

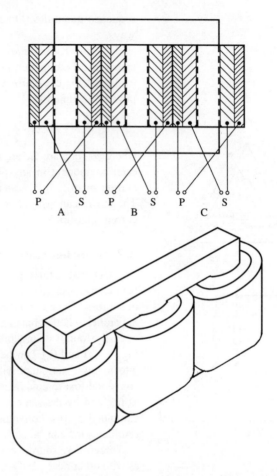

Fig. 6.10

A three-phase transformer core.

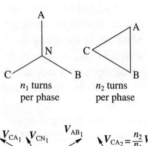

n_1 turns
per phase

n_2 turns
per phase

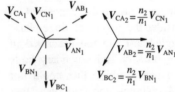

Fig. 6.11

Phase shift from star to delta.

but physically separate single-phase transformers can also be used, but usually less economically. Whatever the construction, the transformation process takes place from phase to phase, so that any one phase voltage of the primary and the corresponding phase voltage of the secondary are, in the ideal approximation, either in phase or in antiphase according to the directions specified. If primary and secondary are both connected in star, or both in delta, then corresponding line voltages are likewise in phase or antiphase. But it is quite possible, and in practice not uncommon, for one side to be in star and the other in delta: in that case there will be a phase difference of $\pi/6$ (or $5\pi/6$) between corresponding line voltages on primary and secondary, as shown in Fig. 6.11 for the case of a star-connected primary with a delta-connected secondary.

It follows from this that when the primary lines of two different three-phase transformers are connected in parallel the line voltages of their secondary windings may be out of phase with each other by $\pi/6$. This would not necessarily matter if the secondaries were to be kept separate, or at any rate not connected directly in parallel; but in practice it is often required to connect the secondaries in parallel as well as the primaries,

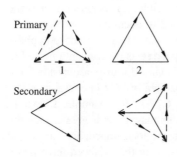

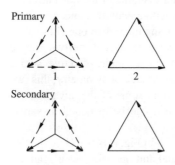

(a) star-delta ‖ delta-star

Primary

Secondary

(b) star-star ‖ delta-delta

Fig. 6.12
Allowed parallel combinations.

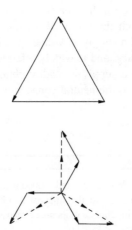

Fig. 6.13
The zig-zag connection.

and in that case only transformers having the same phase shift between primary and secondary line voltages may be used. Identical transformers can, of course, always be connected in parallel, but there are other allowed combinations as shown in Fig. 6.12 (in which for clarity the voltage phasors are so drawn as to indicate the mode of connection).

Naturally the line voltages of windings to be connected in parallel must also have the same magnitude. This condition will evidently be satisfied for transformers having similar primary and secondary connections and the same turns ratio (Section 4.8); the latter condition will also allow a star–star in parallel with a delta–delta, since for both the line voltage ratio is (ideally) equal to the turns ratio. But a star–delta transformer with n_1 primary turns and n_2 secondary turns on each phase has a secondary-to-primary line voltage ratio given by

$$V_{\ell 2}/V_{\ell 1} = n_2/\sqrt{3}n_1$$

while a delta–star has the ratio

$$V_{\ell 2}/V_{\ell 1} = \sqrt{3}n_2/n_1.$$

It follows that to connect a star–delta transformer in parallel with a delta–star the turns ratio n_2/n_1 of the former must be three times that of the latter.

The choice between star and delta connections for three-phase devices, and especially for transformers, depends on a number of factors including, for example, the treatment of the harmonics which tend to corrupt sinusoidal waveforms in practice (Section 9.1.4). Transformers are sometimes made with a third (*tertiary*) winding, the purpose of which may be to provide a neutral point for a delta-connected secondary or to provide a delta connection to short-circuit harmonics. The advantages of the star–delta or delta–star arrangement without the $\pi/6$ phase shift can obtained by using the **zig-zag** connection shown in Fig. 6.13 instead of a star. In this each of the three phase windings is split into two sections; pairs of sections from different phases are connected in series to form three new windings which can be connected in star. The zig-zag connection also has advantages in respect of harmonics and in reducing the effect of unbalanced loads on primary currents; it has the disadvantage of somewhat reduced voltages, which are $\sqrt{3}/2$ of the equivalent star values (its line voltage is therefore 1.5 times the voltage of an undivided phase).

More generally, the splitting of each phase of a transformer secondary winding into several sections which can be reconnected allows not only adjustment of the phase shift between primary and secondary but also reconnection into more than three new phases. This is a convenient way of obtaining an output of six or more phases when that is required for a particular purpose.

6.8 Two-phase systems

Strictly, according to our earlier definitions, a set of balanced two-phase voltages or currents would comprise simply two equal and opposite

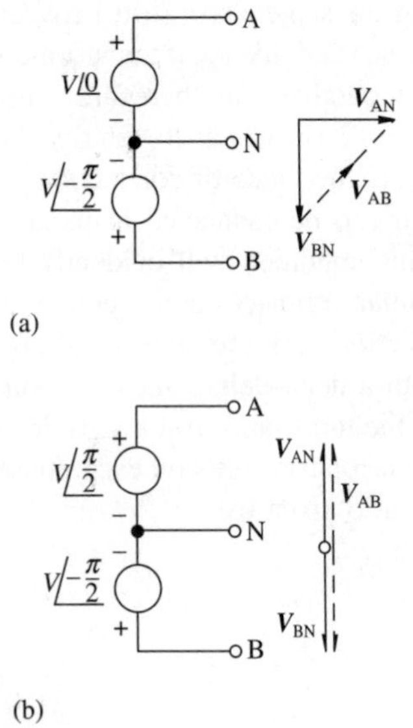

(a)

(b)

Fig. 6.14

(a) Two-phase voltage sources;
(b) single-phase three-wire
sources.

phasors. Such an arrangement is quite possible: it is better known as the single-phase three-wire system mentioned in Section 6.3 and we return to it below. However it lacks some of the advantages of polyphase systems generally, and the term **two-phase** is by convention used to describe what is actually one half of a balanced four-phase system: that is, two phasors of equal magnitude separated by $\pi/2$ in phase, as given by the voltage sources shown in Fig. 6.14(a) for example. Evidently there is only one line voltage here, and its magnitude is greater than the line-to-neutral by the factor $\sqrt{2}$. When the load impedances are equal the line currents have equal magnitudes, as in the balanced three-phase system, but do not now add to zero; indeed their sum, which is the neutral current, has magnitude greater than that of the line current by the factor $\sqrt{2}$. The arrangement shown, with a neutral connection, is the only one possible if the advantages of having more than one phase are to be realized; for if only the two lines were connected the two sources would simply combine in series to form a single effective source, supplying a single load. Despite its apparent lack of symmetry, this two-phase system is occasionally used; it offers some of the advantages of polyphase operation from a supply which can be provided in various ways on a relatively small scale.

A true two-phase arrangement is shown in Fig. 6.14(b). Here the line-to-neutral, or phase, voltages, being equal and opposite, once again add to zero; and the line voltage has magnitude V_ℓ twice that of each phase. For equal load impedances the line currents also add to zero, giving zero current in the neutral. The neutral wire could therefore be omitted without apparent loss, and the arrangement would then be indistinguishable from a single-phase system with voltage V_ℓ. But the neutral connection offers important advantages: one is that of safety, for each line has a voltage of only $V_\ell/2$ with respect to the neutral point; another, that individual loads can either be designed for the lower voltage (and in aggregate shared between the two phases so as to achieve an overall balance, as discussed for the three-phase case in Section 6.3) or for the higher. The arrangement in fact offers both the safety of a supply at voltage $V_\ell/2$ and the relative economy of a supply at voltage V_ℓ. It is used in North America for domestic supplies, and is widely referred to as a single-phase three-wire system. Such a supply is readily obtained from a larger three-phase system by using as the neutral connection the mid-point of a transformer secondary phase.

Problems

6.1 Find (a) the line voltage of a domestic supply providing the standard line-to-neutral voltage of 240 V, (b) the voltage to neutral of each line of a 275 kV transmission system.

6.2 A balanced set of delta-connected impedances of value $20|45°$ Ω is connected to the lines of a 3.3 kV supply. Find the current in a line, and its phase relative to the line-to-neutral voltage for that line.

6.3 A star-connected set of impedances of value $15|30°$ Ω is connected to the lines of Problem 6.2, with the delta set still connected. What is now the total current in a line?

6.4 A certain three-phase motor has a power factor of 0.8 lagging and takes a line current of 20 A from a 2200 V, 50 Hz supply. Find the value of delta-connected capacitors which would make the overall power factor unity. [Hint: assume that the motor is also delta-connected and consider phase currents.]

6.5 If one of the capacitors in Problem 6.4 becomes open-circuit, confirm that the three line currents still add to zero and find their magnitudes.

6.6 Three adjacent houses take electricity from the three successive phases of a 415 V supply. Find the resultant neutral current when they consume, in order of phase sequence, 35 A at 0.9 lagging, 18 A at 0.8 lagging, and 5 A at 0.85 lagging.

6.7 A 5 kW motor is star-connected to a 415 V supply and runs at a power factor of 0.75 lagging. Find its line current.

6.8 Find the total power consumed in the houses of Problem 6.6.

6.9 A short circuit between two lines of a three-phase supply reduces one line voltage to zero while the other two become equal and in phase opposition. Find the symmetrical components of the three line voltages when the non-zero voltages are taken to be $\pm V$.

6.10 Find the symmetrical component magnitudes of the line currents in Problem 6.6, and confirm that the zero-sequence current is one-third of the neutral current.

7 Transient response

7.1 Introduction

In earlier chapters we have dealt almost exclusively with the steady state, in which a linear circuit energized by sinusoidal sources at a certain frequency has all of its currents and voltages varying sinusoidally at the same frequency (d.c. being simply the case of zero frequency). It has been assumed that this is an ultimate condition which is bound to be reached sooner or later and can be regarded as the normal state of the circuit when energized by these sources. But we have not yet asked how the steady state is attained from some different starting point: for example, when the sources are first connected the currents and voltages are likely to start from zero at the same instant, although in the steady state they may end up out of phase; and the sources may for some reason change in value, as for that matter may the circuit components. To answer this question, we must consider the response of a circuit to any newly applied stimulus or disturbance (the connection or disconnection of sources, or any change in the circuit elements) in a very general way. We shall find that the general or **total response** is made up of two parts: the **steady state**, which lasts indefinitely, and the **transient response**, sometimes called the **natural response**, which in most cases decays to zero leaving only the steady state (but which, as we shall see, in certain amplifying circuits may, by design or by default, grow until it is limited by some physical change). Although our everyday experience indicates that the apparent durations of electrical transients are short by common standards (many devices appear to reach a steady state immediately after they are switched on), they may be quite long by the standards of electrical engineering, in which a microsecond can be a long interval.

The behaviour of linear circuits in all circumstances is governed by Kirchhoff's laws (Section 2.2) and by the general voltage–current relations of the passive elements R, L, and C (eqns 1.7). If we combine the two for a circuit we arrive at one or more differential equations with constant coefficients, and it is the solution of these which gives the total response. The fact that we are dealing with a new state of affairs following some change or disturbance is taken into account by **initial conditions**: the equations to be solved are for the new state, but the initial conditions depend on the old state. If there has in fact been no change in the state of the circuit we should find, as we might expect, that the total response is merely the steady state; there is therefore no need to set up differential equations in that case, for we already have the perfectly adequate methods described in Chapters 2 and 3 for dealing with the steady state alone.

7.2 The differential equations of linear circuits

7.2.1 First-order equations

We shall consider two examples which are familiar, namely the charging and discharging of a capacitor through a resistance. Figure 7.1(a) shows the circuit in the former case: the capacitor is initially uncharged and at some instant which we take to be the origin of time t, that is when $t = 0$, the switch is closed to connect it to the ideal voltage source of d.c. voltage V. Suppose now that we wish to know the voltage v_C which subsequently appears across the capacitor. Once the switch is closed, Kirchhoff's second law gives

$$iR + v_C = V$$

which, since the current i is given by

$$i = C \frac{dv_C}{dt}$$

becomes

$$CR \frac{dv_C}{dt} + v_C = V \tag{7.1}$$

which is a first-order linear equation with constant coefficients. It is quite easy to solve in several ways, but we shall use for the time being the standard procedure which can always be applied directly to equations of any order in this category. The voltage V represents what is generally known as the **driving function** or **forcing function**: it is in fact the stimulus, or source, which is responsible for the final steady-state form of the unknown v_C; in this case it is a constant, but in general it could be any function of time (or zero, if the stimulus happened to be removal of sources instead of, as in this example, their connection to the circuit). We now set the driving function to zero, which gives

$$CR \frac{dv_C}{dt} + v_C = 0 \tag{7.2}$$

and assume for this a solution of the form, which always arises in such equations,

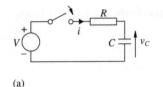

(a)

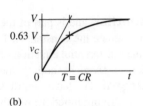

(b)

(c)

(d)

Fig. 7.1
(a) Capacitor charging circuit;
(b) charging voltage;
(c) discharging circuit;
(d) discharging voltage.

$$v_C = Ae^{mt}$$

where A is an unknown constant; it follows that we must then also have

$$\frac{dv_C}{dt} = mAe^{mt}$$

and eqn 7.2 therefore requires that

$$CRmAe^{mt} + Ae^{mt} = 0 \qquad (7.3)$$

which is often called the **auxiliary equation**. It has the solution

$$m = -1/CR$$

and the solution to eqn 7.2 must therefore have the form

$$v_C = Ae^{-t/CR} \qquad (7.4)$$

which is known as the **complementary function** and is part of the total solution we seek, provided that we can evaluate the constant A. It can be shown that the general solution to a linear differential equation of the kind now considered, whatever its order, is given by adding the complementary function the **particular integral**. The latter signifies a solution to the original equation (including the driving function) which contains no arbitrary constant like A; it is in fact the steady-state solution, a point to which we shall return. Inspection of eqn 7.1 shows that it is satisfied by the value

$$v_C = V,$$

so the complete general solution must be

$$v_C = V + Ae^{-t/CR} \qquad (7.5)$$

in which A is still be evaluated. To find A we take into account the initial condition of the circuit: in this case it has been specified that the capacitor is initially uncharged, which means that $v_C = 0$ at $t = 0$. Substituting these values in eqn 7.5 gives

$$0 = V + A$$

from which we find that $A = -V$ and eqn 7.5 becomes finally

$$v_C = V(1 - e^{-t/CR}). \qquad \textbf{(7.6)}$$

This variation of v_C with time is shown in Fig.7.1(b). It is the familiar charging curve of a capacitor and is characterized by the **time constant** CR, a quantity already encountered in connection with steady-state frequency response in Chapter 5. It is easy, and always advisable, to check that such a solution satisfies not only the initial condition (by setting $t = 0$) but also the ultimate steady state, by letting $t \to \infty$; the latter gives $v_C = V$, as expected from the nature of the circuit. We may also note a useful property of the time constant: differentiating eqn 7.6 shows that v_C has an initial rate of change V/CR, so CR is the time which v_C would take to reach its steady-state value V were that rate maintained. The value actually reached by v_C after a time CR is $V(1 - 1/e)$, or about 0.63 of its final value, a figure characteristic of any curve having this rising exponential form.

The second example concerns the circuit shown in Fig. 7.1(c), in which there is no source and the capacitor is initially charged to a voltage V. After the switch is closed, Kirchhoff's second law gives the equation

$$iR + v_C = 0$$

which becomes

$$CR\frac{dv_C}{dt} + v_C = 0 \tag{7.7}$$

the same as eqn 7.1 except that the driving function is now zero, which makes it identical to eqn 7.2. Note that the assumed direction for the current i is the same as before, although it does not have to be and we should now expect current actually to flow the other way. The complementary function is clearly the same as before, and given by eqn 7.4, but the particular integral is now zero (which, again, is simply the steady-state solution) and the complete general solution is therefore simply

$$v_C = Ae^{-t/CR}.$$

This time the initial condition is $v_C = V$ at $t = 0$, and it follows that $A = V$; so the final solution is

$$v_C = Ve^{-t/CR}$$

as shown in Fig. 7.1(d). It is in this case an exponential decay with the same time constant as before; it is a simple matter to confirm that it gives the correct initial and final values, and that CR is again the time in which v_C would reach its steady-state value (in this case zero) at its initial rate of decay. In the time CR it actually falls to $1/e$ or 0.37 of its initial value, a figure characteristic of all such exponential decays.

One or other of the two forms of solution found in the two examples above can arise in the total response of any first-order linear circuit to a constant or zero driving function. Here 'first-order' means a circuit containing only one reactance, because that leads to a first-order equation; the circuit may be a series or parallel arrangement, and the 'response' in question a current or voltage due to a driving function which may itself be either current or voltage. If we were to investigate

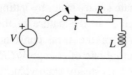

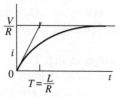

(a)

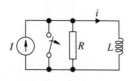

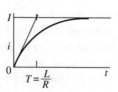

(b)

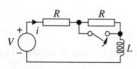

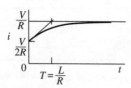

(c)

Fig. 7.2
Transient currents in an
inductance: (a) series circuit,
applied voltage; (b) parallel
circuit, applied current;
(c) following a circuit change.

the response of a circuit containing only inductance L and resistance R to a d.c. stimulus (or the source-free decay of any initial current in L, analogous to the discharge of capacitance) we should be likely to find one of the same two forms, but with the time constant L/R instead of CR. Thus the growth of current from zero in a series LR circuit after a d.c. voltage source is connected to it is shown in Fig. 7.2(a). Similarly, the current in L following the application of a d.c. current to L and R in parallel is shown in (b); in each case the voltage across L decays to zero from its initial value, i.e. from V in case (a) and from IR in (b), with the same time constant.

Example 7.1   An inductor has inductance 1 mH and resistance 10 Ω. Find the current which flows after the application of 10 V d.c. to the inductor at an instant when it carries no current.

After the voltage is applied Kirchhoff's voltage law for the circuit gives

$$L\frac{di}{dt} + iR = 10.$$

Setting the right hand side to zero and assuming a solution

$$i = Ae^{mt}$$

gives the auxiliary equation

$$(mL + R)Ae^{mt} = 0$$

so

$$m = -R/L$$

and the complementary function is therefore

$$i = Ae^{-Rt/L}.$$

The particular integral is the steady-state current in the inductor, which is V/R or 1 A.

The complete solution is then

$$i = 1 + Ae^{-Rt/L}.$$

Since the initial current is zero, we require

$$0 = 1 + A$$

so $A = -1$ and the required solution is

$$i = 1 - e^{-Rt/L} = 1 - e^{-10^4 t} \text{ A}$$

with t in seconds.

In many cases, since we know that the response must take one of two standard forms with a known time constant, it can immediately be deduced without further analysis simply by noting the initial and final conditions of the circuit. In Fig. 7.2(b), for example, the entire source current must flow in R initially, since by eqn 1.6 a sudden rise of current in L would require infinite voltage; we also know that finally no current flows in R because L is a short circuit for d.c. As it happens, in all the examples met so far either the initial value or the final value of the required variable was zero, but this is by no means always the case. Consider the circuit shown in Fig. 7.2(c), in which the d.c. voltage source is permanently connected but one of the two resistors is suddenly short-circuited at $t = 0$. Here the initial current is clearly $V/2R$, the final current is V/R, and the time constant is L/R (since it refers to the nature of the circuit *after* the change); it follows without more ado that the current in the circuit varies as shown in the figure, between two non-zero values.

When one response of a circuit has been found for given conditions it is usually simple to derive any other from it without much more work. For example if we wish to know the charging current of the capacitance in the first example above, it is easily found from the voltage given by eqn 7.6. Thus, we have

$$i = C\frac{\mathrm{d}v_C}{\mathrm{d}t} = \frac{V}{R}\mathrm{e}^{-t/CR}. \tag{7.8}$$

Sometimes the equation which the circuit yields for the chosen variable contains an integrated term. Had we started with the intention to find the current i in that same example we could have written, from Kirchhoff's second law,

$$iR + \frac{1}{C}\int i\mathrm{d}t = V. \tag{7.9}$$

This can readily be reduced to a differential equation by substituting as dependent variable the charge on the capacitor, say q. Since the current i is $\mathrm{d}q/\mathrm{d}t$ eqn 7.9 becomes

$$CR\frac{\mathrm{d}q}{\mathrm{d}t} + q = VC$$

which can be solved for q exactly like eqn 7.1 and i then found by differentiation.

It has been pointed out that the particular integral which is needed for a complete formal solution to the differential equation is simply the steady-state value of the variable, which depends on the driving function. Since the only driving function to appear in the examples above was a d.c. quantity the particular integral was easily deduced by inspection. In other cases it can be found by any of the methods described in earlier chapters, which all applied to the steady state. Suppose, for example, that we want the voltage v_C in the same CR

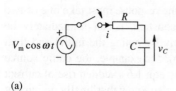

(a)

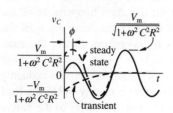

(b)

Fig. 7.3
Capacitor charging from a sinu-soidal source.

circuit as before when it is energized, as shown in Fig. 7.3(a), by an a.c. source of voltage $V_m \cos \omega t$ or $V_m \underline{|0}$. We may conveniently use peak values here, even for phasor notation, because in the end we aim for an actual function of time. By voltage division we may write the (peak) phasor voltage across the capacitance as

$$V_{Cm} = (V_m + j0)\frac{1/j\omega C}{R + 1/j\omega C} = \frac{V_m}{1 + j\omega CR}$$

$$= \frac{V_m}{\sqrt{1 + \omega^2 C^2 R^2}}\underline{|-\phi}$$

where $\phi = \tan^{-1}(\omega CR)$. From this, the steady-state voltage can immediately be written as the function of time

$$v_C = \frac{V_m}{\sqrt{1 + \omega^2 C^2 R^2}} \cos(\omega t - \phi)$$

and this is the particular integral we seek. The complementary function has the same form as before, given by eqn 7.4, so the total solution is

$$v_C = Ae^{-t/RC} + \frac{V_m}{\sqrt{1 + \omega^2 C^2 R^2}} \cos(\omega t - \phi).$$

To satisfy the initial condition that $v_C = 0$ at $t = 0$, we require that

$$A = -\frac{V_m}{\sqrt{1 + \omega^2 R^2 C^2}} \cos(-\phi)$$

and since, from the definition of ϕ above,

$$\cos(-\phi) = \cos \phi = \frac{1}{\sqrt{1 + \omega^2 C^2 R^2}}$$

then

$$A = -\frac{V_m}{1 + \omega^2 C^2 R^2}$$

and the complete solution is therefore

$$v_C = -\frac{V_m}{1 + \omega^2 C^2 R^2} e^{-t/RC} + \frac{V_m}{\sqrt{1 + \omega^2 C^2 R^2}} \cos(\omega t - \phi) \quad (7.10)$$

which is illustrated in Fig. 7.3(b). Here too the transient component is a simple exponential decay superimposed on the steady-state solution, and having such an amplitude as to give the specified initial condition (in this case, $v_C = 0$).

Example 7.2 Find the transient component of current which flows in the series combination of resistance 10 Ω and inductance 100 mH following the sudden application of a voltage $2 \sin 100t$ at $t = 0$, the initial current being zero. Find also the transient voltage across the inductance.

The differential equation of the circuit is similar to that of Example 7.1 and the complementary function is again given by

$$i = Ae^{-Rt/L}$$

which in this case is

$$i = Ae^{-100t}.$$

To find the constant A we need the particular integral, i.e. the steady-state solution. The reactance of the inductance at angular frequency 100 rad s^{-1} is 10 Ω, numerically equal to the resistance, so the impedance is $10\sqrt{2}\lfloor\pi/4$ Ω. The steady-state current is therefore given by

$$i = \frac{2}{10\sqrt{2}} \sin(100t - \pi/4) \text{ A}$$

which at $t = 0$ would have the value

$$\frac{2}{10\sqrt{2}} \sin(-\pi/4) = -0.1 \text{ A}.$$

For the zero initial current given, the transient current must exactly cancel this at $t = 0$, so A must have the value 0.1 A and the transient current is therefore

$$i = 0.1e^{-100t} \text{ A}.$$

The transient voltage across the inductance is most easily found by differentiating the transient current. This gives

$$v_L = L\,di/dt = 0.1 \times 0.1 \times (-100)e^{-100t} = -e^{-100t} \text{ V}$$

and it may be noted that this exactly cancels the transient voltage drop in the resistance, as it must because the two must add to the sinusoidal voltage applied.

7.2.2 General properties

There are important features, which are quite general, to be seen in the two first-order equations we have solved. It may first be noted that the transient and steady-state components of the responses are easily identified: the former is an exponential decay, negative in one case; the latter is merely what would be calculated from steady-state theory (d.c. or a.c.) and may be zero, in which case the entire response is transient. It is also evident that the transient component is the complementary function of the solution, dependent on the initial conditions but not on the source; while, as already pointed out, the steady-state component is the particular integral, dependent on the source but not on the initial conditions. In the case of the transient component we may go further: although its magnitude depends on the initial conditions, its form does not; it would be exponential with time constant CR or L/R for *any* initial condition. This is a very significant fact, for it means that the transient or natural response of a passive circuit takes a form which is characteristic of the circuit itself; it depends neither on the sources connected to it nor on the initial state it happens to be in. But the actual amplitude of the transient is always dictated by the requirement that its initial value added to the steady-state value must yield the specified initial condition of the circuit.

This last requirement leads to an interesting special case: if the initial value of any quantity happens to be the steady-state value, then its response may not have a transient component. Thus, in the example with a CR circuit and an a.c. source in the last section there would be no transient if the initial condition happened to be

$$v_C(0) = V_m/(1 + \omega^2 C^2 R^2)$$

since that is the steady-state value of v_C at $t = 0$. Equally, in the more likely event that the initial value of v_C is zero, the transient could be avoided by connecting the a.c. source at an instant when the steady-state value of v_C would also be zero; in the example this would mean any t such that $(\omega t - \phi)$ is an odd multiple of $\pi/2$. It follows that in connecting an a.c. source to a first-order circuit a transient can be avoided, in theory, by making the connection at the instant when its steady-state value would equal whatever its initial value happens to be. In practice this would require some ingenuity in timing and there is not usually any reason to attempt it (in the minority of cases where transients are likely to be harmful their effects can be mitigated by other means). In circuits of higher order, with more than one reactive element, some kind of transient is almost inevitable: for one thing, it is most unlikely that the initial condition in every reactance would be equal to the corresponding steady-state value.

The question of initial conditions is an important one and will be discussed further in Section 7.4 in connection with the Laplace transform. To find the complete solution for the response of a circuit we need to know the initial voltage in every capacitance and the initial current in every inductance, and the number of arbitrary constants to be found (only one in the above examples) is equal to the number of such voltages and currents which can be independently specified. In some cases the possibility arises of a discontinuity in one of these quantities at $t = 0$. In relation to Fig. 7.2(b) above, for example, it was noted that a sudden change in the current through L was not possible because it would require infinite voltage, and we could equally well have said that it would mean a corresponding change in stored energy and would therefore need infinite power; on either argument it is clearly not possible in the given circuit, and we could therefore take the current to start from zero after the switch operates. Similar arguments apply to many transient problems, and in any case changes which are mathematically 'sudden' could not happen in practice. Nevertheless, it is possible in principle to have infinite values of voltage, current, or power and consequent discontinuities: they can all be provided by *impulse functions*. These have an infinite magnitude for a vanishingly small time, and will be considered further in Section 7.4.

Although impulses are evidently an idealization which is impossible to achieve in practice, they represent an important concept in electrical engineering and are frequently encountered. We can even provide an example from the previous section. In Fig. 7.1, if R is zero then so is the time constant and we are left with a capacitance which is instantaneously charged on connecting the d.c. voltage. If the source is ideal then an infinite current flows for an infinitely short time and this impulse immediately provides the necessary charge CV and the stored energy $\frac{1}{2}CV^2$. In this case the solution is trivial, but in general where impulses are possible we need to decide whether a discontinuity in an initial condition can occur and if so which of the two values to take, just before or just after the change. We return to this question in Section 7.4.

Any purely resistive circuit has an instantaneous response with no transient component, and we may note that such a circuit can have no 'initial condition' in the sense we have used (that is, current in L or voltage on C). Of course, a resistive circuit may have initial currents and voltages when some change takes place, but these have no effect on what subsequently happens: the final steady state is immediately attained in any case. But it must be emphasized that this applies *only* to circuits having no reactive elements; although a reactive circuit appears to be purely resistive to steady-state d.c. this is not true of its transient response.

The method of solution given for first-order equations in the previous section can be applied to most ordinary linear differential equations having constant coefficients, i.e. to the kind arising from linear circuits. For an equation of order n the complementary function requires n constants to be evaluated and that number of initial conditions must therefore be specified. The usual procedure for a solution can be summarized thus:

(1) set the driving function(s) to zero;

(2) assume a solution of the form Ae^{mt};

(3) by substitution deduce the auxiliary equation;

(4) the roots m_1, m_2, ... of the auxiliary equation give the complementary function $A_1e^{m_1t} + A_2e^{m_2t} + \dots$, provided that there are no repeated roots (the latter case is considered in the next section);

(5) add the particular integral which satisfies the original equation, i.e. the steady-state solution;

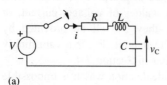

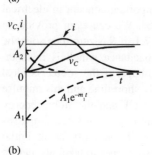

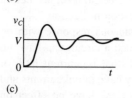

(a)

(b)

(c)

Fig. 7.4
Second-order transients:
(a) circuit; (b) overdamped solution; (c) under-damped solution.

(6)　evaluate constants $A_1, A_2, \ldots$, from the initial conditions.

7.2.3　The second-order transient

A second-order differential equation always arises from a circuit containing both an inductance and a capacitance. Consider the RLC series circuit shown in Fig. 7.4(a) and suppose that we wish to investigate the voltage v_C which follows the connection to it of a d.c. voltage V, the initial conditions being zero. Kirchhoff's second law gives

$$L\frac{di}{dt} + Ri + v_C = V$$

but since i can be written as Cdv_C/dt this gives

$$LC\frac{d^2v_C}{dt^2} + RC\frac{dv_C}{dt} + v_C = V. \tag{7.11}$$

Setting V to zero, assuming the solution of the form Ae^{mt} and substituting, we find the auxiliary equation

$$m^2LC + mRC + 1 = 0 \tag{7.12}$$

which has the solution

$$m = -\frac{R}{2L} \pm \sqrt{\frac{R^2}{4L^2} - \frac{1}{LC}}. \tag{7.13}$$

We may note in passing the common features between eqn 7.12 and eqn 5.10 of Section 5.3.3, which related to resonance in this same circuit. We return to this comparison later, but for the present let us use the general notation introduced in that section to write eqn 7.13 in the form

$$m = \omega_0\left(-\zeta \pm \sqrt{\zeta^2 - 1}\right) \tag{7.14}$$

in which, as before,

$$\omega_0 = \frac{1}{\sqrt{LC}}$$

and

$$\zeta = \frac{R}{2}\sqrt{\frac{C}{L}}$$

According to the value of ζ, there are evidently three possible forms for the roots m of the auxiliary equation and hence for the complementary function. We consider them in turn.

(i)　$\zeta > 1, R > 2\sqrt{L/C}$

In this case eqn 7.14 gives two negative real values for m, say $-m_1$ and

$-m_2$. The complementary function is therefore

$$v_C = A_1 e^{-m_1 t} + A_2 e^{-m_2 t} \qquad (7.15)$$

and since the particular integral, i.e. the steady-state value of v_C, is simply

$$v_C = V$$

the complete solution can be written

$$v_C = A_1 e^{-m_1 t} + A_2 e^{-m_2 t} + V. \qquad (7.16)$$

We now invoke the initial conditions, namely $v_C = 0$ and $i = 0$, of which the second implies that $\mathrm{d}v_C/\mathrm{d}t = 0$. Making these substitutions in eqn 7.16 and its derivative, for $t = 0$, we find that

$$A_1 + A_2 = -V$$

and

$$-m_1 A_1 - m_2 A_2 = 0.$$

These give, with the two values of m from eqn 7.14, the two constants

$$A_1, A_2 = -\frac{V}{2}\left(1 \pm \frac{\zeta}{\sqrt{\zeta^2 - 1}}\right)$$

of which one is positive and the other negative; the former has the smaller magnitude and is associated with the larger m and hence the more rapid decay. It follows that the transient component of the response is the difference of two exponential decays, is always negative, and when added to the particular integral V gives the complete response for v_C shown in Fig. 7.4(b). The figure also shows the form of the current i, which is easily found as $C\mathrm{d}v_C/\mathrm{d}t$; it is entirely transient (as we should expect) and it is readily confirmed to be again the difference between two exponentials, but this time of equal initial amplitude. (The rise from zero to a peak before a falling back to zero is characteristic of the difference between a pair of exponentials with the same end states.)

Example 7.3 An inductance L of 250 μH is connected in series with a capacitance C of 1 nF and a resistance of 1.2 kΩ. Find the time constants of the complementary function for the voltage v_C across C and the initial amplitude of the two exponential current components which flow after a d.c. voltage V of 10 V is applied to the circuit with initial conditions zero.

For this circuit the damping factor is

$$\zeta = \frac{R}{2}\sqrt{\frac{C}{L}} = 600\sqrt{\frac{10^{-9}}{25 \times 10^{-5}}} = 1.2$$

which confirms that we are dealing with case (i) above. The time constants are the reciprocals of m_1 and m_2 in eqn 7.15. From eqn 7.14 we have

$$m_1, m_2 = -\frac{1}{\sqrt{LC}}(-\zeta \pm \sqrt{\zeta^2 - 1}) = -2 \times 10^6(-1.2 \pm \sqrt{0.44})$$

$$= 1.073, \ 3.727 \times 10^6 \ \text{s}^{-1}$$

so the time constants in microseconds are $1/1.073$ and $1/3.727$, i.e. **932** and **268 ns**.

The current is given by $C\mathrm{d}v_C/\mathrm{d}t$, so from eqn 7.16 the initial amplitude of its two components is $-Cm_1A_1$ or $-Cm_2A_2$. From the expression given for A_1 and A_2, allocating the positive sign to subscript 1 gives

$$A_1 = -\frac{V}{2}\left(1 + \frac{1.2}{\sqrt{0.44}}\right) = -5 \times 2.809 = -14.05$$

so that

$$-Cm_1A_1 = -10^{-9}(1.073 \times 10^6)(-14.05) = \mathbf{15.1 \ mA}.$$

The other component can be confirmed to be negative with the same initial amplitude.

(ii) $\zeta = 1$, $R = 2\sqrt{(L/C)}$.

This particular case, corresponding to what we found in Section 5.3.3 (eqn 5.11) to be a critical value of R so far as resonance is concerned, is a special one in that the complementary function does not take the standard form of eqn 7.15. The roots of the auxiliary equation are now, by eqn 7.14, both given by

$$m = -\omega_0$$

and it can be shown that the complementary function in this case must take the form

$$v_C = e^{-\omega_0 t}(A_1 t + A_2).$$

When this is added to the particular integral V, the same initial conditions as before require that

$$A_2 + V = 0,$$
$$-\omega_0 A_2 + A_1 = 0$$

and hence that

$$A_2 = -V,$$
$$A_1 = -\omega_0 V$$

which gives the transient response

$$v_C = -Ve^{-\omega_0 t}(1 + \omega_0 t)$$

and the complete solution

$$v_C = V\{1 - e^{-\omega_0 t}(1 + \omega_0 t)\}. \tag{7.17}$$

This, although mathematically different from the solution of case (i), appears quite similar to it in form; and, again, its derivative, which is a measure of the current response, rises from zero to a maximum before decaying back to zero.

(iii) $0 < \zeta < 1, R < 2\sqrt{L/C}$.

In this case eqn 7.14 shows that the auxiliary equation has two complex conjugate roots; it can be written

$$m = -\zeta\omega_0 \pm j\omega_0\sqrt{1 - \zeta^2}$$
$$= -\alpha \pm j\omega_n$$

in which the notation α for $\zeta\omega_0$ and ω_n for $\omega_0\sqrt{1 - \zeta^2}$ has been introduced for convenience. The complementary function is now

$$v_C = A_1 e^{(-\alpha+j\omega_n)t} + A_2 e^{(-\alpha-j\omega_n)t} \tag{7.18}$$

which can be rearranged as

$$v_C = e^{-\alpha t}\left\{(A_1 + A_2)\frac{e^{j\omega_n t} + e^{-j\omega_n t}}{2} + j(A_1 - A_2)\frac{e^{j\omega_n t} - e^{-j\omega_n t}}{2j}\right\}$$
$$= e^{-\alpha t}(B_1 \cos \omega_n t + B_2 \sin \omega_n t) \tag{7.19}$$

where B_1 and B_2 are new constants to be determined instead of A_1 and A_2. It is worth noting that B_1 and B_2 must both be real since the other factors in the equation are real, despite the fact that B_2 represents $j(A_1 - A_2)$: this requirement can be satisfied because A_1 and A_2 are necessarily complex conjugates in order that eqn 7.18 should give real v_C (we must bear in mind that we are not now considering phasors, nor is an exponential quantity to be taken here as implying only its real part; we are dealing for the present with actual functions of time).

Since the particular integral is V as before, the complete response in this case is thus

$$v_C = e^{-\alpha t}(B_1 \cos \omega_n t + B_2 \sin \omega_n t) + V. \tag{7.20}$$

Inserting the initial conditions, we find that

$$B_1 + V = 0,$$
$$\omega_n B_2 - \alpha B_1 = 0$$

from which

$$B_1 = -V,$$
$$B_2 = -\alpha V / \omega_n$$

and the final solution is therefore

$$v_C = V \left\{ 1 - e^{-\alpha t} (\cos \omega_n t + \frac{\alpha}{\omega_n} \sin \omega_n t) \right\} \qquad (7.21)$$

This time we have a very different kind of solution. The transient response is sinusoidal at the frequency ω_n but its amplitude decays with time constant $1/\alpha$ or $1/\zeta\omega_0$; it is said to be a **damped oscillation**. The solution is illustrated in Fig. 7.4(c). Sometimes α is called a **damping constant**, as distinct from ζ, which was defined in Chapter 5 as a damping factor or damping ratio. The frequency ω_n is the **natural frequency** of the circuit, although the term is sometimes also applied to ω_0, which strictly speaking is the **undamped** natural frequency (as well as the resonance frequency encountered in Section 5.3); the two are very close when $\zeta \ll 1$. The idea of critical damping, for which $\zeta = 1$, was first encountered in Chapter 5 in connection with the frequency response of resonant circuits, but it can now be seen to have wider significance than that: *if the damping is less than critical the transient response is oscillatory, otherwise it is not.*

The responses we have deduced above are those of the capacitor voltage v_C when a d.c. voltage V is applied to a series circuit; we should have found the same auxiliary equation for any other response such as current, or the voltage across the inductance, to an applied voltage whether d.c. or some other waveform. An applied *current* would be a different matter in this circuit: in that case the voltage response of each series element could be found separately and that of the whole found by addition; the results would not resemble those obtained above.

We shall find later in this chapter methods of rapidly predicting the form of any response to any kind of driving function in a given circuit without the exertion of setting up and solving a differential equation in each case.

7.3 Transient response and frequency response

The similarity between the auxiliary equation which governed the characteristic transient response of the *RLC* series circuit (eqn 7.12), and the quadratic factor which governed the frequency response of the same circuit (eqn 5.10) has already been pointed out. It seems to indicate a clear connection between frequency response, which relates to different frequencies in the steady state, and transient response, which relates to a function of time. The same connection can be found in first-order circuits, and we shall see later that it is quite general, but it is most evident in the second-order circuit. The quadratic polynomial which gives an oscillatory transient in the response of the series circuit to a voltage stimulus has the same form as that which gives a peak in the steady-state current taken by the same circuit from a voltage of varying frequency, provided in each case that ζ is small enough; the natural frequency of the transient, defined above by

$$\omega_n = \omega_0\sqrt{1 - \zeta^2},$$

approximates at small ζ to ω_0, the resonance frequency in the steady state, as did the other characteristic frequencies encountered in Section 5.3. It should be no surprise to find that the frequency at which a circuit is markedly more responsive to a steady-state driving function is also that at which energy naturally tends to oscillate between its reactive elements (which is what happens during a transient oscillation), nor that both effects are extinguished by the excessive dissipation implied by too large a value of ζ. It would be quite possible to express transient behaviour in terms of Q, as in Section 5.3, rather than ζ; but by convention the use of Q is normally confined to the steady state.

The transient behaviour of a circuit or other system has been described by functions of time, found by solving differential equations, while the steady-state behaviour investigated in Chapter 5, found essentially from the same equations, was expressed by functions of frequency. Hence it is sometimes said that a given system can be characterized by its behaviour in either the **time domain** or the **frequency domain**. That there is a mathematical connection between the two will become clearer in the following sections. At this point we need only note the form taken by the roots of the auxiliary equation which determines the transient response: in eqn 7.18 they were written as

$$m = -\alpha \pm j\omega_n,$$

in which the magnitude ω_n of the imaginary part defines the natural frequency and α, that of the real part, the rate of decay. Because we may regard these complex values of m as defining 'frequencies' m/j, i.e. $\pm\omega_n(1 \pm j\alpha/\omega_n)$, the term **complex frequency** is sometimes used for such exponents. We shall meet this idea again in the following section.

7.4 The Laplace transformation

7.4.1 Introduction

The method outlined in Section 7.2, based on what may loosely be called classical methods of solving linear differential equations, can be used to solve any transient problem, although in a circuit of appreciable complication the process may be laborious. In fact, a different method is often preferred for the general investigation of transient behaviour; it is based on the **Laplace transformation**, one of a number of mathematical transformations which convert functions from one variable to another in such a way as to result in convenient properties (another, closely related to the Laplace, is the Fourier transformation which we shall meet in Chapter 9). We use the Laplace transformation to change a function of time to a different function of a complex variable s; like all such transformations it has an inverse, which converts functions of s back into functions of time. The variable s is closely related to the quantity m, introduced above for the roots of the auxiliary equation, which in the previous section was seen to be a kind of complex frequency. The advantages of this method are twofold. First, the Laplace transformation has the property that the term-by-term transformation of a differential equation in time t changes it into a purely algebraic equation in s; the latter easily yields an expression for the unknown quantity which can then be transformed back ('inverted') into the required function of t. In this way the actual solution may be less laborious: the transformation and its inverse are usually found from a table of standard cases such as appears in Appendix E.

The second advantage relates to a property which to some extent applies to both the Laplace or the classical method of solution but is more easily extracted from the former. It is this: for linear systems in general the complete solution of differential equations is often made unnecessary by the fact that the general behaviour of a system, transient as well as steady-state, can be readily predicted if its system functions are expressed in terms of s rather than $j\omega$, a change which turns out to be very easily achieved. This more general prediction is merely an extension of the facility afforded by the a.c. theory of Chapter 3, by which we were able to deduce the steady-state behaviour of a circuit without reference to its differential equations.

The **Laplace transform** of a function $f(t)$, t a real variable, is defined as a function $F(s)$, s a complex variable, given by[†]

$$F(s) = \int_0^\infty e^{-st} f(t) \mathrm{d}t \tag{7.22}$$

and the **inverse Laplace transform** is defined by

$$f(t) = \frac{1}{2\pi \mathrm{j}} \int_{\sigma - \mathrm{j}\infty}^{\sigma + \mathrm{j}\infty} e^{st} F(s) \mathrm{d}s \tag{7.23}$$

in which σ is a constant real part of s chosen so as to ensure convergence of the integral. Two functions $f(t)$ and $F(s)$ related by eqns 7.22 and

---

[†] Strictly eqn 7.22 defines the **one-sided** Laplace transformation, a reference to the fact that the integration covers only the range $0 < t < \infty$, that being appropriate to our requirements for circuit analysis.

7.23 are said to form a (Laplace) **transform pair**. The symbols $\mathscr{L}$ and $\mathscr{L}^{-1}$ are used to indicate transformation, thus:

$$\mathscr{L}f(t) = F(s); \quad \mathscr{L}^{-1}F(s) = f(t).$$

There is no need for us to consider the justification of these relations and the formal restrictions to which they are subject; nor, usually, need we actually carry out the integrations, since the tables commonly available suffice for the majority of cases which arise in circuit analysis. It should be noted here that a transformed function of time, although it is in general a complex quantity, is conventionally denoted, as in $F(s)$, by an upper-case symbol which, unlike the usual symbols for phasors and for functions of $j\omega$, is not emboldened.

Example 7.4 From the tables in Appendix E, find the Laplace transform of $10 \sin (20t + 30°)$.

The first entry of Appendix E.1 indicates that the constant multiplier 10 simply carries over into the transform. In the Table E.2 there are transforms for $\sin at$ and $\cos at$ but none for the form $\sin(at + \phi)$. However we can write

$$10 \sin(20t + 30°) = 10(\sin 20t \cos 30° + \cos 20t \sin 30°)$$
$$= 5\sqrt{3} \sin 20t + 5 \cos 20t.$$

The second entry of E.1 confirms the linearity property that the transform of a sum is the sum of the transforms, so we can now use the entries for $\sin at$ and $\cos at$ to write the required transform as

$$\mathscr{L}10 \sin(20t + 30°) = 5\sqrt{3}\frac{20}{s^2 + 400} + 5\frac{s}{s^2 + 400} = \frac{5(s + 20\sqrt{3})}{s^2 + 400}.$$

Distinct from particular pairs $f(t)$ and $F(s)$ are the **properties** of the Laplace transformation consequent on the nature of the integrals. The two properties which underlie the advantages outlined above can be expressed thus:

$$\mathscr{L}\frac{\mathrm{d}}{\mathrm{d}t}f(t) = sF(s) - f(0+) \tag{7.24}$$

and

$$\mathscr{L}\int_0^t f(\tau)\mathrm{d}\tau = \frac{F(s)}{s}. \tag{7.25}$$

The term $f(0+)$ in eqn 7.24 indicates an initial value, to be taken just after $t = 0$ in the event of a discontinuity there. We return to this point below. That apart, these equations show that differentiation with respect to time is equivalent to multiplication by s, and that integration is equivalent to division by s. It is through these that differential (or integral) equations transform into algebraic equations.

Consider again the first-order problem we solved in Section 7.2.1, the charging of a capacitance through a resistance. The differential equation to be solved, eqn 7.1, was

$$CR\frac{dv_C}{dt} + v_C = V,$$

in which v_C is the unknown function of time and V a constant. Since we take the initial condition to be $v_C = 0$, eqn 7.24 (also given in Appendix E.1) shows that here

$$\mathcal{L}\frac{dv_C}{dt} = sV_C(s).$$

The first entry of E.2 shows that

$$\mathcal{L}V = \frac{V}{s}$$

so we can now transform eqn 7.1, above, term by term to give

$$CRsV_C(s) + V_C(s) = \frac{V}{s}$$

and from this the transform of the unknown voltage v_C can be written

$$V_C(s) = \frac{V}{s(1 + sCR)} = \frac{V}{CRs(s + 1/CR)}.$$

The form of this expression appears as an entry in Appendix E.2, and from that we can write its inverse transform immediately as

$$v(t) = \frac{V}{CR}\frac{1 - e^{-t/CR}}{1/CR} = V(1 - e^{-t/CR})$$

in exact agreement with the solution obtained before in eqn 7.6.

As a second example let us find the transform of the voltage v_C across the capacitance in the RLC circuit of Fig. 7.4(a) following the connection of some given voltage $v(t)$ at $t = 0$, with initial conditions zero. The differential equation for v_C is given by eqn 7.11 when the unspecified driving function $v(t)$ is substituted for the d.c. voltage V; thus

$$LC\frac{d^2v_C}{dt^2} + RC\frac{dv_C}{dt} = v_C = v(t). \tag{7.26}$$

Confirming from Appendix E.1 that, as we should expect, double differentiation is simply equivalent to multiplication by s^2 for zero initial conditions, we can transform eqn 7.26 to give

$$LCs^2V_C(s) + RCsV_C(s) + V_C(s) = V(s).$$

Hence

$$V_C(s) = \frac{V(s)}{LCs^2 + RCs + 1}.\tag{7.27}$$

By inserting the appropriate transform $V(s)$ for whatever voltage $v(t)$ is applied, we find the transform $V_C(s)$ which must then be inverted to give the response $v_C(t)$. The results may be available more or less directly from a table of transforms, but if need be the inversion can be simplified by the method of partial fractions outlined in Section 7.4.5 and Appendix E.3.

7.4.2 Generalized system functions

Equation 7.27 has a familiar look about it, because the polynomial in the denominator has the same form as the auxiliary eqn 7.12 which was obtained in solving the same problem. This is hardly surprising: our original method was to assume a complementary function of the form e^{mt}, for which differentiation is clearly equivalent to multiplication by m, just as we now have it equivalent to multiplication by s (if for the time being we continue to assume that initial conditions are zero). What turns out to be more significant is the similarity between the polynomial in s and that of eqn 5.10 in the variable $j\omega$. Here too we might expect it, for we saw in Chapter 3 that the complex functions of a.c. analysis arose because differentiation of a sinusoid is equivalent to the multiplication of its phasor by $j\omega$.

This introduces a very potent analogy. The equivalence of the operation d/dt to multiplication by $j\omega$ led to the reactances $j\omega L$ and $1/j\omega C$ and thence to the whole idea of complex functions which can relate phasor quantities in the steady state; by the same token, the equivalence of d/dt to multiplication by s in a transformed equation leads us to conclude that the substitution of s for $j\omega$ in any system function will correctly relate the corresponding transformed quantities *whatever the time variation they represent*. This means we now have a much more general idea of a system function (including, for example, impedance and admittance) which is not restricted to the sinusoidal steady state. Such a function can be set up either from the steady-state version, by the simple substitution of s for $j\omega$, or by writing the reactances of L and C as sL and $1/sC$ respectively and constructing the

Table 7.1 *Relations between voltage and current*

	Actual	Phasors	Transforms
	$v(t)$, $i(t)$	V, I	$V(s)$, $I(s)$
Resistance	$v = Ri$	$V = RI$	$V = RI$
Inductance	$v = L\,di/dt$	$V = j\omega LI$	$V = sLI$
Capacitance	$i = C\,dv/dt$	$I = j\omega CV$	$I = sCV$

function by the usual rules; the result may be called a **generalized system function**. The basic relations between voltage and current in passive elements may be summarized as in Table 7.1.

To illustrate the use of the transform relations in setting up a generalized function directly, let us apply them to the circuit of Fig. 7.4(a) for which we deduced eqn 7.27 relating the transform $V_C(s)$ for the capacitance voltage to that for the input voltage, $V(s)$. With the relations of Table 7.1 we can use potential division to write the transfer function

$$\frac{V_C(s)}{V(s)} = \frac{1/sC}{sL + R + 1/sC} \tag{7.28}$$

from which eqn 7.27 is obtained immediately. There is no more need to set up a differential equation like eqn 7.26 than there would be if we wanted the steady-state equivalent of eqn 7.28, i.e. the same transfer function in terms of $j\omega$.

Example 7.5 Find the response $v_L(t)$ of the voltage across the inductance L in Fig. 5.1(a) after the connection of a d.c. voltage V to the input, the initial conditions being zero.

We start by finding the transfer function $V_L(s)/V(s)$. By the normal combination rules the input impedance of the circuit is

$$Z_{in}(s) = R_1 + \frac{R_2(R_3 + sL)}{R_2 + R_3 + sL} = \frac{R_1(R_2 + R_3 + sL) + R_2(R_3 + sL)}{R_2 + R_3 + sL}.$$

The input current transform is therefore

$$I_1(s) = \frac{V(s)}{Z_{in}(s)} = \frac{V(s)(R_2 + R_3 + sL)}{R_1(R_2 + R_3 + sL) + R_2(R_3 + sL)}$$

and by current division the current in L is then

$$I(s) = I_1(s)\frac{R_2}{R_2 + R_3 + sL} = \frac{V(s)R_2}{R_1(R_2 + R_3 + sL) + R_2(R_3 + sL)}$$

(compare eqn 5.1). Finally, the voltage $V_L(s)$ is $sLI(s)$, which gives

$$\frac{V_L(s)}{V(s)} = \frac{sLR_2}{R_1(R_2 + R_3 + sL) + R_2(R_3 + sL)}.$$

We should of course obtain the same result by finding the steady-state transfer function and then substituting s for $j\omega$. Now we need only put $V(s)$ equal to V/s, the transform of the d.c. voltage V, and we find

$$V_L(s) = \frac{VLR_2}{R_1(R_2 + R_3 + sL) + R_2(R_3 + sL)} = \frac{VR_2}{(R_1 + R_2)(s + a)}$$

where a is $(R_1R_2 + R_2R_3 + R_3R_1)/L(R_1 + R_2)$. This expression corresponds to a standard case in Appendix E.2 and inverts to

$$v_L(t) = \frac{VR_2}{R_1 + R_2} e^{-at}.$$

It is a helpful exercise to confirm that the initial and final values of this response agree with physical reasoning about the circuit.

There is a residual problem: we have so far assumed zero initial conditions. To be quite general, according to eqn 7.24 we should replace the voltage–current transform relations in L and C with the following:

$$V = \mathscr{L}L\mathrm{d}i/\mathrm{d}t = L\{sI - i(0+)\} = sLI - Li(0+)$$

$$I = \mathscr{L}C\mathrm{d}v/\mathrm{d}t = C\{sV - v(0+)\} = sCV - Cv(0+). \qquad (7.29)$$

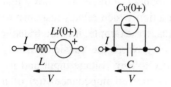

From these it appears that the effect of an initial current $i(0+)$ in L is to introduce an effective voltage source having the transform $Li(0+)$, in series with L, acting in the direction specified for current flow (since its sign shows that it acts so as to augment the applied voltage V); and that the effect of an initial voltage $v(0+)$ on C is to introduce a current source having the transform $Cv(0+)$, in parallel with C, acting in the direction to augment I. The two cases are illustrated in Fig. 7.5. (The voltage source can of course be changed to a current source and vice versa, by the usual equivalence of sources described in Section 1.6.3, if that should be more convenient.) It thus turns out that if initial conditions are regarded as effective sources to be added to the circuit in this way, the required relations can still be constructed directly, without reference to the differential equations.

Fig. 7.5

Equivalent sources to represent initial conditions.

Example 7.6 Use the representation in Fig. 7.5 to deduce the voltage of the discharging capacitance in Fig. 7.1(c), having an initial voltage V.

The current source $Cv(0+)$ in Fig. 7.5 is here CV and must be added in parallel with C in Fig. 7.1(c). The transformed current CV therefore flows into the impedance of C and R in parallel, i.e. into

$$Z(s) = \frac{R/sC}{R + 1/sC} = \frac{1}{C(s + 1/CR)}$$

and the resulting voltage transform is

$$V_C(s) = CVZ(s) = \frac{V}{s + 1/CR}.$$

The inverse of this is quickly seen from Appendix E.2 to be

$$v_C(t) = Ve^{-t/CR}$$

which is precisely the solution found by classical methods in Section 7.2.1.

In fact, the usefulness of the general notation goes far beyond the solving of transient problems to find an actual time response, and the need to find solutions with non-zero initial conditions does not arise very frequently. What is much more commonly required is the *form* of the transient response, which, as we have seen, does not even need a knowledge of the driving function. Although in describing the transform method of solution there has been no indication that the transient and steady-state components are as easily separated as in the classical method of Section 7.3, it must still be true that the form of the transient depends neither on the driving function (i.e. the sources or other stimuli) nor on the initial conditions, which fix only the arbitrary constants and hence the magnitude of the transient component. It therefore follows that its form depends only on the system function which relates response to stimulus, whatever the latter may be: thus, for example, it is the transfer function of eqn 7.29, and only that, which governs the transient form of v_C in the circuit of Fig. 7.4(a) in response to *any* voltage input and for any initial conditions. Similarly, the generalized impedance $Z(s)$ of a circuit, that is $V(s)/I(s)$, is enough to tell us how its voltage will react to any change in the input current; a generalized voltage gain for a four-terminal network tells us how its output voltage will react to any change of input voltage; and so on. We shall see in the following sections how this information can be extracted from a system function without going to the trouble of finding an actual response. In those cases where a solution is required for a particular circumstance, the transform of the driving function together with the appropriate system function gives immediately the transform of the required response, for example $V_C(s)$ in eqn 7.27.

If a system function is already known for the steady state in terms of $j\omega$, say $G(j\omega)$, it is a trivial matter to find $G(s)$ by substituting s for $j\omega$. By the same token, the steady-state function is easily recovered from $G(s)$ by rewriting it as $G(j\omega)$. It must be remembered that s is in general a complex quantity having, like frequency, the dimension of reciprocal time. Regarded as a variable we cannot in general specify its real and imaginary parts; but the functions we deal with yield characteristic values of s which are completely specified (much as steady-state functions like those of Chapter 5 give particular values of frequency such as $1/CR$, $1/\sqrt{LC}$, etc.). These values correspond to the roots of the auxiliary equation in the classical method of finding solutions, and they tell us, quite often, all we need to know about transient behaviour.

7.4.3 Driving functions

In those cases where the complete response of a system to a particular type of stimulus is wanted, the transform of the stimulus or driving function is needed. In circuits this will normally be due to the application of a voltage or current. Where the stimulus comes from a different kind of change or disturbance, for example the removal of a source or a change in the circuit configuration, the transform of any sources which remain in the circuit will still be required (if no sources remain, it is essentially the initial conditions which cause the transient). Generally, therefore, if we should want to know more than the basic form of a transient we are likely to need the transforms of such functions of time as are likely to arise as driving functions: there are one or two which are important apart from the obvious constant and sinusoid to represent d.c. and a.c. sources.

First we consider the constant or d.c. driving function. Because very often the response of interest arises from the sudden connection of a source, the idea of a **step function** is used to describe a quantity which is zero for all time before some instant and thereafter a finite constant. A **unit step** is a step function in which the constant is unity, and if the change occurs at $t = 0$ it is written $u(t)$; thus

$$u(t) = 0 \qquad (t < 0)$$
$$= 1 \qquad (t \geq 0). \tag{7.30}$$

A step which changes from zero to the constant value A at the instant $t = T$, say, is then $Au(t - T)$. The instant at which a connection or other change takes place is often, of course, an arbitrary choice and it is then simplest to take $t = 0$; but sometimes another choice is needed, for example to deal with successive changes at certain intervals.

In fact, the step function is not strictly necessary for our purposes, at least in the common case when $T = 0$, for the Laplace transform is defined only for $t > 0$ and it follows that the transform of $Au(t)$ can be no different from that of the constant A: both are actually given by A/s (Appendix E). For this reason we have been able to solve some of our examples by treating a d.c. voltage as a constant, although it would be more usual to describe its sudden connection as a step function. The difference in physical effect between a constant quantity which already exists and one which has been zero until $t = 0$ is accounted for solely by the initial conditions to which the former gives rise. Nevertheless the step function, sometimes called simply a 'step', is a useful and common description; it can be used to indicate the sudden application of *any* driving function which has until then been zero. Thus, for example, the quantity $A \cos \omega t$ applied at $t = 0$ is represented by the function

$$f(t) = Au(t) \cos \omega t.$$

While this notation is sometimes observed punctiliously, it is not universal and the factor $u(t)$ is often omitted; this, of course, makes no difference to the transform.

The unit step is more useful to indicate the sudden application of any driving function at a later instant $t = T$. If the function $f(t)$ is delayed, i.e. shifted in time, by T it can be written $f(t - T)$ so its delayed application can be represented by

$$u(t - T)f(t - T).$$

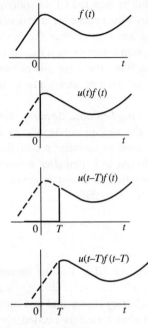

Fig. 7.6
A delayed function.

The transform of this can be obtained by the **time-shift** or **delay theorem** (Appendix E) which applies to any delayed function and gives

$$\mathscr{L}u(t-T)f(t-T) = e^{-sT}F(s). \tag{7.31}$$

There is an important difference between the delayed function $u(t-T)f(t-T)$ and the function which results when $f(t)$ retains its original origin but is not applied until $t=T$; the latter can be written $u(t-T)f(t)$ and in general it does not transform according to eqn 7.31. Figure 7.6 illustrates the two cases for an arbitrary function. If $f(t)$ is a constant there is of course no difference, since then $f(t)$ is the same as $f(t-T)$ and both expressions simply represent a delayed step.

Another important function is the **impulse**, which was encountered in Section 7.2.2. By definition an impulse is a function which has infinite amplitude over an infinitesimal interval and is otherwise zero. Its magnitude is defined as its integral, and can be assigned any value. An impulse of magnitude A can be regarded as the limiting form of a rectangular pulse having duration τ and height A/τ, as $\tau \to 0$. The **unit impulse** has magnitude unity, is also known as the **Dirac** or **delta function** and is written $\delta(t)$ when it occurs at $t=0$; if it occurs at the instant $t=T$ it is then $\delta(t-T)$ and may be formally defined by the statements

$$\delta(t-T) = 0 \qquad (t \neq T)$$
$$\int_0^\infty \delta(t-T)\mathrm{d}t = 1. \tag{7.32}$$

These two statements together require that the amplitude be infinite at $t=T$. Not the least important property of $\delta(t)$, the unit impulse at $t=0$, is the fact that *its Laplace transform is unity*. It follows from this that the response of any system to a unit impulse is given directly by simply inverting the appropriate system function; hence the **impulse response** is a way of specifying characteristic, i.e. transient, behaviour. Of course in practice a perfect impulse is not attainable, but a pulse of very short duration (on some appropriate scale of time) can be a good approximation. The height of a pulse of given duration is then a measure of its magnitude as an impulse.

In Section 7.2.2 the significance of the impulse for discontinuities in initial conditions was mentioned, and eqn 7.24 introduced the convention that the condition just *after* a discontinuity was to be taken. To find that condition, any applied impulse must be taken into account, by physical reasoning or otherwise. In fact it is equally valid to treat the impulse as a driving function applied to the system in the condition just *before* the change, at $t=0-$, and this is exactly how an impulse response is found simply by inverting a system function. The latter choice has the effect of including the impulse in the transformation integral, while the former excludes it and accounts for it in a different initial condition. The equivalence of the two can be seen from a simple example. Suppose we want to know the voltage response when a current impulse is applied to an uncharged capacitance C in parallel with a resistance R, that is to an impedance $Z(s)$ given by $1/C(s1/CR)$, as in Example 7.6 above. If the magnitude of the impulse (which has the dimensions of charge) is Q, then its transform is also Q and the voltage is given by

$$V(s) = QZ(s) = \frac{Q/C}{s + 1/CR}$$

since the initial voltage, just before the impulse, is zero.

Now consider the other view, taking the initial condition just after the impulse to be $v(0+) = Q/C$, on the grounds that the charge Q must be instantaneously deposited on C in its entirety because there can be no discharge through R in an infinitesimal time. We now ignore the impulse, but incorporate the initial condition, as in Fig. 7.5, as a current source having the transform $Cv(0+)$, which in this case is $C(Q/C)$, i.e. Q. The voltage produced by this current has the transform $QZ(s)$, exactly as before. Inverting this transform gives the voltage of a discharging capacitance which was found in Section 7.2.1 and again in Example 7.6. In these cases we took the capacitance to have an initial voltage; the current impulse we have just considered is merely one way of producing that condition.

The convention of choosing initial conditions at $t = 0+$, as in eqn 7.24, can be defended on the grounds that classical solutions in effect take this origin and that these are also the values given by the initial-value theorem (Section 7.4.4). In practice there is rarely any problem about choice, if only because in most cases initial conditions show no discontinuity. It should be borne in mind, however, that discontinuities commonly arise in quantities other than voltage on C or current in L: the current into a charging capacitance, for example.

The remaining basic function which is sometimes encountered is the **ramp**, a descriptive name given to a linear increase starting from zero and having no upper limit. By definition the gradient of a ramp is taken to be its magnitude; thus a ramp of magnitude A which starts at $t = 0$ can be written

$$Atu(t)$$

while a similar ramp starting at $t = T$ becomes

$$A(t - T)u(t - T).$$

A **unit ramp** starting at the origin, $t = 0$, is simply $tu(t)$ and has the formal definition

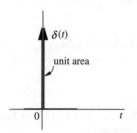

(a)

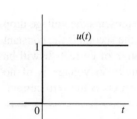

(b)

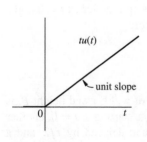

(c)

Fig. 7.7
Driving functions: (a) unit impulse or delta function; (b) unit step; (c) unit ramp.

$$\begin{aligned} tu(t) &= 0 \quad (t < 0) \\ &= t \quad (t \geq 0) \end{aligned} \tag{7.33}$$

It is readily shown from the transformation integral, eqn 7.24, that the transform of this unit ramp is $1/s^2$.

Figure 7.7 shows the unit impulse, step, and ramp functions $\delta(t)$, $u(t)$ and $tu(t)$, and it will be noticed that the last two represent successive integration of the first; it is therefore quite consistent to find that their transforms are respectively 1, $1/s$, and $1/s^2$. The sequence may be extended to any function having the form $t^n u(t)$, as shown in Appendix E.

Example 7.7   Find the voltage which appears across a series circuit of inductance L and resistance R in response to a ramp of current of magnitude 2 As^{-1}, there being no initial current.

The voltage transform is given by $I(s)Z(s)$, which here gives

$$V(s) = \frac{2}{s^2}(R + sL) = \frac{2R}{s^2} + \frac{2L}{s}.$$

From the description above, or from Appendix E.2, it is clear that the inverse of this is the sum of a ramp and a step, namely

$$v(t) = (2Rt + 2L)u(t),$$

in which the first term on the right-hand side represents the voltage drop in R due to a current increasing at 2 As^{-1}, and the second is the constant voltage across L due to that same rate of change of current. It will be noted that this is a case where the discontinuity in voltage is of no significance: the only initial condition which matters is the zero current.

Certain driving functions, especially those in the form of pulses, can be regarded as the superposition of two or more of these basic functions. For example, the rectangular pulse of width τ and height A shown in Fig. 7.8(a) is the superposition of a positive step function at $t = 0$ and a negative step at $t = \tau$; its transform is therefore

$$F_1(s) = A\left(\frac{1}{s} - \frac{e^{-s\tau}}{s}\right).$$

Similarly, the triangular pulse shown in (b), of width τ and height B, is the superposition of a ramp having slope $2B/\tau$ starting at $t = 0$, another having a negative slope of twice as much and delayed by $\tau/2$, and a third which again has slope $2B/\tau$ but starts at $t = \tau$. Summing the transforms of these three components gives the transform of the whole pulse to be

$$F_2(s) = \frac{2B}{\tau}\left(\frac{1}{s^2} - \frac{2e^{-s\tau/2}}{s^2} + \frac{e^{-s\tau}}{s^2}\right).$$

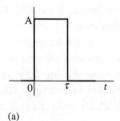

(a)

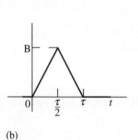

(b)

Fig. 7.8
Synthesis of (a) rectangular and (b) triangular pulses.

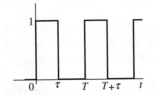

Fig. 7.9
Synthesis of a periodic pulse train.

It is also possible to transform periodic functions other than sinusoids by means of superposition. For example, the train of pulses shown in Fig. 7.9 with pulse width τ and period T can be written

$$f(t) = u(t) - u(t - \tau) + u(t - T) - u(t - T - \tau) + \ldots$$

and therefore has the transform

$$F(s) = \frac{1}{s}\{(1 - e^{-s\tau}) + e^{-sT}(1 - e^{-s\tau}) + \ldots\}$$

$$= \frac{1}{s}(1 - e^{-s\tau})(1 + e^{-sT} + e^{-2sT} + \ldots)$$

$$= \frac{1}{s}\left(\frac{1 - e^{-s\tau}}{1 - e^{-sT}}\right)$$

which can be generalized to

$$F(s) = \frac{F_1(s)}{1 - e^{-sT}} \tag{7.35}$$

for any periodic function of period T whose first period begins at $t = 0$ and has the transform $F_1(s)$.

7.4.4 Initial- and final-value theorems

The initial value of any quantity as a transient variation begins is usually clear from physical reasoning; likewise the final or steady-state value (given by the particular integral in the 'classical' method of solving the governing differential equation) can be quite quickly found from steady-state theory. However, if the Laplace transform of the quantity concerned is known then both values can also be found formally by means of the following two theorems, and these can serve as useful checks when a complete solution is not required.

Initial-value theorem

$$\lim_{t \to 0+} f(t) = \lim_{s \to 0} sF(s) \tag{7.36}$$

Final-value theorem

If $f(t)$ approaches a unique limit at large t, then

$$\lim_{t \to \infty} f(t) = \lim_{s \to 0} sF(s). \tag{7.37}$$

The qualification to the second theorem is necessary because even those functions which do not diverge may not approach a final limit: a sinusoid, for example.

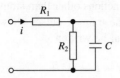

Fig. 7.10
The circuit for Example 7.8.

Example 7.8 Find the current transform $I(s)$ for the circuit shown in Fig. 7.10 when a d.c. voltage V is applied to it at $t = 0$, the capacitor being then uncharged, and confirm the initial and final values of i.

We may first decide the two limiting values of i by inspection. The voltage on C cannot change suddenly since infinite current is not possible here, hence the initial voltage across C and R_2 at $t = 0+$ remains at zero; the initial current i should therefore be V/R_1. Ultimately, since the source is d.c., the capacitor carries no current so the expected final value of i is $V/(R_1 + R_2)$.

To find the required transform we first need the generalized impedance of the circuit, which is

$$Z(s) = R_1 + \frac{R_2/sC}{R_2 + 1/sC} = \frac{R_1(1 + sCR_2) + R_2}{1 + sCR_2}.$$

The applied d.c. voltage is in effect a step function and its transform $V(s)$ is therefore V/s. Hence the current transform is

$$I(s) = \frac{V(s)}{Z(s)} = \frac{V(1 + sCR_2)}{s(R_1 + R_2 + sCR_1R_2)}.$$

The initial value of i according to the theorem is therefore

$$i(0+) = \lim_{s \to \infty} sI(s) = \lim_{s \to \infty} \frac{V(1 + sCR_2)}{R_1 + R_2 + sCR_1R_2} = \frac{V}{R_1}$$

as expected from our previous reasoning. The final value of i, according to the theorem, is

$$i_\infty = \lim_{s \to 0} sI(s) = \frac{V}{R_1 + R_2}$$

which again agrees with the expected value.

7.4.5 Inversion

It was pointed out in Chapter 5 that a steady-state system function is rational, so that it can be written as a ratio of polynomials in the variable $j\omega$; it follows that the generalized system functions which have been discussed in connection with transients must be ratios of polynomials in s. The response transform obtained when such a function is multiplied by one of the driving function transforms described above will also comprise such a ratio, save for a possible factor e^{-sT} due to any delayed stimulus; we can ignore the latter for the present, for it can only signify a corresponding delay in the final response.

We may therefore represent the basic form of any response we are likely to want by the transform

$$F(s) = K\frac{(s+z_1)(s+z_2)\ldots}{(s+p_1)(s+p_2)\ldots} \tag{7.38}$$

in which K is a constant and the z and p represent real or complex conjugate values which depend on the parameters of the circuit and on the driving function; some of them may be zero, in which case $F(s)$ will contain a factor s^r, where r is a positive or negative integer. In general, then, the actual total response $f(t)$ in a transient problem is given by the inverse transform of an expression like eqn 7.38. To find it, we often need to reduce the latter to the more elementary forms which are commonly tabulated (as in Appendix E).

The reduction is done by the method of **partial fractions**, whereby eqn 7.38 may in many cases be rewritten as

$$F(s) = K\left(\frac{A}{s+p_1} + \frac{B}{s+p_2} + \ldots\right) \tag{7.39}$$

in which the constants A, B,... are found from the requirement that the two versions of $F(s)$ are identically equal for any value of s. (Exceptions to this standard pattern are given in Appendix E.) The solution is thus reduced to the inversion of terms which, apart from constants, have the form $1/(s + p)$ which on inversion yields the exponential e^{-pt}. This should be no surprise, since all the transients encountered so far have been exponential in form (even oscillatory responses, such as eqn 7.19, arise from conjugate pairs of exponentials); this feature is in fact characteristic of all linear systems. If one of the terms turns out to have $p = 0$ then its inverse becomes a constant, which can be regarded as merely a special case of the exponential. The most significant thing to notice about eqn 7.39, however, is the disappearance of all z. They are of course contained in the constants A, B,... and therefore must affect the composition of the solution, but they do not affect the *form* of each component part. We shall return to this point.

Example 7.9 Find the current response $i(t)$ for the circuit shown in Fig. 7.10 when the d.c. voltage V is applied, the capacitor being uncharged.

The current is given, as in Example 7.8, by

$$\begin{aligned}
I(s) &= \frac{V(1 + sCR_2)}{s(R_1 + R_2 + sCR_1R_2)} \\
&= \frac{V}{R_1}\frac{s+z_1}{s(s+p_1)}
\end{aligned}$$

where $z_1 = 1/CR_2$ and $p_1 = (R_1 + R_2)/CR_1R_2$.

Now let

$$\frac{s + z_1}{s(s + p_1)} = \frac{A}{s} + \frac{B}{s + p_1}$$

$$= \frac{A(s + p_1) + Bs}{s(s + p_1)}$$

For this to be true at $s = -p_1$, we require

$$-p_1 + z_1 = -Bp_1$$

or

$$B = (p_1 - z_1)/p_1.$$

Similarly, for the identity to hold at $s = 0$ we require

$$z_1 = Ap_1$$

or

$$A = z_1/p_1.$$

Hence we find

$$I(s) = \frac{V}{R_1}\left\{\frac{z_1}{p_1 s} + \left(1 - \frac{z_1}{p_1}\right)\frac{1}{s + p_1}\right\}$$

which inverts immediately to

$$i(t) = \frac{V}{R_1}\left\{\frac{z_1}{p_1} + \left(1 - \frac{z_1}{p_1}\right)e^{-p_1 t}\right\}$$

$$= \frac{V}{R_1}\left(\frac{R_1}{R_1 + R_2} + \frac{R_2}{R_1 + R_2}e^{-t/T}\right)$$

in which

$$T = \frac{1}{p_1} = \frac{CR_1R_2}{R_1 + R_2}.$$

It can easily be confirmed that this result gives initial and final values for $i(t)$ which agree with those already found in Example 7.8.

Example 7.10   Use the Laplace transform to confirm eqn 7.18 for the capacitor voltage in the circuit of Fig. 7.4(a) following the application of a d.c. voltage V, the initial conditions being zero.

The required transfer function has already been found as eqn 7.27, which for $V(s) = V/s$ gives

$$V_C(s) = \frac{V}{s(LCs^2 + RCs + 1)}$$

The case we are now asked to consider is that having $\zeta < 1$, that is case (iii) of Section 7.2.3, for which the quadratic expression has complex conjugate factors. We therefore write

$$V_C(s) = \frac{V}{LC} \frac{1}{s(s + \alpha + j\omega_n)(s + \alpha - j\omega_n)}$$

where, in our previous notation,

$$\alpha = \zeta\omega_0 = R/2L$$

and

$$\omega_n = \omega_0\sqrt{1 - \zeta^2} = \sqrt{\frac{1}{LC} - \frac{R^2}{4L^2}}.$$

To cast $V_C(s)$ into partial fractions, we could write

$$\frac{1}{s(s + \alpha + j\omega_n)(s + \alpha - j\omega_n)} = \frac{A}{s} + \frac{B}{s + \alpha + j\omega_n} + \frac{C}{s + \alpha - j\omega_n}$$

$$= \frac{A\{(s + \alpha)^2 + \omega_n^2\} + Bs(s + \alpha - j\omega_n) + Cs(s + \alpha + j\omega_n)}{s\{(s + \alpha)^2 + \omega_n^2\}}.$$

Then, for $(s + \alpha) = j\omega_n$, we require

$$1 = C(j\omega_n - \alpha)(2j\omega_n)$$

which gives

$$C = 1/2j\omega_n(j\omega_n - \alpha)$$

and similarly for $(s + \alpha) = -j\omega_n$ we require

$$1 = B(-j\omega_n - \alpha)(-2j\omega_n)$$

from which

$$B = 1/2j\omega_n(j\omega_n + \alpha).$$

Finally, for $s = 0$ we require

$$1 = A(\alpha^2 + \omega_n^2)$$

so

$$A = 1/(\alpha^2 + \omega_n^2) = 1/\omega_0^2.$$

From these we may now write

$$V_C(s) = \frac{V}{LC}\left\{\frac{1}{\omega_0^2 s} + \frac{1}{2j\omega_n(j\omega_n + \alpha)(s + \alpha + j\omega_n)} + \frac{1}{2j\omega_n(j\omega_n - \alpha)(s + \alpha - j\omega_n)}\right\}$$

and this inverts to

$$v_C(t) = V + \frac{V\omega_0^2}{2j\omega_n}\left\{\frac{(-j\omega_n + \alpha)e^{-(\alpha + j\omega_n)t}}{(j\omega_n + \alpha)} - \frac{(j\omega_n + \alpha)e^{-(\alpha - j\omega_n)t}}{-j\omega_n + \alpha}\right\}$$

which when rearranged becomes

$$v_C(t) = V\left\{1 - e^{-\alpha t}\left(\cos \omega_n t + \frac{\alpha}{\omega_n}\sin \omega_n t\right)\right\}$$

in agreement with eqn 7.21.

From the second of these examples it is clear enough that the process of inversion is hardly less laborious than the 'classical' method of solution we used earlier for this second-order problem. Solution by transform methods can be preferable in more demanding cases, but for the most part the advantages of the Laplace transform lie in the formal notation and the ease of manipulation which it affords in the s domain. Because the variable s has the attributes of a complex frequency (Section 7.3) we can justifiably regard the s domain as a general frequency domain. On this view transformation and inversion mean simply passage from time functions to frequency functions and vice versa; the trouble involved in these two processes is often justified by the great convenience of the frequency domain. In the steady state we found that it was usually much easier to handle functions of $j\omega$, such as a transfer function or an impedance, than sinusoidal functions of time; now we have made that convenience more general by extending the frequency variable from $j\omega$ to s.

7.4.6 Convolution

The transform of the product of two functions of time is not given by the product of their transforms, nor is the inverse of a product given by the product of the inverses. However, the product of two functions in either domain can be transformed into the other by a process known as **convolution**. If $F(s)$ and $H(s)$ are the Laplace transforms of the functions $f(t)$ and $h(t)$, then it can be shown that

$$
\begin{aligned}
\mathcal{L}^{-1}F(s)H(s) &= \int_{-\infty}^{\infty} f(t - \tau)h(\tau)\mathrm{d}\tau \\
&= \int_{-\infty}^{\infty} f(\tau)h(t - \tau)\mathrm{d}\tau.
\end{aligned}
\tag{7.40}
$$

The two integrals are alternative forms of the **convolution integral**,[†] which is also said to be the convolution of $f(t)$ and $h(t)$ and is often written in the notation $f(t)^*h(t)$ or even simply f^*h. The dual of eqn 7.40 is also true, that is the Laplace transform of the product $f(t)h(t)$ is the convolution of $F(s)$ and $H(s)$, defined by integrals corresponding to those above and written $F(s)^*H(s)$ or simply F^*H.

The significance of this convolution theorem in circuit analysis needs a little consideration, for there is no reason to suppose that the function $F(s)H(s)$ could not be inverted when necessary by the normal method described in the previous section. Indeed we have already done this in the examples, for in each case it was the product of a system function with the transform of a driving function which we inverted to obtain the final solution. What we may now say, according to eqn 7.40, is that if $H(s)$ is a system function and $F(s)$ the transform of a driving function then the response of the former to the latter is given by the convolution of $f(t)$ and $h(t)$; of these, $f(t)$ is the actual driving function and $h(t)$ is the **impulse response** of the system (which, as we saw in Section 7.4.3, is merely the inverse of the system function). Hence if the impulse response of a system is known then its response to any driving function can be obtained as the convolution of the two. In other words we have a means of obtaining an actual response *without* working in the s domain (and, for that matter, without formally solving a differential equation) if for any reason that should be preferred. In practice this approach is useful in calculating responses to inputs which are not analytical functions but for which the convolution integral is readily evaluated numerically. In effect, convolution treats the input $f(t)$ as the superposition of successive impulses of magnitude $f\,d\tau$.

The integration in eqn 7.40 can be illustrated diagrammatically, as in the following very simple example.

Example 7.11 Find by convolution the response of a system having the transfer function $1/(s+a)$ to the rectangular pulse shown in Fig. 7.11(a).

In this case $f(\tau)$ has the form of the pulse shown in the figure, with the variable changed to the dummy variable τ. The impulse response $h(t)$ for the given system is simply the inverse of $1/(s+a)$, which is e^{-at}. Choosing the second version of eqn 7.40, we now need $h(t-\tau)$, which can be written $e^{a(\tau-t)}$; as a function of τ this has the form of $h(t)$ reflected about the vertical axis (with the variable changed to τ) and shifted by an amount t, as shown in (b).

To integrate the product of $f(\tau)$ and $h(t-\tau)$ to get the required response according to eqn 7.40, we can consider the three ranges of t shown in (c). For $t < 0$ the integral is clearly zero, so (as we expect) there

[†] The limits of integration in eqn 7.40 are often shown as 0, t since $f(t)$ and $h(t)$ are here zero for $t < 0$.

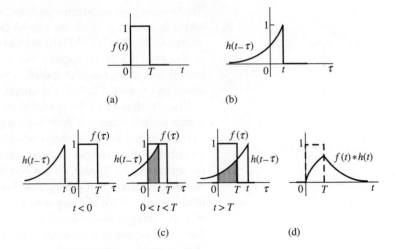

Fig. 7.11

Convolution in Example 7.11.

is no response before $t = 0$. For $0 < t < T$, the integral becomes

$$\int_0^t e^{a(\tau - t)}d\tau = \left| \frac{1}{a}e^{a\tau}e^{-at} \right|_0^t = \frac{1}{a}(1 - e^{-at})$$

which is the required response for the interval from t to T; it is recognizably the response of a first-order system to a step function. The third range of t to consider is $t > T$; from the figure this gives the integral

$$\int_0^T e^{a(\tau - t)}d\tau = \left| \frac{1}{a}e^{a\tau}e^{-at} \right|_0^T = \frac{1}{a}e^{-at}(e^{aT} - 1)$$

which is an exponential decay and completes the total response shown in (d). In this particular case the same result could easily be obtained by inverting the product of the given $H(s)$ and the transform $F(s)$ of the given pulse, regarded as the superposition of a unit positive step and a unit negative step delayed by T.

7.5 Poles and zeros

7.5.1 The general system function

It has already been observed, in Section 7.4.5, that the general system function is the ratio of two polynomials in s. It may be written

$$H(s) = Ks^r \frac{(s + z_1)(s + z_2)\ldots}{(s + p_1)(s + p_2)\ldots} \tag{7.41}$$

in which r as before is a positive or negative integer. Clearly if s takes any of the values $-z_1, -z_2,\ldots$ then the numerator and hence $H(s)$ itself will be zero; these values of s are therefore called **zeros** of $H(s)$.

Similarly if s takes any of the values $-p_1$, $-p_2$, ... then the denominator will be zero and $H(s)$ infinite; these values of s are called **poles** of $H(s)$. In addition, the value zero for s is also a zero of $H(s)$ if r is positive and a pole of $H(s)$ if r is negative.

Now in Section 7.4.5 it became clear that it was the values of p, rather than those of z, which characterized the exponential forms in the response to any stimulus, because the partial fractions produced terms like $A/(s+p)$ which inverts to Ae^{-pt}. Although $H(s)$ is not the transform of the complete response (unless the driving function happens to be a unit impulse), we know that it alone governs the form of the *transient* part of the response whatever the stimulus; it follows that the transient response has a form which depends on the values of p, that is *on the poles of the system function*. Thus, for example, a factor $(s+p)$ in which p is positive and real gives a negative real pole at $s = -p$ and an exponential decay in the response with time constant $1/p$; and a complex conjugate pair of poles with negative real parts gives a decaying oscillation. This ability to predict the form of the response simply by examining the poles of the appropriate system function is one of the most useful and powerful features of circuit analysis in the s domain.

If we write a system function as

$$H(s) = P(s)/Q(s)$$

in which the $P(s)$ and $Q(s)$ are both polynomials, then the poles of $H(s)$ are those values of s which make $Q(s)$ zero; hence they are the roots of the equation

$$Q(s) = 0$$

which is called the **characteristic equation**. Strictly speaking, it is characteristic only of the particular system function considered, since several such functions can be defined for a given system. (The same is true of poles and zeros, of course: they relate to a particular function and are not a unique set for the system. For example, the poles of an impedance become the zeros of the corresponding admittance and vice versa.) It may be noted by comparing eqns 7.27 and 7.12 that the characteristic equation in s is exactly the same as the auxiliary equation in m for the same response. The information conveyed is the same in each, but the analysis of the circuit in the notation of the Laplace transform makes it more easily accessible and can be readily used for a complete solution whenever that is necessary.

7.5.2 Pole–zero diagrams

A very convenient way of presenting and interpreting the values of the poles, and of the zeros if need be, is to plot their positions as an Argand diagram on the s plane, which may be regarded as a complex frequency plane. Figure 7.12 shows an example in which there are three poles and one zero, marked in the conventional way with a cross to indicate a pole

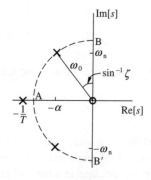

Fig. 7.12

A pole–zero diagram on the s plane.

and a circle to indicate a zero. Two of the poles are complex conjugate and have the values

$$s = -\alpha \pm j\omega_n$$

in the notation used for the second-order circuit previously investigated; the third pole is (negative) real and is indicated by

$$s = -1/T$$

because we have previously used a time constant T to describe a first-order circuit and the same notation is convenient for any pole which is not one of a conjugate pair. The zero is at the origin and therefore indicates the factor s in the numerator of the function. (Note that a factor s^r would have meant r zeros or poles at the origin; if $|r| > 1$ in such a case, the convention is to indicate on the diagram the multiplicity r of zeros, or poles, at the origin. The same would be done for multiplicity at any other point.) We deduce that this diagram represents a transfer function which can be written as

$$H(s) = K \frac{s}{(s + 1/T)(s + \alpha + j\omega_n)(s + \alpha - j\omega_n)}$$

where K is an unknown constant. Simply by interpreting the diagram we can say that the transient response will contain a damped oscillation at the frequency ω_n decaying in amplitude as $e^{-\alpha t}$, as for a second-order circuit, together with a separate but superimposed exponential decay of the form $e^{-t/T}$; the characteristic equation is of the third order. The choice of a damping factor α to describe the oscillatory decay and of a time constant T to describe the other is merely a matter of convention. The distances α, ω_n and $1/T$ are easily identified on the diagram, but we may also recall the relationship of the first two to the quantities ζ and ω_0 which have been previously used in a standard form for quadratic expressions. In Section 7.2.3 we used the definitions

$$\alpha = \zeta\omega_0$$

and

$$\omega_n = \omega_0\sqrt{1 - \zeta^2}$$

for $\zeta < 1$, that is when two poles are complex conjugate. From these we have

$$\omega_0^2 = \omega_n^2 + \alpha^2$$

so ω_0 is clearly the distance from each of the conjugate poles to the origin, as shown in Fig. 7.12; ζ is then the sine of the angle marked. The position of two conjugate poles can therefore be specified in terms either of α and ω_n or of ζ and ω_0; the advantage of the former is that they directly describe the oscillation, while the latter are algebraically slightly more convenient. All four, of course, are determined by the parameters of the circuit or other system. We should recollect here that the damping of an oscillation, that is the value of α or ζ, depends on the dissipation in the system (i.e. on resistance, in the case of an electrical circuit), while ω_0 does not. Hence, in a second-order system of given ω_0, the poles will clearly describe a circular path of radius ω_0 as the damping is varied. At $\zeta = 1$, the damping is critical and both poles lies on the negative real axis at the point A in the figure, distant ω_0 from the origin. As ζ is reduced to zero, the poles move in their respective arcs until they lie at $\pm \omega_0$ on the imaginary axis (points B and B'); this represents a loss-free undamped oscillation which does not decay, with factors $(s \pm j\omega_0)$ in the characteristic equation giving the quadratic $s^2 + \omega_0^2$. If the circuit parameters are such that $\zeta > 1$, there is, as we saw in Section 7.3.2, no oscillation but instead two separate time constants; the poles then move in opposite directions along the negative real axis as ζ increases, in such a way that their product remains as ω_0^2.

There is another feature of the pole–zero diagram to be remarked. Because the steady-state behaviour of a system can always be found by substituting $j\omega$ for s, it follows that the imaginary axis on the complex s plane can represent actual frequencies. Suppose we wish to evaluate a transfer function, in magnitude and angle, at a certain frequency ω in the steady state. To do so, we may substitute for s in eqn 7.41 the particular value of $j\omega$, to get

$$H(j\omega) = K(j\omega)^r \frac{(j\omega + z_1)(j\omega + z_2)...}{(j\omega + p_1)(j\omega + p_2)...}.$$

In this, except for the constant K the magnitude and angle of each factor is easily identified on the pole–zero diagram: for the magnitude of $(j\omega + z_1)$, for example, is the distance from the point $j\omega$ on the positive imaginary axis to the zero at $s = -z_1$; while the argument of the same factor is the angle which the line joining these two points makes with the positive real direction. This is shown in Fig. 7.13. The same applies to the other factors, including $(j\omega)^r$ which evidently has magnitude ω^r and angle $r\pi/2$. With the magnitude and angle of each factor known, those of the whole function follow if K is also known. Hence we find that the complex value of $H(j\omega)$ depends on the geometrical disposition of its poles and zeros in relation to the given point $j\omega$. Although this is not generally the most efficient method of evaluating the steady-state function, it can be a helpful way of assessing its behaviour along with the methods discussed in Chapter 5.

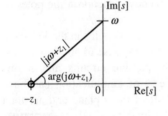

Fig. 7.13
Steady-state response from the pole–zero diagram.

7.5.3 Stability

In the previous section it was assumed that the non-zero poles and zeros were negative if real and had negative real parts if complex, and all the functions met in earlier sections have satisfied that assumption. Positive real values have not arisen, and one reason for this is that systems made up only of linear passive elements cannot have poles which are real and positive, or have positive real parts (zeros of this nature are, however, possible). That this should be so is not too surprising; for, on inverting a transform, a positive real pole would give rise to a positive exponent, that is exponential *growth*, and similarly a conjugate complex pair with positive real parts would give a growing oscillation. But in passive circuits which include resistance and hence dissipation, which in practice means all such circuits, a transient is bound to decay; in the ideal case of a loss-free passive circuit the poles lie on the imaginary axis and the response neither decays nor grows.

Poles to the right of the imaginary axis, which cause growing transients, also imply negative losses and, what is equivalent, negative resistance. While this is impossible in a passive circuit, it can actually happen under certain conditions in amplifying circuits with feedback: it is discussed in the following chapter. Here we may simply observe that while a growing transient cannot in theory lead to a steady state, in practice it must clearly result in an amplitude of response limited by factors not accounted for in our linear theory. A growing transient reaches a kind of steady state quite different from that which the driving function or input source produces; the system is said to be **unstable**, and often such instability is a highly undesirable state which must be avoided. But there is one exception: a transient oscillation which grows until some kind of steady-state limit is reached, in the absence of any independent a.c. source whatever, is the basis of an **oscillator**, a useful form of a.c. generator described in the next chapter (Section 8.4). The output power of such a device comes ultimately, as in other amplifying circuits, from the d.c. sources which are essential to its operation; but the occurrence of the oscillation, and its frequency, depend on the poles of the appropriate transfer function.

In summary then, any function having all of its poles in the left-hand half of the *s* plane is stable and its transient response to any input must decay: the transfer functions of purely passive circuits are in this category. Active circuits with the property of amplification can have functions with poles in the right-hand half of the *s* plane and are then unstable. Any function with poles on the imaginary axis is **marginally unstable** and produces continuous oscillation at a frequency determined by the transfer function. No actual passive circuit can quite reach this state; its attainment in an active circuit is discussed in Section 8.4.

Problems

7.1 An inductance of 1 mH, carrying no current, is connected suddenly to a 20 V d.c. source through a total resistance of 1 kΩ. Find the time taken, from the instant of connection, for the current to reach 99% of its final value. If the source is then short-circuited, how much longer would it take for the voltage across the resistance to fall to 1 V?

7.2 A sinusoidal voltage of peak value 30 V and frequency 50 kHz is connected to the same circuit as in Problem 7.1 at the instant when its instantaneous value is

24 V positive and increasing. Find an expression for the transient current which follows.

7.3 A capacitance of 100 nF is connected in series with the inductance and resistance of Problem 7.1. Find the time at which the current reaches a peak, and the value of this peak, following the sudden connection of the 20 V d.c. source.

7.4 In the circuit of Problem 7.3, the resistance is reduced to the critical value at which $\zeta = 1$. Find the new values of resistance, time to peak current, and peak current when the 20 V source is connected as before.

7.5 In the circuit of Problem 7.4, the resistance is further reduced to 50 Ω. Find the damping constant α, the natural frequency of oscillation, and the ratio of two successive positive peaks in the oscillation, of the transient response to an applied voltage.

7.6 Confirm from eqn 7.22 that the Laplace transform of e^{-at} is $(s+a)^{-1}$.

7.7 By writing the generalized impedance of the series circuit of Problem 7.3 above, find its voltage response to a unit step of current at $t = 0$.

7.8 Find the current response of the series combination of R and L to an applied voltage impulse of magnitude 2 Vs.

7.9 Find the Laplace transform of a triangular voltage pulse which rises from zero to 10 V in 1 μs, and then falls instantaneously back to zero. [Hint: superimpose two ramps and a step.]

7.10 Find the Laplace transform of a saw-tooth waveform which repeats the pulse of Problem 7.8 at a frequency of 1 MHz.

7.11 Two coils have mutual inductance 10 mH. One has self-inductance 10 mH and is connected to a resistance of 100 Ω. The other has self-inductance 20 mH and is connected to an input voltage. Find the generalized input admittance.

7.12 A potential divider comprises the parallel combination of R_1 and C_1 in series with the parallel combination of R_2 and C_2. Show that is has a zero transient response when $R_1 C_1 = R_2 C_2$.

7.13 Find the inverse of the transform $1/(s+2)(s+3)$.

7.14 Find the inverse of the transform $1/s(s+2)(s+3)$.

7.15 Find the inverse of $1/(s^2 + 4s + 6)$.

7.16 Find the inverse of $(s+1)/(s+3)$. [Hint: write the function in the form $1 + F(s)$.]

7.17 Find the characteristic features in the transient response of the system function $(s+2)/(s^2 + 6s + 100)$.

7.18 From the convolution integral confirm the current response in Problem 7.1. [Hint: use the impulse response found in Problem 7.8.]

7.19 A coil with inductance 1 mH and resistance 20 Ω is connected in parallel with a capacitance of 1 μF. Find the poles and zeros of the total admittance.

7.20 A transfer function has three poles which lie in the left-hand s-plane on unit radii at angles 0, $\pm 60°$ to the negative real axis. Verify that the function represents a Butterworth low-pass filter of order 3 and unit cut-off frequency, according to eqn 5.35, and show that its transient response includes an oscillation with damping factor 1/2.

Feedback circuits

8.1 General

The characteristic behaviour of an amplifier, or more formally any four-terminal active network containing one or more controlled sources, can be markedly altered by connecting its output terminals back to its input terminals in some fashion, by way of a second four-terminal network or otherwise. The output of the amplifying network may, as usual, be defined as either a voltage or a current, related to its input voltage or current by a transfer function $H(s)$ say, and this output quantity, while it remains as the output of the whole circuit, also provides the input of the feedback network having transfer function $F(s)$ say. The output from the latter is the feedback quantity which is then added to or subtracted from the original input, and it is their resultant which forms the final input to $H(s)$. Thus the output of $H(s)$ and the input of $F(s)$ are always *common*; but the input to $H(s)$ is always the *sum or difference* of the original input and the feedback quantity. There are several ways in which the connections may be made, according as the input and output of $H(s)$ are defined as voltage or current: a series connection is required for a common current, and for adding or subtracting voltages (depending on how their directions are defined), while a parallel connection has a common voltage but adds or subtracts currents. Figure 8.1 shows all four possibilities, with the connections so made that for the directions indicated the feedback quantity is subtracted from the input, giving what is known as **negative feedback**; simply reversing the output connections of $F(s)$ would add its output to the external input, to give **positive feedback**. Whether the two quantities actually reinforce or oppose each other depends not only on the connections but also on any phase changes which may be caused by $H(s)$ and $F(s)$.

If $H(s)$ and $F(s)$ both represented entirely passive linear networks there would be little point in showing them separate as in Fig. 8.1, for the whole circuit would simply form a new four-terminal network, different in its parameters from its constituents but having the same fundamental properties; like them it would be entirely passive and bilateral, unable to amplify in the proper sense of the word (Section 4.4)

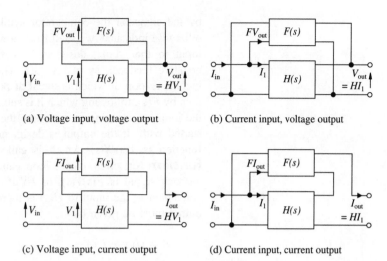

Fig. 8.1

The standard types of feedback in circuits.

(a) Voltage input, voltage output   (b) Current input, voltage output

(c) Voltage input, current output   (d) Current input, current output

and always stable (Section 7.5.3). But if $H(s)$ represents an active amplifying circuit then there are more interesting possibilities. (There is no need for $F(s)$ to be active also, although it may be.)

It has been pointed out in earlier chapters that transfer functions of four-terminal networks depend on the impedance connected to their output terminals. In what follows we shall assume that $H(s)$ and $F(s)$ take account of terminal conditions: that is $H(s)$ is defined with the input impedance of $F(s)$, as well as any external load, connected to its output; while $F(s)$ likewise takes into account whatever impedance is presented to its output terminals. Although it is possible to analyse a complete feedback circuit from first principles without identifying $H(s)$ and $F(s)$ separately, it is more informative at this point to use these functions to investigate the general behaviour of such circuits.

8.2 The closed-loop transfer function

A feedback circuit represented by an arrangement like those of Fig. 8.1 can be regarded as a loop; this is more easily appreciated by drawing it

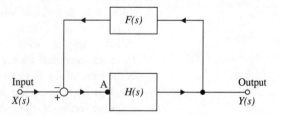

Fig. 8.2

A general diagram for feedback.

not as a conventional electrical circuit but as a single-line diagram, as in Fig. 8.2. Here arrows are used to indicate not current or voltages but the directions in which transfer functions are defined, from input to output; the junction at the output indicates the commonality between the output of $H(s)$ and the input of $F(s)$, while the sign of the feedback is indicated by the signs on the comparator symbol at the input to $H(s)$. Unless otherwise indicated, we shall from now use 'input' to mean the external input to the whole, designated by the transform $X(s)$, and assume negative feedback as in Fig. 8.2. The whole system can be seen to include a loop in which a signal at point A is multiplied by $H(s)$ and then by $F(s)$, following which it is subtracted from the input to complete the loop by returning to A, where the resultant must be the signal we started with. If the output is designated $Y(s)$ then the overall transfer function is $Y(s)/X(s)$, which is called the **closed-loop transfer function** (or, often, closed-loop gain). It is easily deduced, thus: the fedback signal is $F(s)Y(s)$, or FY if we use the simpler notation for convenience; the input to $H(s)$ is therefore $X - FY$. It follows that the output must be given by

$$Y = H(X - FY)$$

from which rearrangement gives the closed-loop transfer function as

$$G(s) = \frac{Y}{X} = \frac{H}{1 + HF}. \tag{8.1}$$

8.3 Stability

8.3.1 The Barkhausen criterion

Equation 8.1 for the closed-loop transfer function allows us to investigate the general behaviour of a circuit with feedback. That there are now more possibilities than we have met before becomes obvious if we consider first the steady state, in which s becomes $j\omega$, and suppose that the functions H and F are so chosen that

$$H(j\omega)F(j\omega) = -1 \tag{8.2}$$

at some value of ω[†]. Equation 8.1 indicates that the circuit would then have infinite gain; this immediately suggests instability, for an output of indeterminate amplitude would be obtained for zero input. It is not difficult to appreciate why this possibility should arise. The function $H(j\omega)F(j\omega)$ is the steady-state **open-loop transfer function**, often called simply the loop transfer function or loop gain because it arises from one complete tour around the loop, from any starting point, and it is the ratio of the fedback signal to that which starts off as the resultant input to H, at the point A in Fig. 8.2. Because eqn 8.1 is derived for negative feedback, the value -1 for this ratio means that the feedback is actually positive, having the same direction and the same magnitude as the signal which started at A: in other words, the circuit can sustain a signal of any postulated magnitude without the need of an external input. The system can be described as having positive feedback with an

[†] In this chapter and the next we again drop the bold notation for functions of $j\omega$.

open-loop gain of unity. If the idea that a circuit can provide its own input seems to be physically unlikely, it is helpful to remember that the required condition of eqn 8.2 could not be attained with only passive circuits: $H(s)$ at least must represent an active, amplifying circuit; and that means its output power is provided from sources other than its input signal (Section 7.5.3). Equation 8.2 is sometimes known as the **Barkhausen criterion**, and can be considered to represent an amplitude condition, namely a gain of unity around the loop, and a phase condition, namely a phase shift of $180°$. These modest-seeming requirements cannot in fact be achieved without amplification at some point.

The above argument shows that a given combination of H and F may produce instability in the form of a (theoretically) unlimited output which is at such a frequency as to give the steady-state open-loop gain $H(j\omega)F(j\omega)$ the value -1 and is quite independent of any input or driving function. To investigate this in more detail we return to the consideration of poles and zeros as in Section 7.5. There it emerged that instability in a given transfer function would occur unless its poles were confined to the left-hand half of the s plane. Now the poles of the closed-loop transfer function, eqn 8.1, are obviously the zeros of its denominator, that is the roots of the characteristic equation

$$H(s)F(s) + 1 = 0 \qquad (8.3)$$

or in other words the values of s which satisfy the condition

$$H(s)F(s) = -1.$$

If these values of s, which are the poles of the closed-loop transfer function $G(s)$, lie in the left-hand half plane, the system is stable; otherwise it is unstable. This is the generalized criterion corresponding to eqn 8.2. If the latter can be satisfied for some value of ω, it means that $G(s)$ has poles on the imaginary axis and the system therefore has steady-state instability (sometimes called **marginal** instability); but even if there are no poles on the axis, so that eqn 8.2 cannot be satisfied, poles of $G(s)$ in the right-hand half plane will still produce instability for the reasons discussed in Section 7.5. Hence the general condition for instability in a feedback system is that eqn 8.3 has roots either on the imaginary axis *or* in the right-hand half of the s plane. In practice, of course, it is impossible to know the position of poles so precisely in the first instance as to assert that they are exactly on the axis.

If we recollect that $H(s)$ and $F(s)$ can each be expected, like any transfer function for a linear system, to be the ratio of two polynomials, then we may write

$$H(s) = P(s)/Q(s)$$

and

$$F(s) = R(s)/S(s)$$

so that eqn 8.3 becomes

$$\frac{P(s)R(s)}{Q(s)S(s)} + 1 = 0$$

or

$$P(s)R(s) + Q(s)S(s) = 0 \tag{8.4}$$

which is another version of the characteristic equation. It may be noted that the poles of $H(s)$ and $F(s)$ taken separately would set only the second term to zero.

8.3.2 The Routh criterion

The characteristic eqn 8.3 or 8.4 reduces to the general form

$$a_n s^n + a_{n-1} s^{n-1} + a_{n-2} s^{n-2} + \ldots + a_1 s + a_0 = 0 \tag{8.5}$$

for any linear system, and the poles are given by the roots of this equation. There is an algebraic criterion for deciding whether all of its roots lie in the left-hand half plane (i.e. whether they are real and negative, or have negative real parts) which allows stability to be checked without having to solve the equation. It is due to Routh in the first instance, but is more often known as the **Routh-Hurwitz** criterion. In the case of a quadratic equation, $n = 2$, it simplifies to the requirement that for stability all three coefficients should be non-zero and of the same sign (easily confirmed from the general quadratic solution). It is interesting to refer here to the characteristic or auxiliary equation of a typical second-order passive circuit, such as eqn 7.27 or eqn 7.12; comparison shows that the stability criterion in this case amounts to a requirement that there should be resistance in the circuit. Zero resistance would give marginal instability, and it may be concluded by extension that the ability of positive feedback to produce instability implies that it can produce the effect of negative resistance. This has, of course, nothing to do with any physical resistance, but can be taken to mean that overall the circuit can *generate* power rather than dissipate it. It is this condition which is actually realized in oscillator circuits (Section 8.4).

In its general form the Routh criterion sets conditions on an array of elements constructed from the coefficients $a_0, \ldots, a_n$. For many purposes it is sufficient to quote, in addition to the quadratic case above, the results for the cubic and quartic cases. They are as follows:

(i) $n = 3$: all coefficients must be non-zero and carry the same sign, which must also be the sign of $(a_2 a_1 - a_3 a_0)/a_2$.

(ii) $n = 4$: all coefficients must be non-zero and carry the same sign, which must also be the sign of $(a_3 a_2 - a_4 a_1)/a_3$ and of

$$a_1 - \frac{a_3^2 a_0}{a_3 a_2 - a_4 a_1}.$$

While the Routh criterion gives a quick means of deciding whether or not a system is stable, it gives no further information about the position of poles or the nature of the response.

Example 8.1 Show that a feedback system with an open-loop transfer function of the form

$$H(s)F(s) = K/(s + a)(s + b),$$

in which K, a, and b are real constants, and a and b are positive, is always stable for positive K.

The characteristic equation for the closed-loop system is

$$1 + K/(s + a)(s + b) = 0$$

or

$$(s + a)(s + b) + K = 0.$$

This gives a quadratic equation in which, if K is positive, all coefficients have the same sign and therefore, by the Routh criterion, the system is stable. Instability is evidently possible for large enough negative K, which is hardly surprising since then the negative feedback connection stipulated for eqn 8.1 would tend to cause positive feedback. (This statement is necessarily imprecise, for the steady-state open-loop transfer function is not in general a real quantity.)

Example 8.2 Show that a feedback system with an open-loop transfer function of the form

$$H(s)F(s) = K/(s + a)(s + b)(s + c),$$

with a, b, c positive, may be unstable for both positive and negative K.

The overall characteristic equation is

$$1 + K/(s + a)(s + b)(s + c) = 0$$

which becomes

$$s^3 + s^2(a + b + c) + s(ab + bc + ca) + abc + K = 0.$$

The coefficient $abc + K$ is clearly negative, and the system therefore unstable, if K is negative and sufficiently large. If K is positive, all coefficients have the same sign, but by the Routh criterion the system will be stable only if

$$(a + b + c)(ab + bc + ca) > abc + K$$

that is, if

$$(b + c)(a^2 + ab + bc + ca) > K$$

which is obviously not true for positive K exceeding a critical value.

8.3.3 The Nyquist criterion

A second criterion for stability can be deduced from the mathematical behaviour of a function of a complex variable. If $w(s)$ is a function of the kind met in the present context, namely a rational function of polynomials, any closed path or **contour** on the complex s plane maps into another closed path on the complex plane representing $w(s)$, provided that the contour in s does not pass through any pole of $w(s)$. The following statement is then true: if the contour in s encloses P poles and Z zeros of $w(s)$ and is traced in a clockwise sense, then its map in the plane of $w(s)$ encircles the origin $Z - P$ times in the same sense.

Now consider the function $w(s)$ to be $1 + H(s)F(s)$, the denominator of the closed-loop transfer function $G(s)$ as in eqn 8.1. The poles of $G(s)$, which are our chief concern, are clearly the zeros of the function $1 + H(s)F(s)$. Let there be N of them in the right-hand half plane. As for the poles of $1 + H(s)F(s)$, we may assume that none is in the right-hand half, because if that were not so one or both of the two component networks must be unstable to start with. (The case of poles on the $j\omega$ axis is considered below.) It follows, then, that a contour of s enclosing the whole right-hand half plane and traced clockwise would map into a curve of $1 + H(s)F(s)$ encircling the origin N times clockwise. But such a contour of s consists of the entire $j\omega$ axis from $-\infty$ to $+\infty$, and an infinite semi-circle to close the contour as shown in Fig. 8.3; hence its map is simply the steady-state polar diagram of the function $1 + H(j\omega)F(j\omega)$, but including negative ω as well as positive. And if this diagram encircles the origin N times, a simple shift of origin means that a diagram of $H(j\omega)F(j\omega)$ would encircle the point $(-1, j0)$ N times.

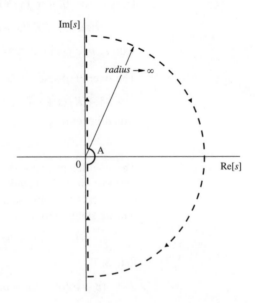

Fig. 8.3
The s contour for all $j\omega$.

Thus the statement we started from has led to the following.

If the function

$$G(s) = \frac{H(s)}{1 + H(s)F(s)}$$

has N poles in the right-hand half of the s plane, then the polar diagram of $H(j\omega)F(j\omega)$ encircles the point $(-1, j0)$ N times clockwise as ω increases.

For stability, therefore, the curve of the steady-state polar diagram must not enclose this point, a condition known as the **Nyquist criterion**. It may be noted that a curve which just passes through the critical point satisfies the Barkhausen criterion, eqn 8.2, at the corresponding frequency. The usefulness of the Nyquist criterion lies in the fact that $H(j\omega)F(j\omega)$, the steady-state open-loop transfer function, can if necessary be found in practice by direct measurements.

There are one or two formal points to be mentioned. One is that the contour of s must not pass through any poles of $w(s)$. In our case this can happen if $H(s)F(s)$ has one or more poles at the origin, due to a factor s^{-r} (r positive) in either $H(s)$ or $F(s)$, and in such event the contour of s is imagined to exclude the origin by an infinitesimal detour to the right, as shown in Fig 8.3. This allows us to deduce how to close the polar diagram of $H(j\omega)F(j\omega)$, which tends to infinite magnitude as s nears the origin: see Example 8.4. (A similar device is used in the unlikely event that either H or F has a pole on the imaginary axis, which would mean that one of these was itself marginally unstable.) On the other hand, if there are zeros of $(1 + H(s)F(s))$, i.e. roots of the characteristic equation and poles of $G(s)$, on the imaginary axis then the whole system has marginal instability and the curve of $H(j\omega)F(j\omega)$ passes through the point $(-1, j0)$.

For feedback systems the polar (or Argand) diagram of $H(j\omega)F(j\omega)$, because of its significance for the Nyquist criterion, is sometimes referred to as the **Nyquist diagram**. Because the s contour embraces both negative and positive halves of the $j\omega$ axis, the complete Nyquist diagram must include both negative and positive ω; however, since the value of $H(j\omega)F(j\omega)$ for negative ω is the complex conjugate of its value for the same positive ω, the diagram is symmetrical about the real axis. It is conventional, therefore, to draw only the half for positive ω, the other half being understood to be its mirror image in the real axis.

Example 8.3 Find by the Nyquist diagram the critical value of positive K for the closed-loop system having the open-loop transfer function

$$H(s)F(s) = \frac{K}{(s + 2)^3}.$$

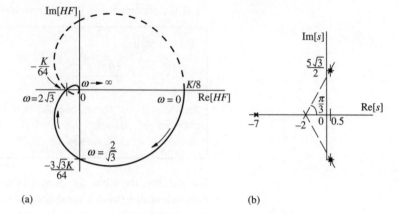

Fig. 8.4
Diagrams for Example 8.3:
(a) Nyquist diagram; (b) closed-
loop poles.

The steady-state open-loop function is

$$H(j\omega)F(j\omega) = \frac{K}{(j\omega + 2)^3} = K\frac{8 - 6\omega^2 - j\omega(12 - \omega^2)}{(4 + \omega^2)^3}.$$

The polar diagram for this function with K positive is shown in
Fig. 8.4(a), values for negative ω being indicated by a broken line. The
curve crosses the negative real axis at $\omega = 2\sqrt{3}$ and has then the value
$-K/64$. Hence the closed path, which is traced clockwise for increasing
ω (i.e. for a clockwise circuit of the s contour in Fig. 8.3), encircles the
point $(-1, j0)$ only when K exceeds the critical value **64**. There is then a
double encirclement, indicating that instability is caused by two roots of
the characteristic equation in the right-hand half s plane.

We may confirm this result algebraically. Suppose that $K = 125$, say.
The characteristic equation

$$1 + H(s)F(s) = 0$$

becomes

$$(s + 2)^3 + 125 = 0$$

or

$$s + 2 = \sqrt[3]{(-125)}$$

of which the solution corresponding to the real cube root is

$$s = -7$$

and evidently lies in the left-hand half plane. The other two solutions
come from the complex cube roots of -125, which are most easily
expressed in terms of the operator 'a' (Section 6.6) as $-5a$ and $-5a^2$;
these are conjugate and each has the real part 2.5. Hence the remaining
two solutions for s have the real part $(2.5 - 2)$, or 0.5, and therefore lie in
the right-hand half plane, as expected. The three solutions, which give
the positions of the poles of the closed-loop transfer function, are shown
on the s plane in Fig. 8.4(b).

It is interesting to note also that for negative K the diagram of
Fig. 8.4(a) would simply be rotated by π, as would the disposition of the

cube roots shown in (b). Only one encirclement could then occur, for $K < -8$, corresponding to the single root which would lie in the right-hand half-plane; and the critical value -8 is in evident agreement with the Routh criterion as applied in Example 8.2 when, as here, $a = b = c = 2$.

Example 8.4 Confirm by the Nyquist diagram that a closed-loop system whose open-loop transfer function is

$$H(s)F(s) = K/s(s + a)$$

is always stable for positive real a and K.

The steady-state open-loop transfer function is

$$H(j\omega)F(j\omega) = \frac{K}{j\omega(j\omega + a)} = K\frac{-\omega - ja}{\omega(a^2 + \omega^2)}$$

$$= -K\left\{\frac{1}{a^2 + \omega^2} + j\frac{a}{\omega(a^2 + \omega^2)}\right\}$$

which has the form shown in Fig. 8.5. The given function has a pole at the origin, which makes the Nyquist curve tend to the limits $\pm j\infty$ for small values of ω. This is a case where the small detour of the s contour around the pole, shown in Fig. 8.3, must be considered. At the point A on that diagram, where s has a small positive real value, say ϵ, the value

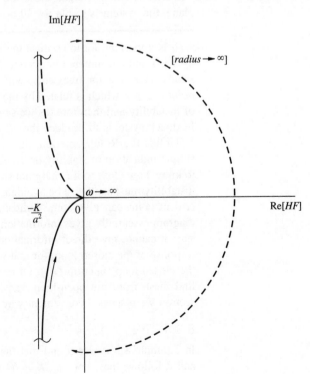

Fig. 8.5
The Nyquist diagram for Example 8.4.

of the given expression is

$$H(\epsilon)F(\epsilon) = K/\epsilon(\epsilon + a)$$

which is large, real and positive. Hence the Nyquist diagram, Fig. 8.5, must be closed at infinite radius in the *right-hand* half plane, as is shown. Evidently the curve can never encircle the point (−1, j0) whatever the (positive) values of a and K. Were K negative, however, the whole curve would be mirrored in the imaginary axis and would then encircle the critical point once for any positive a and negative K.

This again may be confirmed algebraically. The characteristic equation is

$$1 + K/s(s + a) = 0$$

or

$$s^2 + as + K = 0$$

of which the solution is

$$s = \frac{-a \pm \sqrt{a^2 - 4K}}{2}.$$

For negative K, this gives two real values of s of which one is always positive, conforming instability. For positive K, the two roots are negative and real for $K \le a^2/4$, and complex conjugates with real part $-a/2$ for $K > a^2/4$; since these are all in the left-hand half of the s plane, the system is stable for all positive K.

It is important not to confuse the idea of critical *damping*, described by $\zeta = 1$ and encountered in Section 5.3.6 and 7.2.3 respectively for the steady state and for transients, with what we have called the critical *point* (−1, j0), which is related by the Nyquist criterion to the possibility of instability and therefore (as we saw in Section 7.5.3) to the ability of feedback systems to produce the effect of zero damping, i.e. $\zeta = 0$.

Unlike the Routh criterion, the Nyquist diagram gives more than a simple indication of stability or instability. Very often a designer wants to know how close to the marginal state a system will be, for example if instability must at all cost be avoided; and the relation of the curve to the critical point can give some indication of this. However, because the diagram essentially gives information about the open loop in the steady state it cannot give direct information about, for example, the transient response of the closed loop; for that we need to know about the poles of the closed-loop function $G(s)$. Of course it is a fairly simple matter to find these from the open-loop transfer function, but in the following section we discuss a convenient way in which to deal with them.

8.3.4 The root locus

In Example 8.3 it was found that the system was unstable for $K > 64$, and it follows that choosing $K = 64$ will give marginal instability with poles of $G(s)$ just on the $j\omega$ axis. This result suggests that it would be useful to know more generally how the poles of a closed-loop function change their positions if parameters of the system change in value; of these an obvious one to examine is the constant which has been denoted K in the above examples, because as a factor in the open-loop transfer function, it can represent the magnitude of the gain provided by the amplifier which is necessarily part of the system. In this section we consider how to visualize the effect of changing the value of K (which is sometimes called the loop gain, although it is strictly only the 'constant' part of that).

Let us return to the system of Example 8.3 with K positive. The characteristic equation was

$$(s+2)^3 + K = 0$$

and its roots are given by

$$s = \sqrt[3]{-K} - 2,$$

a three-valued expression whose values can again be conveniently written in terms of a, the operator defined in Section 6.6, as

$$s = -\left|\sqrt[3]{-K}\right| - 2, \quad -\mathrm{a}\left|\sqrt[3]{-K}\right| - 2, \quad \text{and} - \mathrm{a}^2\left|\sqrt[3]{-K}\right| - 2.$$

These are symmetrically equidistant from the point $(-2, \mathrm{j}0)$ as shown in Fig. 8.6, the distance increasing with K. Figure 8.4(b) already showed the positions for the particular case $K = 125$; what we now see is how these positions would change with K, in other words their locus. A

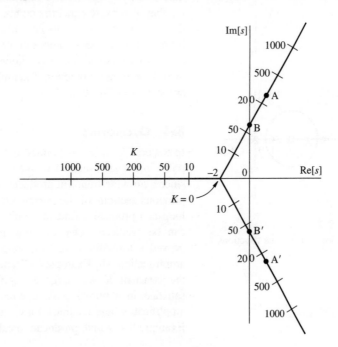

Fig. 8.6
The root locus for Example 8.3.

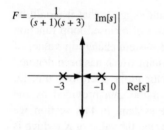

$$F = \frac{1}{(s+1)(s+3)}$$

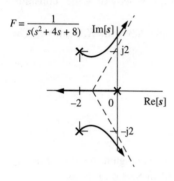

$$F = \frac{1}{s(s^2+4s+8)}$$

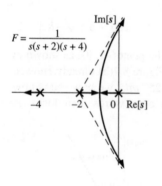

$$F = \frac{1}{s(s+2)(s+4)}$$

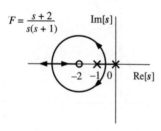

$$F = \frac{s+2}{s(s+1)}$$

Fig. 8.7

Root loci for various functions.

diagram of this kind is called a **root locus**. It gives yet another way of assessing stability, for if any part of the locus enters the right-hand half-plane then instability will occur at some value of K (in our example, at $K = 64$). But it also shows the range of possible pole positions as K is varied, and hence indicates how the closed loop can behave not only in the matter of instability but in general.

In the example above, the root locus was quite simply deduced as a precise solution of the characteristic equation. It is not always so, but the tedium of obtaining it can be avoided by a few rules which allow it to be sketched very simply. It is not appropriate to rehearse these here (they are used mostly in the design of control systems) but one in particular is readily appreciated. If K is taken to be a separate positive factor multiplying the open-loop transfer function then the characteristic equation can be written in the form

$$1 + KH(s)F(s) = 0. \tag{8.6}$$

Now suppose that the locus of its solution is to cover the whole possible range of K, from zero to infinity. For $K = 0$, the equation can be satisfied only when $H(s)F(s) \to \infty$, that is for the poles of the *open-loop* transfer function; on the other hand, for $K \to \infty$, it can be satisfied only by the zeros of that function. Hence we see immediately that any branch of the root locus must begin at a pole, and end at a zero, of the open-loop transfer function. Conversely, each pole is the starting point of a branch, and each zero is an end-point. It follows (at first sight improbably) that the numbers of poles and of zeros must be equal, a requirement which is always met when zeros at infinity are included: it is easily confirmed that, for any linear transfer function, zeros at infinity make up the total number of zeros to equal the number of poles. In the example of Fig. 8.6 the three poles happen to be coincident and the three zeros are all at infinity. Taken with a number of other rules, identifying the terminations of the root locus in this way allows it to be sketched directly from the open-loop transfer function. Examples of loci for a number of functions are shown in Fig. 8.7.

8.4 Oscillators

In Section 7.5.3 it was pointed out that the instability consequent upon poles in the right-hand half plane can form the basis of an oscillator, since a growing transient produces an output which requires no input. In previous sections of the present chapter we have seen how a feedback loop can produce a function with such poles, and how their existence can be predicted. One essential requirement, although we have not proved it formally, is that the loop must contain an active circuit with amplification. (In Examples 8.2 and 8.3 above, instability arose only if the constant K was large enough, a condition which could not be satisfied in a purely passive circuit.) Of course, the combination of amplification and feedback may cause instability without oscillation: in Example 8.4, which produced instability for all negative K, the solution of the characteristic equation showed that the pole of the closed-loop transfer function which then lay in the right-hand half-plane was real; this pole would not cause oscillation, because that can only result from a pair of complex conjugate poles. Such a pair of poles can arise only if the system is frequency dependent, and this too is a requirement for an oscillator based on the principles discussed here.

There is one more point to consider. Since in practice an oscillator must normally operate in the steady state, it follows that to produce a sinusoidal output at a given frequency the relevant poles must lie exactly on the imaginary axis, making the system marginally unstable. But this seems to impose some practical difficulty: on the one hand, the slightest inaccuracy in an attempt to arrange this might put the poles just to the left, and any oscillation would then decay; on the other hand, poles just to the right of the axis would give a growing oscillation. However it was pointed out in Section 7.5.3 that a growing oscillation must inevitably reach some limit which will yield a steady state, and we must now ask how this is related to the original poles. Have they, for instance, somehow moved on to the axis? And will the frequency of oscillation be what they indicated? To answer these questions we need only consider the root locus of our earlier example, shown in Fig. 8.6. Suppose the poles lie in the right-hand half-plane at the points A, A$'$ in the figure corresponding to a chosen value of the constant K. Now the gain of any amplifying device ultimately falls as its output amplitude increases, so a growing oscillation will inevitably cause the gain of the amplifier to fall; K therefore decreases and the poles move along the locus until they reach the axis at B, B$'$. These points correspond to a steady-state oscillation at a frequency ω_B somewhat less, in this case, than the frequency ω_A indicated by the original poles. It can now be seen that to produce a steady-state oscillation there is no need to aim accurately at poles on the axis; it is sufficient, and indeed necessary, to place them to the right of it on the locus which crosses the axis at the required frequency. On this basis, an oscillator could be designed by first choosing a circuit and a value of K which will put the poles on the axis (or, what is exactly equivalent, satisfy the Barkhausen criterion, eqn. 8.2) at the stipulated frequency, and then ensuring that a somewhat larger value of K is available. In practice however, some means of controlling the variation of K with output amplitude is usually incorporated, since the natural fall in the gain of an amplifier occurs only when the output amplitude reaches the limits of linearity and leads to a poor waveform. We return briefly to this point below.

Example 8.5 The circuit shown in Fig. 8.8 is a basic form of a circuit known as a Wien bridge oscillator. Find the value of the voltage gain A_v which will produce oscillation, and at what frequency.

Here $H(s)$ is the gain A_v to be found and, since both input and output are voltages as in Fig. 8.1(a), $F(s)$ is the voltage transfer function of the RC

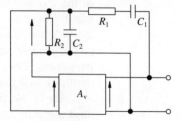

Fig. 8.8
The Wien bridge oscillator (Example 8.5).

network, which must be derived. The impedance of R_2 and C_2 in parallel is

$$Z_2 = \frac{R_2/sC_2}{R_2 + 1/sC_2} = \frac{R_2}{1 + sT_2}$$

if we write T_2 for R_2C_2; and that of R_1 and C_1 in series is

$$Z_1 = R_1 + \frac{1}{sC_1} = \frac{1 + sT_1}{sC_1}$$

with $T_1 = R_1C_1$. Taking Z_1 and Z_2 as a potential divider then gives

$$F(s) = \frac{Z_2}{Z_1 + Z_2} = \frac{R_2}{(1 + sT_2)\left\{\dfrac{(1 + sT_1)(1 + sT_2) + sT_3}{sC_1(1 + sT_2)}\right\}}$$

$$= \frac{sT_3}{1 + s(T_1 + T_2 + T_3) + s^2 T_1 T_2}$$

in which $T_3 = C_1 R_2$. Hence, by setting s to $j\omega$, the steady-state open-loop transfer function becomes

$$H(j\omega)F(j\omega) = \frac{j\omega T_3 A_v}{1 + j\omega(T_1 + T_2 + T_3) - \omega^2 T_1 T_2}. \tag{8.7}$$

The criterion for steady-state or marginal instability, eqn 8.2, requires this expression to satisfy two conditions: that it should be negative real, and that it should have magnitude unity; these are exactly equivalent to the phase and amplitude conditions mentioned in Section 8.3. Inspection shows that HF is real only when

$$\omega^2 T_1 T_2 = 1,$$

that is when

$$\omega = (C_1 C_2 R_1 R_2)^{-1/2} \tag{8.8}$$

and this is therefore the angular frequency at which oscillation may occur provided that the value of HF is then negative with at least unit magnitude. At this frequency the function given by eqn. 8.7 has the value

$$H(j\omega)F(j\omega) = \frac{T_3 A_v}{T_1 + T_2 + T_3} \tag{8.9}$$

and to make this -1 we need

$$A_v = -\frac{T_1 + T_2 + T_3}{T_3} = -\left(\frac{R_1}{R_2} + \frac{C_2}{C_1} + 1\right). \tag{8.10}$$

This is the critical value of A_v, which here includes the constant K used in our earlier discussion; hence a negative value of A_v with a magnitude in excess of that given by eqn 8.10 will cause oscillation. The negative sign can in practice be obtained either from an amplifier with a phase shift of $180°$ or of course, for an amplifier with no phase shift, simply by reversing the connections shown in Fig. 8.8.

The same results can be obtained by writing the condition for the roots of the characteristic equation to be purely imaginary, putting the closed-loop poles on the $j\omega$ axis. The equation is

$$1 + H(s)F(s) = 0$$

which with the H and F given above becomes

$$s^2 T_1 T_2 + s(T_1 + T_2 + T_3 + A_v T_3) + 1 = 0.$$

This has purely imaginary roots when the coefficient of s is zero, a condition which again yields eqn 8.10, and the magnitude of the resulting roots is clearly $1/\sqrt{T_1 T_2}$ which agrees with eqn 8.8 for the frequency of oscillation.

In the above example the frequency dependence which ensures that the phase condition implicit in the Barkhausen criterion can be satisfied only at one frequency was provided by a CR circuit. Another possibility is a tuned, i.e. resonant, circuit, in which case it is common to incorporate that in the amplifier which provides $H(s)$; the feedback circuit $F(s)$ need then provide only a constant ratio less than unity, for which a simple passive network suffices.

Example 8.6   The output current of an amplifier is passed through a tuned circuit having parallel elements R, L, and C, such that its overall voltage gain is AZ_p where Z_p is the impedance of the tuned circuit and A a constant of the amplifier. A fraction β of the output voltage is fed back in the positive direction to the input. Find the value of the product $A\beta$ which will produce oscillation, and the frequency.

The generalized impedance of the parallel circuit is

$$Z_p = \frac{1}{1/R + 1/sL + sC} = \frac{sL}{s^2 LC + sL/R + 1}$$

and the 'forward' transfer function is therefore given by

$$H(s) = AZ_p = \frac{AsL}{s^2 LC + sL/R + 1}.$$

The feedback transfer function $F(s)$ is in this case simply $-\beta$, the negative sign being needed because the feedback has been specified as positive whereas all our original definitions were based on negative feedback. The steady-state open-loop transfer function is therefore

$$H(j\omega)F(j\omega) = -\frac{j\omega A\beta L}{(1 - \omega^2 LC) + j\omega L/R}. \tag{8.11}$$

As above, this must be real to satisfy the Barkhausen criterion, and it evidently is so when $\omega^2 LC = 1$, that is at the resonant frequency $\omega_0 = 1/\sqrt{LC}$. At this frequency we have

$$H(j\omega_0)F(j\omega_0) = -A\beta R;$$

so the critical condition for oscillation is

$$-A\beta R = -1$$

which means that we require

$$A\beta > 1/R. \tag{8.12}$$

The same results could, of course, be obtained by requiring the roots of the characteristic equation to lie on the imaginary axis of the s plane. The equation in this case reduces to

$$s^2 LC + sL(1/R - A\beta) + 1 = 0 \tag{8.13}$$

from which the required conditions leads to eqn 8.12 again, and which by comparison with the standard quadratic form

$$s^2/\omega_0^2 + 2\zeta s/\omega_0 + 1 = 0$$

(Section 5.3.3) shows very clearly how positive feedback has the effect of cancelling the damping in a system; indeed zero damping can be cited as the marginal instability condition, and negative damping as the requirement for oscillation. As expected, the roots of eqn 8.13, with the damping term set to zero, give the oscillation frequency again as ω_0.

The feedback for a tuned-circuit oscillator can be arranged in various ways. In some designs a proportion of the voltage across the tuned circuit is obtained by dividing either L or C into two series parts; in others the feedback is provided by a transformer of which the inductance L forms the primary and whose secondary is connected to the amplifier input; β is then given by the turns ratio (Section 4.9).

A commercial electrical oscillator is, of course, likely to have many refinements over and above a simple feedback circuit operating on the principles outlined above: it will very often, for example, have means for varying at will both the frequency and the amplitude of its output. Oscillator circuits can be designed in which the frequency is varied by means of a d.c. control voltage, rather than by changing component values; they are known as **voltage-controlled** oscillators. In the discussion above output amplitude has hardly been considered, beyond noting the fact that for steady-state oscillation it must reach a value at which the gain is just at the value for marginal instability. In many designs the variation of gain with amplitude is controlled by incorporating a non-linear element in a separate loop with negative feedback (Section 8.5); this leads to a more predictable and satisfactory performance than simply depending on the overall limits of linearity for a basic amplifier.

So far all our theories for linear circuits have depended on the fact that the steady state means sinusoidal waveforms, and in the foregoing discussion we have tacitly assumed that oscillation is sinusoidal. However, not all oscillators produce sinusoidal waveforms: some, for example, have square-wave outputs. These can still be based on positive feedback, used in some cases to drive the amplifying component of the circuit far beyond the limits of linearity. For a transistor the two extremes are saturation (maximum current) and cut-off (zero current), and it is easy to switch it rapidly from one to the other by positive feedback. (Switching between the two is a commonplace in digital circuits, but there it need have nothing to do with oscillation and may not involve feedback.) By the same token, an amplifier as a whole can be driven to either of two extremes of its output voltage, the values of which will depend on its d.c. supplies: one extreme may be positive and the other zero, or they may form a balanced positive and negative pair, for example. If the feedback is so arranged that the amplifier is stable at neither of these limiting states, then it will change rapidly between the two at intervals which depend on the feedback circuit. Such a system is said to be **astable**, rather than unstable. An oscillator of this sort is considered in Section 8.6.5. Variations on the same principle can give circuits which do not oscillate but are **monostable** or **bistable**: the former has one of its two limiting states stable and will always revert to that, but can be changed to the other for a predictable interval by means of a 'trigger' i.e. an input stimulus; a bistable circuit, having both states stable, can be changed from one to the other only by a suitable trigger. The transition between states is achieved by overwhelming positive feedback which puts the loop gain well beyond marginal instability, and it is therefore very rapid. Also known as **flip-flops** or **latches**, such circuits are widely used in digital systems. They have the property of 'memory', since their response to a given input depends on their previous state.

It is worth noting, finally, that the principles of positive feedback, instability and oscillation apply to many systems other than electrical circuits. For example, mechanical clockwork constitutes an oscillator in which positive feedback to the pendulum or hairspring is provided by an escapement and maintains the oscillation at the natural frequency determined by the length of the pendulum or by the stiffness and mass of the hairspring; the weight or mainspring provides the equivalent of the d.c. power required by an electronic amplifier. A quartz oscillator similarly is an electromechanical system with a natural frequency determined by mechanical vibration of the quartz crystal; the feedback loop is completed by an electronic circuit coupled to the crystal by virtue of the piezoelectric effect. A mechanical equivalent for a bistable circuit is any toggle device (like those used for small switches, for example) which can be switched between two stable configurations.

8.5 Negative feedback

The only consequence of feedback considered so far is that of the instability, and in particular oscillation, which can occur when the feedback is actually positive. Negative feedback, however, is used widely in active circuits and offers important advantages. Because the functions $H(s)$ and $F(s)$ were originally defined so as to subtract the feedback signal, it follows that to make the feedback actually negative there must be zero phase shift in the steady-state transfer function $H(j\omega)F(j\omega)$. But unlike the conditions for oscillation, this is not an absolute requirement for any particular effect: a modest amount of phase shift would not alter the tendency of such feedback to reduce the magnitude of the input signal. Nevertheless, for simplicity we shall now assume that $H(j\omega)$ and $F(j\omega)$ are real and positive for all frequencies of interest, and write these simply as H and F since we deal now only with the steady state and have no need for the generalized functions $H(s)$ and $F(s)$. In general H can be considered to be the gain of an amplifier and F the fraction of its output which is subtracted from the external input.

By eqn 8.1 the overall gain of the system can be written

$$G = \frac{H}{1 + HF}. \tag{8.14}$$

If the gain H is large then, so long as F is not too small,

$$HF > 1$$

and hence

$$G \approx 1/F. \tag{8.15}$$

In the limiting case when all of the output is fed back (often called unity feedback) we have $F = 1$, and for large H the overall closed-loop gain G is therefore also unity. So one conclusion is clear: negative feedback reduces the amplifier gain, and that drastically if F approaches unity. However an amplifier with unit gain is not necessarily worthless: in Section 4.6.3 we discussed the source and emitter followers, transistor circuits having voltage gain even less than unity, whose merits were those of an impedance transformer, or buffer, namely a high input impedance and low output impedance. The resistance R_s or R_e in these circuits actually provides negative feedback because the resulting voltage drop is subtracted from the input. In general it turns out that reduced gain is the price paid for several advantages to be had from negative feedback. One of these can be seen immediately from eqn 8.15:

if H is large enough the overall gain G is almost independent of the value of H and depends only on F. The merit of this lies in the facts, touched upon in Section 4.5.6, that F can be fixed quite accurately and consistently by passive components, whereas H is almost always an amplifier gain which is liable to vary because some transistor parameters can be both unpredictable and quite strongly temperature dependent. (The positive temperature coefficient of resistance in semiconductors can even lead to thermal instability, since resistance falls as current increases.) The fact that G can be made virtually independent of variations in H can be described as the *stabilizing* effect of negative feedback, if we now use that word in a slightly wider sense than before. The improvement can be expressed a little more precisely by finding the change dG in overall gain due to a change dH in that of the amplifier if F is assumed to be constant. Differentiating eqn 8.14 gives

$$\frac{dG}{dH} = \frac{1 + HF - HF}{(1 + HF)^2} = \frac{1}{(1 + HF)^2}$$

so the fractional change in G is

$$\frac{dG}{G} = \frac{dH}{(1 + HF)^2} \times \frac{1}{G} = \frac{dH}{H(1 + HF)} \tag{8.16}$$

which is less than that in H by the factor $(1 + HF)$: as it happens, precisely the factor by which the negative feedback reduces the gain in the first place.

Example 8.7 Find the change in the overall gain of an amplifier with a negative feedback fraction 0.1 if the forward gain H falls from 100 to 95.

The original overall gain, by eqn 8.14, was

$$G = H(1 + HF) = 100/11 = 9.0909$$

and this becomes

$$G = 95/10.5 = 9.0476$$

so the fractional loss of gain is $0.0433/9.0909 = \mathbf{0.00476}$, or about **0.5%**, which may be compared with the 5% change in H. According to eqn 8.16 these figures should be related by the factor $(1 + HF)$, i.e. by 11 or 10.5 depending on whether we use the old or new value for H; the agreement is of course imperfect not only because of our rounded figures but also because the differential relation in the equation is accurate only for infinitesimal changes. With either choice the figures clearly demonstrate how the feedback renders the overall gain relatively immune from changes in the forward gain of the amplifier.

For general purposes a substantial loss in gain may seem a very high price to pay for a commensurate improvement in another respect, but in fact it is not usually so: very high gains are rather easily achieved, by cascade connection if necessary, while the general stability and predictability of gain is for many applications essential.

There are other advantages to negative feedback, one of which is the reduction of distortion in analogue amplifiers. Distortion results from non-linearity in amplifying devices and means, in essence, that a sinusoidal signal at a given frequency does not produce a purely sinusoidal output. (The waveform is said to contain **harmonics**; these are discussed in Chapter 9.) It can be shown that the proportion of distortion, defined in a suitably precise way which need not concern us here, is reduced by the same factor $(1 + HF)$ as we have already encountered.

Negative feedback also has a pronounced effect on input and output impedances, which depends on the feedback arrangement. Consider for example Fig. 8.1(a), which shows a voltage amplifier with voltage feedback, and suppose that the input impedance of the amplifier itself is Z_{in}, as shown in Fig. 8.9. Because the input current I_{in} to the whole circuit is also that to the amplifier, we may write

$$\frac{V_{in} - FV_{out}}{I_{in}} = \frac{V_1}{I_{in}} = Z_{in}$$

and therefore

Fig. 8.9

Input impedance with voltage feedback.

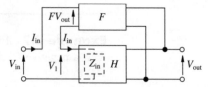

$$\frac{V_{in}}{I_{in}} = Z_{in} + \frac{FV_{out}}{I_{in}}$$
$$= Z_{in} + \frac{FV_1H}{I_{in}} \tag{8.17}$$
$$= Z_{in}(1 + HF)$$

which shows the factor $(1 + HF)$ appearing yet again, as that by which in this case the negative feedback has increased the input impedance. We should find the same results for Fig. 8.1(c), because the feedback voltage FI_{out} can still be written FV_1H where H is now the transfer function I_{out}/V_1. However a comparable calculation for (b) or (d), in both of which the input is current, shows that the input impedance is then reduced by the same factor. Both results are beneficial, since a low input impedance is as desirable for a current input as is a high input impedance for a voltage input.

Output impedances are also changed for the better by negative feedback. It can be shown that if the input source is ideal then the output impedance for cases (a) and (b) of Fig. 8.1, where the output is voltage, is reduced by the factor $(1 + HF)$; while for (c) and (d), with current output, it is increased by the same factor.

There is one other important effect of negative feedback: it increases **bandwidth**. Any practical amplifying device has a finite bandwidth, in the sense that its gain must ultimately decrease with increasing frequency, if only because of unavoidable 'stray' capacitance (it is nearly always capacitance, rather than inductance, which dominates in determining the bandwidth of an electronic circuit). In addition, the gain of some amplifiers decreases, by design, as the frequency falls below a certain range: these so-called **a.c. amplifiers** have some advantages and avoid certain practical problems associated with d.c. amplifiers (by which is meant those designed to amplify all frequencies from d.c. to some upper limit). In consequence the Bode diagram for the gain H of an a.c. amplifier may appear at its simplest as shown in Fig. 8.10. The breakpoint frequencies ω_1 and ω_2, where the asymptotes

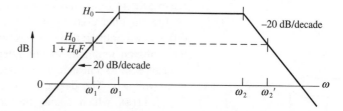

Fig. 8.10

The effect of negative feedback on gain and bandwidth.

are in error by about 3 dB (Section 5.2.2), allow the bandwidth to be defined as $\omega_2 - \omega_1$; these frequencies correspond exactly to the half-power points for resonant circuits, mentioned in Section 5.3.5. The effect of negative feedback, we know, is to reduce gain by the factor $(1 + HF)$ according to eqn 8.1, but on substituting in that equation the appropriate frequency-dependent function for H it turns out that the high- and low-frequency asymptotes for the overall gain $G(j\omega)$ are the same as those for the gain $H(j\omega)$ of the amplifier itself; the result is that the negative feedback simply truncates the Bode diagram for amplitude, as shown by the broken line in the figure. The effect of this is readily shown to be an increase in ω_2 by the factor $(1 + H_0F)$ where H_0 is the maximum or **midband** gain of the amplifier, and an identical decrease in ω_1. For a d.c. amplifier, in which there is no lower breakpoint, the bandwidth is simply ω_2 and is therefore increased by that same factor. The bandwidth of an a.c. amplifier, $\omega_2 - \omega_1$, is evidently increased by rather more than this; but since in many cases $\omega_2 \gg \omega_1$ the factor $(1 + H_0F)$ gives a good approximation. More precisely, we can if need be express the new bandwidth logarithmically: measured in decades, the increase in ω_2 and the decrease in ω_1 are each given by $\log_{10}(1 + H_0F)$; hence the increase in bandwidth is $2\log_{10}(1 + H_0F)$ decades, which is numerically equal to one tenth part of the decibel reduction in midband gain.

This effect of negative feedback on bandwidth illustrates a rather general (but not precise) principle, that in circuit design the product of gain and bandwidth tends to be constant. By feedback (or otherwise) one may be traded against the other; the actual value of the product may be regarded as a figure of merit in system design. The effect is also evident in resonant circuits, for which we saw in Section 5.3.5 that the bandwidth varies inversely as the quality factor Q, which in turn is a measure of peak response.

8.6 Operational amplifiers

8.6.1 General

In Chapter 4 we considered the basic ideas of amplifying circuits by way of their linear small-signal equivalents. Here we consider a large and versatile class of electronic circuits known as **operational amplifiers** whose behaviour is dominated by the feedback chosen and can be investigated on the basis of simple assumptions about the amplifier itself. The name arises from the fact that by choosing suitable feedback arrangements the circuit can be designed to carry out various **operations** on its input signal, beyond that of mere amplification.

The amplifier used, in practice an integrated circuit, is a d.c. voltage difference amplifier: that is, it has two separate input terminals as well as the common terminal, and its internal design is such that the output voltage is in its linear regime proportional to the *difference* in the two input voltages. The device is given the standard symbol indicated in Fig. 8.11(a), which shows the common terminal but none of the connections for the d.c. power supplies and operating adjustments which are required in practice; often the common terminal is omitted, its potential being assumed to be that of earth or some other obvious point in the actual circuit, to leave the symbol shown at (b). The marks – and + are given to what are known as the inverting and non-inverting inputs respectively, enabling the output voltage to be written

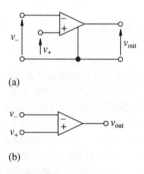

(a)

(b)

Fig. 8.11
The operational amplifier.

$$v_{out} = A_v(v_+ - v_-) \qquad (8.18)$$

where we assume that A_v, the voltage gain, is a positive real quantity which is constant from d.c. up to some breakpoint frequency (we return to the question of frequency response below). Equation 8.18 is not necessarily confined to small signals superimposed on a d.c. operating point, such as we discussed in Section 4.5; the voltages in the equation are potentials with respect to earth, and the power supplies and biassing of the difference amplifier are usually such that v_{out} can take any value between the positive and negative maxima which are determined by the amplifier's saturation limits.

The amplifier is designed to have in high degree the attributes to be looked for in a voltage amplifier: a large gain A_v (of order 10^4 to 10^5 or more at low frequencies), large input impedances (100 kΩ or more) between the two inputs and between each and earth, and a low output impedance (in the order of 100 Ω). For many purposes it can accordingly be assumed that both gain and input impedances are infinite and that the output impedance is negligible; these assumptions define the **ideal** operational amplifier, and they are often used, but in general they should of course be justified in relation to the other parameters of the circuit, failing which actual values should be used in the first instance.

8.6.2 The inverting amplifier

For some applications the non-inverting input terminal is connected to earth, so that by eqn 8.18

$$v_{\text{out}} = -A_{\text{v}}v_{-} \tag{8.19}$$

if A_{v} is again positive. This is the basis of the common arrangement shown in Fig. 8.12(a) which has the feedback provided by the two impedances Z_1 and Z_2; these we take to be generalized functions of the variable s so as not to restrict the input and output waveforms, but we may also if need be take them to be complex quantities with certain magnitude and angle at a given frequency, as in the steady state.

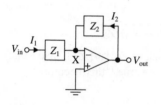

(a)

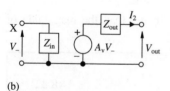

(b)

Fig. 8.12
An inverting amplifier.

We now consider this circuit in detail, assuming at this stage only that the amplifier is linear with open-circuit gain A_{v} and input and output impedances Z_{in} and Z_{out}. In a case like this, with one input earthed, Z_{in} is simply the resultant internal impedance between the amplifier inputs; it is shown in Fig. 8.12(b), which also shows the Thévenin equivalent of the amplifier itself. The output of the whole we assume to be open-circuit. If the voltage at the input node X is written in transform notation as V_{-} then the current in Z_2, which is also the output current of the amplifier, must be

$$I_2 = \frac{V_{\text{out}} - V_{-}}{Z_2}$$

in the direction shown. The output voltage, which for I_2 zero would be $-A_{\text{v}}V_{-}$ according to eqn 8.19, is therefore now

$$V_{\text{out}} = -A_{\text{v}}V_{-} - I_2 Z_{\text{out}} = -A_{\text{v}}V_{-} - Z_{\text{out}}\frac{V_{\text{out}} - V_{-}}{Z_2}$$

or

$$V_{\text{out}} = V_{-}\frac{-A_{\text{v}} + Z_{\text{out}}/Z_2}{1 + Z_{\text{out}}/Z_2}. \tag{8.20}$$

This shows in passing that for the whole circuit V_{out} can still be taken as $-A_{\text{v}}V_{-}$ provided that the magnitude of Z_{out} is much less than that of Z_2 for all frequencies of interest.

The input current to the whole circuit is the current in Z_1, which can be written

$$I_1 = \frac{V_{in} - V_-}{Z_1},$$

while the actual input current to the amplifier is V_-/Z_{in}. Kirchhoff's first law at the node X then gives

$$\frac{V_-}{Z_{in}} = I_1 + I_2 = \frac{V_{in} - V_-}{Z_1} + \frac{V_{out} - V_-}{Z_2}$$

or

$$V_-\left(\frac{1}{Z_{in}} + \frac{1}{Z_1} + \frac{1}{Z_2}\right) = \frac{V_{in}}{Z_1} + \frac{V_{out}}{Z_2}.$$

In this, V_- can be written in terms of V_{out} from eqn 8.20 and rearrangement then gives an overall gain

$$\frac{V_{out}}{V_{in}} = -\frac{A_v Z_2 - Z_{out}}{Z_1(1 + A_v) + (Z_2 + Z_{out})(Z_1 + Z_{in})/Z_{in}}.$$

If now we make the assumptions that $|Z_{in}| \gg |Z_1|$ and that $|Z_{out}| \ll |Z_2|$, both of which are likely to be true,[†] this becomes

$$\frac{V_{out}}{V_{in}} = -\frac{A_v Z_2}{Z_1(1 + A_v) + Z_2} = -\frac{Z_2}{Z_1 + \dfrac{Z_2 + Z_1}{A_v}}. \qquad (8.21)$$

Finally, if we assume that A_v is much greater than the magnitude of $(1 + Z_2/Z_1)$ then eqn 8.21 reduces to

$$\frac{V_{out}}{V_{in}} = -\frac{Z_2}{Z_1}. \qquad \mathbf{(8.22)}$$

This simple and useful result could have been produced much more easily had we analysed the circuit by taking the amplifier to be ideal at the outset, so that A_v and Z_{in} were infinite and Z_{out} negligible; but by starting with the general case we have also shown that the required conditions are respectively

$$A_v \gg |1 + Z_2/Z_1|$$
$$|Z_{in}| \gg |Z_1| \qquad\qquad\qquad (8.23)$$
$$|Z_{out}| \ll |Z_2|.$$

In many applications of circuits having the form of Fig. 8.12(a) all three of these conditions are likely to be satisfied, the amplifier can be assumed to be ideal, and eqn 8.22 can be invoked accordingly.

[†] In this chapter the notation Z has usually been used to mean the generalized impedance $Z(s)$. By $|Z|$ we mean here the magnitude of the steady-state impedance $Z(j\omega)$ at any frequency which may be concerned.

The above assumptions for the ideal amplifier lead directly to two consequences which are often used to simplify the analysis of any operational amplifier circuit for which they can be justified. First, if the open-circuit gain A_v of the amplifier itself is very high and its output impedance low it follows that its actual gain is also high and the potential difference $v_+ - v_-$ accordingly very small so long as the amplifier operates in its linear region with its output below saturation level; it is therefore common to assume that $v_+ \approx v_-$. In a circuit such as the inverting amplifier just considered, where the non-inverting input is earthed so that $v_+ = 0$, this means that $v_- \approx 0$; in such cases the inverting input (X in Fig. 8.12) is therefore commonly assumed to be at zero potential and is then called a **virtual earth**. In an amplifier with both inputs active their voltage difference is likewise assumed to be zero, and the two therefore at the same potential. Secondly when the input impedance is assumed to be very high the current into each amplifier input terminal can be taken to be zero and Kirchhoff's first law applied accordingly at node X or at each input node, as the case may be. These two conditions allow the rapid deduction of an overall voltage transfer function for any operational amplifier circuit in the linear regime.

There is clearly similarity between eqns 8.21 and 8.22 and the more general pair eqns 8.14 and 8.15: $-A_v$ corresponds to the forward gain G and it appears that F has become $-Z_1/Z_2$, although on this basis the correspondence between eqn 8.21 and eqn 8.14 is not perfect. In fact the circuit just considered cannot very conveniently be related to the general forms shown in Fig. 8.1, partly because its output is not fed back directly to the input of the circuit as in those cases. Comparison of the two figures shows that while the circuit of Fig. 8.12 essentially uses current feedback as in Fig. 8.1(b), its input is actually a voltage and its forward amplifier is specified by a voltage gain. In particular cases such as this it is often simpler to deal with operational amplifier circuits from first principles, as here, than by attempting to adapt eqn 8.14 for the purpose. Nevertheless the above results confirm that for this circuit, as in the general case, the effect of negative feedback with a large forward gain is to produce an overall gain which is dependent only on the feedback network, in this case the impedances Z_1 and Z_2. In the case when these are both resistances, R_1 and R_2, the overall voltage gain is ideally $-R_2/R_1$, so the output waveform is an inverted replica of the input (if that is sinusoidal they differ in phase by π) and the circuit is simply an **inverting amplifier**.

The feedback used in an operational amplifier circuit has a marked effect also on the input and output impedances of the circuit as a whole, as might be expected from the more general discussion of Section 8.5. In the case of the inverting amplifier discussed above, the input impedance is easily deduced by inspection from Fig. 8.12(a): since the node X is a virtual earth, the input current is simply V_{in}/Z_1 and the input impedance is therefore Z_1. This is typically much less in magnitude than Z_{in} for the amplifier, so the feedback in this case has an adverse effect. The output impedance of the circuit can be found by any of the methods described in earlier chapters, when the amplifier is represented by its Thévenin equivalent as in Fig. 8.12(b); it can be shown to be

$$Z'_{\text{out}} = \frac{Z_{\text{out}}(Z_1 + Z_2)}{Z_{\text{out}} + Z_2 + (1 + A_v)Z_1}$$

when Z_{in} is assumed to be large. This result shows that the output impedance of the circuit is less than Z_{out}, the output impedance of the amplifier itself; in this respect the feedback is again beneficial, as it is in the general cases shown in Fig. 8.1.

8.6.3 The non-inverting amplifier

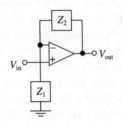

(a)

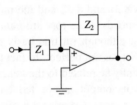

(b)

Fig. 8.13
(a) The non-inverting amplifier;
(b) an unstable circuit.

A gain similar to that of the inverting amplifier described above, but without the inversion, can be had from the circuit shown in Fig. 8.13(a), as is easily confirmed by analysing the circuit on the ideal assumptions outlined in the previous section. Thus, taking the two inputs to be at the same potential gives

$$v_- = v_+ = v_{\text{in}}$$

and assuming no current into either amplifier terminal gives, by Kirchhoff's first law at the inverting input, and in our previous notation,

$$V_-/Z_1 = (V_{\text{out}} - V_-)/Z_2.$$

Combining these relations gives

$$\frac{V_{\text{out}}}{Z_2} = V_-\left(\frac{1}{Z_1} + \frac{1}{Z_2}\right) = V_{\text{in}}\frac{Z_1 + Z_2}{Z_1 Z_2}$$

and so

$$\frac{V_{\text{out}}}{V_{\text{in}}} = \frac{Z_1 + Z_2}{Z_1} \tag{8.24}$$

which is to be compared with the gain $(-Z_2/Z_1)$ of the inverting amplifier of Fig. 8.12.

In this case the similarity with the result of eqn 8.15 is clear, for the ratio $Z_1/(Z_1 + Z_2)$ is here, by potential division, the fraction of output voltage fed back to the inverting input and therefore corresponds precisely to the (negative) feedback function F in the general notation. In the case the two impedances are actually resistances R_1 and R_2, the output reproduces the input waveform with a gain given by eqn 8.24 and without inversion; the circuit is a **non-inverting amplifier**.

It may be observed here that the non-inverting property is not achieved, as one might think possible, by simply taking the inverting circuit 8.12 and interchanging the + and − inputs as shown in Fig. 8.13(b). This circuit, in fact, cannot act as an amplifier at all, for its feedback is positive and it is in consequence unstable. This may be deduced by inspection: if the input potential v_+ for any reason rises above earth potential then the output voltage must become large and positive, and by potential division this must increase v_+ and hence the output still more, so that the amplifier rapidly attains its positive saturated output voltage whatever the value of v_{in}; similarly, if v_+ happens to fall below earth potential, the output voltage will become negative, further lowering v_+ and the output until the latter reaches its negative saturated limit. Although it cannot be used as an amplifier, the ability of this and similar positive-feedback circuits to switch very rapidly between extremes of output voltage can be put to use in various ways (some already mentioned in Section 8.4; see also Section 8.6.5).

The input impedance of the non-inverting circuit is very different from that of the inverting circuit, which we saw to be approximately Z_1. Inspection of Fig. 8.13(a) shows that the input impedance of the circuit is simply that between the amplifier's non-inverting input and earth, so it may be assumed to be very large; in this respect the circuit is superior to its inverting counterpart. Its output impedance, like that for the inverting circuit, can be shown to be much smaller in magnitude than Z_{out} for the amplifier itself when Z_{in} and A_v are both large.

8.6.4 Limitations in performance

Apart from the question of whether or not the conditions of eqn 8.13 are satisfied in a particular case there are several imperfections in actual amplifiers which are not accounted for in the foregoing outline. One is the fall in the gain A_v with increasing frequency (since difference amplifiers are normally d.c. amplifiers there is no difficulty at low frequencies). The breakpoint frequency f_0 at which A_v begins to fall with frequency is typically designed, for reasons to which we return below, to be no more than about 10 Hz, above which it decays at 20 dB per decade or perhaps more. This low bandwidth is not in itself serious because an operational amplifier is always used with feedback and, as was discussed in Section 8.5 and illustrated in Fig. 8.10, negative feedback increases bandwidth. A non-inverting amplifier, for example, with gain $(R_1 + R_2)/R_1$ according to eqn 8.24, might have the frequency response shown in Fig. 8.14(a). In practice the decline in gain with frequency may exceed 20 dB per decade, with a corresponding increase in phase lag, at the higher frequencies. The frequency f_1 at which the gain falls to unity or 0 dB is known as the **unity-gain bandwidth**; its value evidently gives the (constant) gain–bandwidth product, which is of course also given by $A_v f_0$.

The phase lag at high frequency can give rise to problems of instability: if the total lag around the amplifier-feedback loop should reach π, either because the gain falls off at 40 dB per decade or because of unwanted phase shift in the feedback connections, then what was intended to be negative feedback becomes positive and oscillation may result according to the Barkhausen criterion (eqn 8.2). An amplifier with a fall of 40 dB per decade can be improved in this respect by **frequency compensation**. By means of capacitance, either external or internal to the integrated circuit, this extends to higher frequencies and lower gains the range over which the fall is only 20 dB per decade, at the expense of

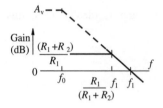

(a)

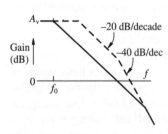

(b)

Fig. 8.14
Frequency response with negative feedback: (a) simplest case; (b) compensated.

reducing the breakpoint f_0, as shown in Fig. 8.14(b). It is for this reason that the typical value of f_0 is as low as 10 Hz, as in (a).

The frequency response of Fig. 8.14 should be regarded as applicable to small signals. The response to relatively large signals can be affected by the fact that the rate of change which the output voltage can achieve has a limiting value known as the **slew rate**, typically in the order of 1 V/μs. Its value reflects the maximum current levels which are available to charge and discharge the internal capacitances of the amplifier. At a given frequency the peak rate of change of a voltage increases with its amplitude, so the slew rate limitation can cause distortion of a large signal at frequencies for which a small signal would be amplified normally.

Another imperfection not yet accounted for arises when the amplifier output, which should depend only on the difference $v_+ - v_-$, depends also to some small extent on the average of the two inputs. The total output voltage can in that case be written

$$v_{out} = A_v(v_+ - v_-) + A_{cm}(v_+ + v_-)/2$$

where A_{cm} is the **common-mode** gain and is ideally zero; this expression can also be written as

$$v_{out} = A_1 v_+ - A_2 v_-$$

where

$$A_1, A_2 = A_v \pm \frac{1}{2} A_{cm}.$$

The ratio of A_v to A_{cm} is the **common-mode rejection ratio**, a figure of merit which should ideally be infinite and in a good difference amplifier is at least 10^4.

The above expressions indicate that when the two inputs are precisely equal, so that $v_+ = v_- = v$, say, the output voltage is $A_{cm}v$; even with finite A_{cm} this should be zero when $v = 0$, i.e. when both inputs are earthed. This too is not quite true in an actual amplifier: imperfect balance of its component parameters results in a small output voltage, an imperfection which can be described by an **offset voltage**. By this is meant the voltage which must be applied to one input, the other being earthed, in order to produce zero output; it is usually no more than one or two millivolts, but is somewhat temperature dependent. For analysing the performance of a circuit it can simply be represented as a voltage source connected to one input in addition to any other applied voltage. The term 'offset voltage' is also used to describe the voltage, ideally zero, which small but unbalanced input bias currents, flowing in the amplifier's input impedances, cause to appear between the input terminals in the absence of any signal; this is often described in terms of an **offset current**, which can also be represented by a source at the input.

8.6.5 Applications of operational amplifiers

In Sections 8.6.2 and 8.6.3 we have already discussed the inverting and non-inverting configurations in which an operational amplifier with

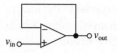

Fig. 8.15
The voltage follower.

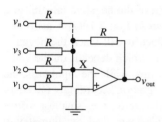

Fig. 8.16
An adding circuit.

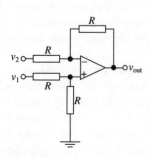

Fig. 8.17
A subtracting circuit.

purely resistive feedback gives overall gain $-R_2/R_1$ or $(R_1 + R_2)/R_1$ as the case may be, and the several advantages of these circuits as amplifiers have been pointed out. The very high input impedance and low output impedance of the non-inverting circuit make it useful as a buffer or impedance transformer. If its gain is required to be unity for this purpose we need small R_2 and large R_1; the circuit then takes the limiting form shown in Fig. 8.15, which is known as a **voltage follower**. Inspection confirms that $v_{\text{out}} = v_{\text{in}}$ to the degree of accuracy that $v_- = v_+$, and the arrangement is an example of unity feedback, $F = 1$, as mentioned in Section 8.5. It is superior to the simple emitter or source follower (Section 4.6.3) in that its voltage gain is unity to a very close approximation. The inverting circuit (Section 8.6.2), although less good as a buffer because of its relatively low input impedance, is also sometimes given unit gain (by choosing $R_2 = R_1$) because it then serves as an **inverter**. This function simply represents a change of sign, but can be particularly useful in cancelling an unwanted inversion elsewhere in a system (in steady-state terms, the total phase shift then becomes 2π, equivalent to zero).

An **adding circuit** is a variation of the inverting amplifier made by connecting together at a single amplifier input currents proportional to the several voltages to be added, as shown in Fig. 8.16. Since the input X is a virtual earth, Kirchhoff's first law there gives

$$v_1/R + v_2/R + \ldots + v_n/R = -v_{\text{out}}/R$$

from which

$$v_{\text{out}} = -\sum_{i=1}^{n} v_i. \tag{8.25}$$

The negative sign can be removed if required by an inverter of unit gain. By allocating different values to the input resistances R the input voltages can be added in any required proportions to give a weighted sum. A different purpose is served by setting the feedback resistance to R/n, for the circuit then gives the **average** of the inputs. **Subtraction** can be achieved by first inverting the appropriate inputs to the adding circuit or, alternatively, by using both amplifier inputs as shown in the example of Fig. 8.17, for which it is readily shown that

$$v_{\text{out}} = v_1 - v_2 \tag{8.26}$$

These circuits are of course analogue applications, and have little in common with the digital circuits which perform similar functions in binary arithmetic.

Integrating and **differentiating** circuits can be made with a suitable choice of Z_1 and Z_2 in Fig. 8.12. For example the circuit of Fig. 8.18(a) has $Z_1 = R$, $Z_2 = 1/Cs$, so eqn 8.22 gives the gain or transfer function

$$v_{\text{out}}/v_{\text{in}} = -1/RCs$$

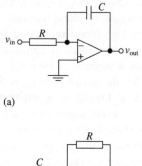

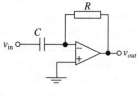

(a)

(b)

Fig. 8.18
(a) An integrating circuit; (b) a differentiating circuit.

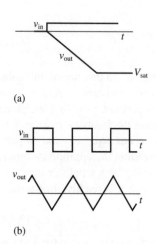

(a)

(b)

Fig. 8.19
Integration: (a) of a d.c. input; (b) of a periodic input.

which means that the output voltage is proportional to the integral (with respect to time) of the input, again with a change of sign. Similarly the circuit shown in Fig. 8.18(b) has the transfer function

$$v_{out}/v_{in} = -RCs$$

which means that the output is proportional to the negative derivative of the input. It is worth noting that in this case the forms of Z_1 and Z_2 mean that the high-gain assumption of eqns 8.23 would not be valid for high-frequency sinusoids (its validity for non-sinusoidal waveforms generally depends on their frequency spectra, a topic which will be considered in Chapter 9). For this and other reasons differentiating circuits are not greatly favoured.

The integrating circuit can illustrate the difficulties which may arise from the limitations of operational amplifiers, including the assumptions and imperfections already discussed as well as the natural limits to the output voltage imposed by saturation and determined by the voltages of its power supplies. Suppose the circuit of Fig. 8.18(a) to have a small positive voltage applied to its input terminal. From its transfer function we may expect the output to take the form of the negative integral of a constant, that is a ramp function as shown in Fig. 8.19(a). Clearly the output cannot maintain this waveform, since it will reach the negative saturation voltage of the amplifier in a time which depends on the value of RC (and on the initial conditions); after this, while the input lasts, the output will remain at this value and the integration accordingly fails. Not only that: even if the input is exactly zero, the smallest offset imperfection will be integrated in the same way and the output will eventually, even if slowly, reach saturation as before, a phenomenon known generally as **drift**. On the other hand the circuit will successfully integrate a square wave, as indicated in (b), provided that the value RC is so chosen in relation to the frequency that the output never reaches a saturation limit.

At this point it could be asked whether integration and differentiation of waveforms by what may be called analogue, as opposed to digital, electrical methods could not be carried out without an amplifier, for example simply by measuring the voltage across a capacitance, which is the integral of the current into it. The answer is not only that a voltage input may be preferable in this example, but also that a combination of high input impedance and low output impedance can be provided by an amplifier circuit, and that the choice of two elements rather than one affords versatility.

Integrating and differentiating circuits could also be produced by using operational amplifiers with combinations of resistance and inductance; but the latter is not a preferred component in electronic circuits, and is often avoided for the practical reason that it cannot be produced as cheaply or as purely as capacitance, and would be especially difficult in an integrated circuit. Indeed, operational amplifiers are used in certain applications where in principle passive circuits would serve as well, mainly because they allow the design of circuits without

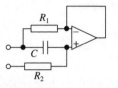

Fig. 8.20
A circuit to simulate inductance.

the inductances which would otherwise be necessary (see Example 8.8 below). This possibility can be simply illustrated by the circuit of Fig. 8.20, which if need be can be used on its own to simulate inductance. It can be shown, on the ideal amplifier assumptions, to have a steady-state input impedance given by

$$Z_{in} = \frac{R_1 + j\omega CR_1R_2}{1 + j\omega CR_1}$$

which on rationalization becomes

$$Z_{in} = \frac{R_1(1 + \omega^2C^2R_1R_2)}{1 + \omega^2C^2R_1^2} + j\omega\frac{R_1C(R_2 - R_1)}{1 + \omega^2C^2R_1^2}$$
$$= R' + j\omega L'.$$

Provided that L' is positive, i.e. that $R_2 > R_1$, the circuit evidently behaves like an inductance in series with a resistance R'; its quality factor is

$$Q = \frac{\omega L'}{R'} = \frac{\omega C(R_2 - R_1)}{1 + \omega^2C^2R_1R_2}.$$

Active filters are operational amplifier circuits designed to fulfil the functions described in Section 5.4, having substantial gain for frequencies within a prescribed band and otherwise negligible gain. They are no different in principle from filters which use passive elements only, but the active form is for many purposes superior: it can offer high input and low output impedances, which allows high-order filters to be made by cascading low-order sections; it can provide a pass-band gain higher than unity if required; and it avoids any need for inductance.

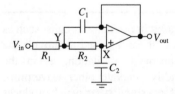

(a)

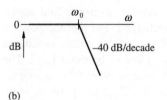

(b)

Fig. 8.21
Example 8.8: (a) an active filter circuit; (b) its Bode diagram.

Example 8.8 Find the steady-state transfer function and frequency response of the circuit shown in Fig. 8.21(a).

For this circuit, assuming an ideal amplifier, we have

$$V_+ = V_- = V_{out}$$

and if V_1 is the phasor voltage at the node Y the nodal equation there is

$$(V_{in} - V_1)/R_1 + (V_{out} - V_1)/R_2 + (V_{out} - V_1)j\omega C_1 = 0$$

while at node X we have

$$(V_1 - V_{out})/R_2 = V_{out}j\omega C_2,$$

which gives

$$V_1 = V_{out}(1 + j\omega C_2R_2).$$

The equation above for node Y may then be written

$$\frac{V_{in}}{R_1} + V_{out}\left(\frac{1}{R_2} + j\omega C_1\right) = V_1\left(\frac{1}{R_1} + \frac{1}{R_2} + j\omega C_1\right)$$

$$= V_{out}(1 + j\omega C_2 R_2)\left(\frac{1}{R_1} + \frac{1}{R_2} + j\omega C_1\right)$$

which, rearranged, gives the required transfer function

$$\frac{V_{out}}{V_{in}} = \frac{1}{1 + j\omega C_2(R_1 + R_2) + (j\omega)^2 C_1 C_2 R_1 R_2}. \tag{8.27}$$

The quadratic term in this is different from those encountered in previous chapters, which related to RLC circuits, but we may still find a resonant frequency ω_0 and damping ratio ζ by comparing it with the expression 5.21. This gives

$$\omega_0 = \frac{1}{\sqrt{C_1 C_2 R_1 R_2}}$$

$$\zeta = \frac{1}{2}\sqrt{\frac{C_2}{C_1}}\left(\sqrt{\frac{R_1}{R_2}} + \sqrt{\frac{R_2}{R_1}}\right). \tag{8.28}$$

The Bode diagram corresponding to eqn 8.27 is shown in Fig. 8.21(b). The gain decreases with frequency at 40 dB per decade above the breakpoint ω_0, and the circuit evidently acts as a second-order low-pass filter with cut-off frequency ω_0. The precise form of the response curve for a given breakpoint can be adjusted to give a Butterworth or other desired characteristic (Section 5.4.2) by changing the values of the parameters so as to vary ζ but not ω_0. (With four parameters to choose it is also possible to vary the input and output impedances; but the latter is in any case small.)

The operational amplifier is widely used in oscillator circuits, such as were discussed generally in Section 8.4. It can be used to provide the amplification needed to produce sinusoidal oscillation, as in the examples given there, but it can readily generate other waveforms. The circuit of Fig. 8.22(a) is a **relaxation oscillator** which produces a square-wave output. It is easily seen that the circuit has no stable state; for any positive v_{out} would tend to make v_- a higher potential than v_+, so v_{out} would then be driven negative and the reverse would apply. Suppose the capacitance C is initially uncharged, and that the positive feedback through the two resistances R causes the output rapidly to switch to the positive saturation value V_m so that v_+ becomes $V_m/2$. The capacitance then charges through R_1 with time constant CR_1, raising v_- accordingly until it just exceeds the value $V_m/2$; at this point $v_+ - v_-$ becomes negative and the positive feedback drives the output rapidly to the negative limit $-V_m$ and v_+ to $-V_m/2$. The capacitance now begins to discharge and recharge negatively, until v_- falls just below v_+ and the

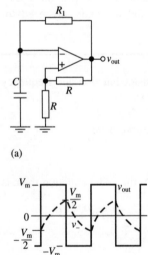

(a)

(b)

Fig. 8.22
A relaxation oscillator: (a) the circuit; (b) its output waveform.

output switches again; the pattern repeats itself, with the output oscillating between the limits $\pm V_m$ as the capacitance voltage v_- varies exponentially between the values $\pm V_m/2$. The resulting waveforms are shown in Fig. 8.22(b). The oscillation frequency can be deduced from the fact that when the capacitance begins to charge from its minimum voltage $-V_m/2$ it is being driven by the output voltage V_m; hence we may write

$$v_- = V_m\left(1 - \frac{3}{2}e^{-t/CR_1}\right)$$

and the next transition occurs when this expression reaches the value $V_m/2$, after the half-period of oscillation $T/2$. We therefore have

$$\frac{1}{2} = 1 - \frac{3}{2}e^{-T/2CR_1}$$

from which it follows that the period is

$$T = 2CR_1 \ln 3.$$

There are many other applications of operational amplifiers, including some which use non-linear feedback elements to produce a specified input/output relationship.

Problems

8.1 An amplifier has a voltage gain given by $A_0/(1 + j\omega T)$. One half of its output voltage is fed back negatively to the input, so that $F = 1/2$. Find the closed-loop voltage gain and bandwidth.

8.2 If the feedback in Problem 8.1 is reconnected as positive feedback, what range of A_0 would cause instability?

8.3 A system has a transfer function given by $1/(1 + 0.1s + 0.04s^2)$. What are its undamped natural frequency and damping factor? Find the new values of these quantities if negative feedback is applied directly between output and input ($F = 1$).

8.4 The open-loop gain of an amplifier is given by

$$HF = \frac{K}{(1 + 0.01s)(1 + 10^{-4}s)(1 + 10^{-5}s)}.$$

Find the critical value of K for stability.

8.5 An oscillator can be made from an amplifier with feedback through a ladder network of three identical RC sections, giving the transfer function

$$F(s) = \frac{s^3 C^3 R^3}{s^3 C^3 R^3 + 6s^2 C^2 R^2 + 5sCR + 1}.$$

Find the minimum gain needed for the amplifier and the frequency of oscillation.

8.6 A d.c. amplifier has a gain of 100 and bandwidth 10 kHz. Find the feedback fraction needed to increase the bandwidth to 100 kHz, and the resulting gain.

8.7 An operational amplifier has input resistance 1 MΩ, output resistance 60 Ω and low-frequency gain 10^4. It is used as an inverting amplifier with resistances $R_1 = 10$ kΩ and $R_2 = 100$ kΩ. Find the error in assuming that the voltage gain has the ideal value -10.

8.8 Find the output impedance of the feedback amplifier in Problem 8.7.

8.9 The amplifier in Problem 8.7 has a slew rate of 1 V μs^{-1}, a unity-gain bandwidth of 200 kHz, and an output saturation level of ± 10 V. Find the maximum frequency of a sinusoidal voltage which can be amplified without distortion when its output amplitude just reaches saturation, and estimate the maximum amplitude of an input signal at 20 kHz for no distortion.

8.10 The operational amplifier in Problem 8.7 gives a d.c. output of 0.5 V when $v_+ = 0.405$ V and $v_- = 0.406$ V, and a d.c. output of 1.25 V when $v_+ = 1.000$ V

and $v_- = 1.001$ V. Estimate (a) the offset voltage, (b) the common-mode rejection ratio.

8.11 An integration circuit like that of Fig. 8.18(a) has $C = 10$ nF and $R = 47$ kΩ. Find the peak-to-peak output voltage when the input is a square-wave at 10 kHz with amplitude ± 2 V.

9 Non-sinusoidal waveforms

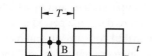

Fig. 9.1

A periodic rectangular waveform.

9.1 Periodic waveforms

9.1.1 Fourier series

In earlier chapters the analysis of circuits has been discussed for
sinusoidal waveforms (of which d.c. can be regarded as a special case)
and for transient waveforms such as arise from the sudden application of
sources or from other disturbances. There are however, many
waveforms common in engineering which are not sinusoidal but which
do not conveniently fit into the class of transients. The rectangular wave
shown in Fig. 9.1 is an example: while it would be possible to represent
it as an infinite series of positive and negative step functions this seems
neither very convenient nor appropriate for a waveform which is
periodic and regular, and therefore seems to describe a steady state in
some sense of that term.

It is a fortunate fact that a periodic waveform, of whatever kind (a
few formal exceptions need not concern us), can be represented
mathematically by a series of sinusoids. We shall also see later that our
good fortune goes further, for even non-periodic waveforms can be
represented in a comparable way. A function may of course be periodic
in any variable, but here, as in Section 3.2, we consider only functions of
time t. Formally, a function $f(t)$ is defined as periodic if there is a
positive real quantity T such that for all t

$$f(t) = f(t+T) = f(t-T). \tag{9.1}$$

The constant T is the period, the interval of time after which the
waveform repeats itself. The reciprocal of T is the **fundamental
frequency**, which we denote f_1, in hertz, corresponding to an angular
frequency ω_1 given by $2\pi/T$ in radians per second; this too is called
simply the fundamental frequency unless there is risk of confusion. Any

function $f(t)$ which satisfies eqn 9.1 can be written as the **Fourier series**

$$f(t) = \sum_{n=0}^{\infty} (a_n \cos n\omega_1 t + b_n \sin n\omega_1 t) \quad (n = 0, 1, 2, ...) \quad \textbf{(9.2)}$$

in which

$$a_0 = \frac{1}{T} \int_t^{t+T} f(u)\,du,$$

$$a_n = \frac{2}{T} \int_t^{t+T} f(u) \cos n\omega_1 u\,du, \qquad (n = 1, 2, 3, ...)$$

$$b_n = \frac{2}{T} \int_t^{t+T} f(u) \sin n\omega_1 u\,du.$$

The series thus comprises the following components: a constant term of value a_0 which in an electrical context is often called the 'd.c.' term, and is simply the average value of the waveform over a whole number of periods or over a long interval; two terms of amplitudes a_1 and b_1 which are sinusoids at the frequency ω_1, have therefore the same period as the waveform itself, and represent its **fundamental** component; and pairs of sinusoids at frequencies which are integer multiples of the fundamental and are known as **harmonics**. The terms in $2\omega_1$ are said to comprise the second harmonic, those in $3\omega_1$ the third harmonic, and so on. The two sinusoids at each frequency, a sine and a cosine, can of course be combined into a single term with some phase angle dependent on the values of their amplitudes a and b: writing the terms in pairs in this way is simply one way of specifying the relative phase of each frequency component, including the fundamental. Thus the nth harmonic, which comprises the two terms

$$a_n \cos n\omega_1 t + b_n \sin n\omega_1 t$$

can equally well be written

$$\sqrt{a_n^2 + b_n^2} \cos(n\omega_1 t - \phi) \tag{9.3}$$

where

$$\phi = \tan^{-1}(b_n/a_n).$$

Evidently the amplitude of the nth harmonic is $\sqrt{a_n^2 + b_n^2}$. It turns out that for many waveforms this amplitude decreases with n (although not necessarily monotonically), so higher order harmonics contribute less to the waveform $f(t)$; it follows that often a given waveform can be represented to a good degree of accuracy by a finite number of terms, and for practical purposes this number may be quite small. The fundamental must always exist, although either a_1 or b_1 may be zero, according to the choice of origin for $f(t)$; but the d.c. component and some harmonics may be zero in a given case. A pure sinusoid, such as we have considered in previous chapters, consists of a fundamental only, having neither d.c. nor harmonic components.

It is sometimes possible to deduce information about the components of a waveform by inspection. For example, the waveform of Fig. 9.1 has obviously a zero average, hence $a_0 = 0$ and there is no d.c. component. If the origin, where $t = 0$, is chosen to be the point A then the waveform has the property

$$f(t) = f(-t) \tag{9.4}$$

for any t and it is then said to be an **even** function; any such function which is also periodic has a Fourier series containing only cosine terms, and so $b_n = 0$ for all n. Similarly, the choice of point B for origin in Fig. 9.1 means that

$$f(t) = -f(-t) \tag{9.5}$$

and such a function is said to be **odd**. An odd periodic function has a Fourier series which contains only sine terms, and $a_n = 0$ for all n. It follows that there may be considerable advantage to be had from a particular choice of origin in calculating the series for a given waveform. However, the following points should be noted:

(1) a function which can be made even or odd by certain choices of origin will have neither property for other choices, such as points between A and B in Fig. 9.1;

(2) some functions can be made neither even nor odd; others can be made one of these but not the other;

(3) no function with a d.c. component can be made odd.

Although the choice of origin will always affect the relative importance of the a and b coefficients, it cannot affect the relative amplitudes, as given by eqn 9.3 for example, of the fundamental, d.c. and harmonic components, nor their phases relative to each other; these must of course be characteristic of the waveform whatever the choice of origin.

In practice, engineers do not very often need to evaluate the integrals given in eqn 9.2, since the coefficients of many common waveforms are readily available; a selection is given in Appendix F. Nevertheless it is instructive to apply the expressions to an actual case, as in the example which follows.

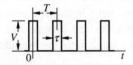

Fig. 9.2
A rectangular pulse train (Example 9.1).

Example 9.1 Find a Fourier series to represent the pulse train shown in Fig. 9.2, with pulse length τ and period T.

If the origin is chosen as shown, then the waveform is an even function and $b_n = 0$. By inspection we may also deduce that a_0, the d.c. or average value, is $V\tau/T$. In the definition of a_n, eqn 9.2, let us take $t = 0$ and note that $f(t)$ takes the following values:

$$f(t) = V \qquad (0 < t < \tau/2)$$
$$f(t) = 0 \qquad (\tau/2 < t < T - \tau/2)$$
$$f(t) = V \qquad (T - \tau/2 < t < T).$$

The coefficient a_n for $n \geq 1$ is therefore given by

$$a_n = \frac{2V}{T}\left\{\int_0^{\tau/2} \cos n\omega_1 u \, du + \int_{T-\tau/2}^{T} \cos n\omega_1 u \, du\right\}$$

$$= \frac{2V}{n\omega_1 T}\left\{\sin\frac{n\omega_1\tau}{2} + \sin n\omega_1 T - \sin n\omega_1 T \cos\frac{n\omega_1\tau}{2}\right.$$

$$\left. + \cos n\omega_1 T \sin\frac{n\omega_1\tau}{2}\right\}$$

but since by definition $\omega_1 = 2\pi/T$ the coefficient becomes

$$a_n = \frac{V}{\pi n}\left(2\sin\frac{n\omega_1\tau}{2}\right) = \frac{2V}{\pi n}\sin\frac{\pi n\tau}{T} = \frac{2V}{\pi n}\sin n\pi r$$

if we write r for τ/T, the ratio of pulse length to period. Inserting this expression, together with the value of a_0, into the general form of Fourier series in eqn 9.2 gives the required series as

$$f(t) = Vr + (2V/\pi)\{\sin \pi r \cos \omega_1 t + \tfrac{1}{2}\sin 2\pi r \cos 2\omega_1 t$$
$$+ \tfrac{1}{3}\sin 3\pi r \cos 3\omega_1 t + \tfrac{1}{4}\sin 4\pi r \cos 4\omega_1 t + \ldots\}$$

and we deduce that the train of rectangular pulses of height V contains, apart from its d.c. component, a fundamental of amplitude $(2V/\pi)\sin \pi r$ and a complete series of harmonics. Had we chosen a different origin we should have found a series which looked different but nevertheless had these same properties.

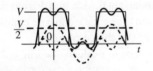

Fig. 9.3

An approximation to a pulse train: d.c., fundamental, and third harmonic only.

In passing, it might be noted that for given r the coefficients a_n in this example vary in proportion to $(\sin x)/x$, where x is here $n\pi r$; this is an important function, to which we shall return. It is also interesting to observe how well only a few terms can represent the waveform. Figure 9.3 illustrates the case $\tau = T/2$ or $r = \tfrac{1}{2}$ in the above example, for which the even-numbered harmonics vanish; only the d.c., fundamental and third-harmonic components have been superimposed in the figure, with the correct relative amplitudes and phases, and the result can be seen to be a passable approximation to the waveform specified.

Periodic functions containing no even-numbered harmonics, nor a d.c. component, form a class said to have **half-wave symmetry**: formally, they have the property that

(a)

(b)

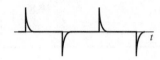

(c)

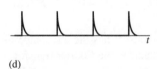

(d)

Fig. 9.4
Symmetry and asymmetry in
periodic waveforms.

$$f(t) = -f(t + T/2) \tag{9.6}$$

which means that a period can be divided into two consecutive halves of which one is the reflection of the other in the time axis. This property is independent of the choice of origin, and of whether the function is odd or even or neither; any waveform which has it contains neither a d.c. component nor even-numbered harmonics. Many waveforms with a high degree of symmetry, including the rectangular wave of Fig. 9.1, have half-wave symmetry and can be odd or even, but less symmetrical waveforms are not uncommon: Fig. 9.4(a) shows a function which can be even but not odd and (b) another which can be odd but not even, neither having half-wave symmetry; (c) shows a function which does have half-wave symmetry but can be neither even nor odd; and the waveform shown in (d) has no symmetry and can be neither even nor odd.

Because the application of Fourier series is not in general confined, as it is here, to functions of time, it is quite common to find the angular variable $\omega_1 t$ expressed simply as θ and the series accordingly written as

$$f(\theta) = \sum_{n=0}^{\infty} (a_n \cos n\theta + b_n \sin n\theta) \qquad (n = 0, 1, 2, ...) \tag{9.7}$$

in which

$$a_0 = \frac{1}{2\pi} \int_{\theta}^{\theta+2\pi} f(\phi)\, d\phi$$

$$a_n = \frac{1}{\pi} \int_{\theta}^{\theta+2\pi} f(\phi) \cos n\phi\, d\phi \qquad (n = 1, 2, 3, \ldots)$$

$$b_n = \frac{1}{\pi} \int_{\theta}^{\theta+2\pi} f(\phi) \sin n\phi\, d\phi.$$

The period of the waveform, and of its fundamental, now corresponds to the angle 2π, and fractions of a period (such as the pulse width τ in Example 9.1) are similarly specified as angles. For the sake of generality this is the form used in the standard cases given in Appendix F; the time-varying form usually needed in electrical engineering is easily recovered by setting θ to $\omega_1 t$ or $2\pi t/T$, where T is the actual period.

Apart from the choice of origin and of variable, there are alternative versions of the Fourier series based on expressing the sine and cosine terms in exponential form. For the nth harmonic we may write

$$a_n \cos n\omega_1 t + b_n \sin n\omega_1 t = \tfrac{1}{2}(a_n - jb_n)e^{jn\omega_1 t} + \tfrac{1}{2}(a_n + jb_n)e^{-jn\omega_1 t}.$$

To make up the series both terms on the right-hand side have to be summed over all positive integers n. However, it happens that the second term is merely the first with the sign of n changed (the definition of b_n in eqn 9.2 shows that it changes sign with n). Hence the series can be got simply by summing the first term over all integers n, negative as well as positive. From the definitions in eqn 9.2 it can also be shown that

$$\frac{1}{T}\int_{t}^{t+T} f(u)e^{-jn\omega_1 u}du = \frac{1}{2}(a_n - jb_n) \qquad (n = \pm 1, \pm 2, \pm 3, ...)$$

$$= a_0 \qquad (n = 0)$$

and it follows that the series can be written

$$f(t) = \sum_{-\infty}^{\infty} c_n e^{jn\omega_1 t} \qquad\qquad (9.8)$$

where

$$c_n = \frac{1}{2}(a_n - jb_n) = \frac{1}{T}\int_{t}^{t+T} f(u)e^{-jn\omega_1 u}du \qquad (n = 0, \pm 1, \pm 2, ...).$$

This is the complex form[†] of the Fourier series. It has the advantage of compactness and of being more readily related to the Fourier transform (Section 9.2). The complex coefficients c_n specify the phase as well as the amplitude of each component; for an even function all the c_n are real, and for an odd function all are imaginary. It should be noted, however, that the fundamental and harmonics are each represented by two terms, one for each sign of n, whose coefficients make a conjugate pair in general and have magnitude $\frac{1}{2}\sqrt{a_n^2 + b_n^2}$. Thus the nth harmonic has amplitude $2|c_n|$ and the fundamental amplitude is $2|c_1|$, but the d.c. component is simply c_0 because there is only one term for $n = 0$. It should also be noted that eqn 9.8 is a correct representation, as it stands, of the real quantity $f(t)$: there is no question of taking the real part of the exponentials, as is done (or understood) in phasor representation.

A Fourier series can be represented diagramatically as a **line spectrum**, made up of vertical lines drawn at equal intervals on a horizontal scale of frequency, with heights proportional to the amplitudes a_0 (or c_0) and $\sqrt{a_n^2 + b_n^2}$ (or $2|c_n|$) of the various components. Figure 9.5(a) shows the line spectrum for the train of pulses analysed in Example 9.1. In that case we had $b_n = 0$, since the function is even, and the amplitudes were given by

$$a_0 = V\tau$$
$$a_n = 2V\tau \ \sin(\pi n\tau)/\pi n\tau. \qquad (n = 1, 2, 3, ...)$$

As we noted earlier, the variation with n of the amplitudes a_n in this example takes the form of the function $(\sin x)/x$. Because it arises in the spectra of rectangular waveforms, this function is important enough to consider in some detail. Since it takes the value zero at $x = \pi, 2\pi, 3\pi, ...$

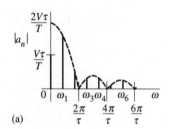

(a)

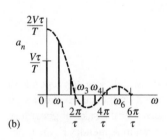

(b)

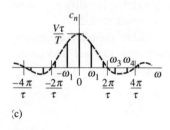

(c)

Fig. 9.5
Different forms of the line spectrum of a pulse train.

[†] For the subject matter of this chapter it is conventional not to use bold notation for complex quantities, even when they are functions of the variable $j\omega$.

it follows that the envelope of the a_n falls to zero at frequencies $n\omega_1$ such that

$$\pi nr = \pi, 2\pi, 3\pi, \ldots$$

or

$$n = \frac{1}{r}, \frac{2}{r}, \frac{3}{r}, \ldots$$

This of course cannot in general be satisfied by an integer n for any r, so the envelope zeros do not generally coincide with the harmonic frequencies. Since $\omega_1 = 2\pi/T = 2\pi r/\tau$ and therefore $n\omega_1 = 2\pi nr/\tau$, the frequencies of the zeros can be written $2\pi/\tau$, $4\pi/\tau$, $6\pi/\tau$, ... as shown in the figure. We should also note that the envelope has *twice* the amplitude a_0 at zero frequency.

Because any even function has, like this example, $b_n = 0$ (and, what is equivalent, c_n real) each component must be simply positive or negative according to the sign of a_n; that is its phase is either zero or π. The line spectra of such functions can therefore be drawn so as to indicate the sign of each component, as shown, again for the same example, in Fig. 9.5(b). Even so, it is common to show only the magnitudes, as in (a), since line spectra in general are restricted to that.

An alternative presentation of a line spectrum shows the amplitudes of the c_n including negative n (i.e. negative frequencies) as in Fig. 9.5(c) for the same example. Because, as we saw above, c_n and c_{-n} each contribute half the amplitude of the nth component, this spectrum has only half the height of that in (a) except for the d.c. component. It has the advantage of a continuous envelope with no discontinuity at $n = 0$, which gives it a plausible appearance, but the inclusion of negative frequencies may seem to separate it slightly from physical reality. The three spectra shown in Fig. 9.5 for the same waveform are, of course, equally correct so long as the information they give is properly specified. It is also possible to show a **phase spectrum** in the general case, indicating the angle of c_n, i.e. $\tan^{-1}(-b_n/a_n)$, for each n.

The series given in Appendix F, which includes most of the common periodic waveforms, are expressed only in terms of a and b coefficients as in eqn 9.2; the complex form of each can be readily deduced if need be from the relations given above.

9.1.2 Circuit analysis by Fourier series

Since the methods of a.c analysis described in Chapter 2 are available for a circuit energized by sinusoidal sources at a given frequency, they can be extended to a circuit energized by periodic sources equivalent to a series of sinusoids. So long as the circuit is, as we habitually assume, made up of linear components then superposition holds and the response to a series of sinusoids is the sum of the responses to each, found in the usual way from the appropriate steady-state system function.

Thus, for example, the steady-state current flowing in an impedance due to an applied periodic voltage expressed as a Fourier series is itself given by a series of which the nth term is

$$I_n = V_n/Z_n \qquad (9.9)$$

in which V_n is the phasor representing the nth voltage component, of frequency $n\omega_1$, and Z_n is the impedance at that frequency. The computation may be done with r.m.s. values, as is common for a.c. analysis; the r.m.s. value of the nth voltage harmonic is $1/\sqrt{2}$ of its peak value, that is

$$V_n = \sqrt{\frac{a_n^2 + b_n^2}{2}} = \sqrt{2}\,|c_n|\,. \qquad (9.10)$$

However, there is some advantage in working with peak values since the series coefficients give these directly. Similarly, while the various terms can be regarded as phasors (at the appropriate frequencies) they can equally well be retained as functions of time, that being the form in which they appear in the series. Each current component I_n has its own phase angle, which depends on the phase angle of V_n and the argument of Z_n.

Once the series for the required quantity has been found, it is not generally possible to sum it mathematically into a single function in any compact way; nor will the waveform which results from the super-position of its terms bear any necessary resemblance to that of the source which produces it, beyond the fact that both must have the same fundamental frequency. However it is always possible to calculate the r.m.s. value of a truncated series, since the mean square of the whole is the sum of the mean squares of its terms; thus the resultant r.m.s. value I of a series of r.m.s. current components I_n is given by

$$I = \sqrt{\Sigma I_n{}^2} \qquad (9.11)$$

whatever the phase angles (or signs) of the individual terms.

Example 9.2   Find the current which results when a train of voltage pulses like those in Fig. 9.2, with $V = 1$ V, $T = 1$ μs and $\tau = 0.5$ μs, is applied to an impedance consisting of a capacitance C of $(1/2\pi)$ μF in series with a resistance R of 1Ω.

Here the ratio r or τ/T is 0.5. Hence, from the Fourier series $f(t)$ deduced in Example 9.1, the voltage waveform in this case, with $V = 1$ V, is

$$v(t) = \frac{1}{2} + \frac{2}{\pi}\left(\cos\omega_1 t - \frac{1}{3}\cos 3\omega_1 t + \frac{1}{5}\cos 5\omega_1 t - \ldots\right)$$

where ω_1 is the fundamental frequency $2\pi/T$ and is in this case $2\pi \times 10^6$ rad s^{-1}. The impedance at frequency $n\omega_1$ is given by

$$Z_n = R + 1/j\omega C = 1 + 2\pi \times 10^6/jn\omega_1 = 1 - j/n$$
$$= \sqrt{1 + 1/n^2}\lfloor -\tan^{-1}(1/n)\ \Omega.$$

The impedance values in ohms corresponding to the terms in the voltage series, given by setting $n = 0, 1, 3, 5, ...$, are then

$$Z_0 \to \infty$$
$$Z_1 = \sqrt{2}\underline{|-45°}$$
$$Z_3 = \sqrt{10}/3\underline{|-18.4°}$$
$$Z_5 = \sqrt{26}/5\underline{|-11.3°}$$

and so on. These may be used with the voltage terms as expressed above, as functions of time, to deduce current terms in the same form by subtracting the argument of Z as a phase shift in each case. Thus we find the current in amperes to be

$$i(t) = \frac{2}{\pi}\left\{\frac{1}{\sqrt{2}}\cos(\omega_1 t + 45°) - \frac{1}{\sqrt{10}}\cos(3\omega_1 t + 18.4°)\right.$$
$$\left. + \frac{1}{\sqrt{26}}\cos(5\omega_1 t + 11.3°) - ...\right\}.$$

As would be expected, there is no d.c. term in this series: Z_0 is infinite because a capacitance is an open circuit for d.c. The r.m.s. value of the current due to the first three terms is readily estimated from eqn 9.11. The r.m.s. values of the individual components are

$$I_1 = 1/\pi$$
$$I_3 = 1/\pi\sqrt{5}$$
$$I_5 = 1/\pi\sqrt{13}$$

and so on; the first three give a resultant r.m.s. value

$$I = \frac{1}{\pi}\sqrt{1 + \frac{1}{5} + \frac{1}{13}} = \mathbf{0.360\ A}.$$

This may be compared with the r.m.s. value of the fundamental alone, namely $1/\pi$ or **0.318 A**.

Example 9.3 Find the error in estimating the r.m.s value of the voltage waveform in Example 9.2 from the first three terms of its Fourier series.

For the given waveform the exact r.m.s. value is easily calculated. The mean square voltage is clearly 0.5 V^2, and the r.m.s. value therefore $1/\sqrt{2}$ V. The r.m.s. values of the first three voltage terms, from the previous example, are

$$V_0 = 0.5$$
$$V_1 = \sqrt{2}/\pi$$
$$V_3 = \sqrt{2}/3\pi$$

and from eqn 9.11 the resultant r.m.s. value due to these three is

$$V = \sqrt{V_0^2 + V_1^2 + V_3^2} = \sqrt{0.25 + 2/\pi^2 + 2/9\pi^2}$$
$$= 0.689 \text{ V}.$$

The fractional error is therefore $(1/\sqrt{2} - 0.689)\sqrt{2} = \mathbf{0.025}$ or $\mathbf{2.5\%}$.

Although a non-sinusoidal waveform can, like any other quantity, be given an overall r.m.s. value, and each component of its Fourier series can, like any sinusoidal function of time, be represented as a phasor at the appropriate frequency, it is important to remember that the waveform itself *cannot* be represented by a single phasor. Phasor methods can be applied to each component of a series, but that is all. Similarly, except in the case of purely resistive circuits no single impedance or other system function for the steady state can be attributed to a circuit in respect of a non-sinusoidal waveform, for such a function in general takes a different value, and the steady-state circuit response differs accordingly, for each component frequency. We may, however, regard a circuit to which any periodic waveform is continuously applied as being in a steady state, even though driving function and response differ in waveform and neither is sinusoidal, simply because its overall behaviour is the superposition of a series of steady states. The transient behaviour of a circuit to any *change* in a periodic driving function, such as its connection or disconnection, can of course be deduced by the same general methods as were discussed in Chapter 7.

9.1.3 Power

The power dissipated in a circuit having periodic but non-sinusoidal voltages and currents can be calculated from the principles encountered in Section 3.9. It is always possible to calculate average power by averaging the instantaneous product of voltage and current, but such a cumbersome procedure is best avoided whenever possible.

In general the instantaneous power represented by a periodic current flowing between terminals having a periodic potential difference must be given by the product of two Fourier series, and the power we usually require is the average of this product over one period, or over a large interval. But it can be shown that

$$\int_t^{t+T} \cos(n\omega_1 u + \phi_n) \cos(p\omega_1 u + \phi_p)du = 0 \qquad (p \neq n)$$

where ω_1 is the frequency $2\pi/T$, n and p are integers, and ϕ_n and ϕ_p are arbitrary phase angles. This means simply that over any interval equal to one period of the fundamental the product of two terms of different frequencies always averages to zero. It follows that in multiplying a voltage series by a current series, both (as is common) having the same fundamental, only the products of *identical frequencies* contribute average power. Hence we need consider only a series of products of the form

$$V_{mn} \cos{(n\omega_1 t + \phi_{vn})} I_{mn} \cos{(n\omega_1 t + \phi_{in})}$$

in which V_{mn} and I_{mn} are the peak voltage and current of the nth harmonic (or, if $n = 1$, the fundamental) and ϕ_{vn}, ϕ_{in} are the respective phase angles. From eqn 3.55 we know immediately the average power represented by such a product: in this case it is

$$P_n = V_n I_n \cos{(\phi_{vn} - \phi_{in})}$$
$$= V_n I_n \cos{\phi_n}$$

where V_n and I_n are the r.m.s. values of the nth harmonic (as given, for example, by eqn 9.10) and ϕ_n is the phase difference between the voltage and current of that harmonic. It is of course possible for either of these to be expressed as a sine instead of a cosine as assumed above, and for either term to carry a negative sign; as in any a.c. circuit the result is still correct so long as ϕ_n is the relative phase angle between the two. (For the significance of sign in a.c. power calculation see Section 3.9.1.) The total power which results from the voltage and current series is the sum of all contributions like the P_n above; only the d.c. contribution is special, in that the r.m.s. values are simply the d.c. values and the relative phase zero. With that understanding we may write the total average power as

$$P = \sum_{n=0}^{\infty} P_n = \sum_{n=0}^{\infty} V_n I_n \cos{\phi_n}. \qquad (9.12)$$

Equation 9.12 may be used to calculate the total power to any desired approximation by truncating the series accordingly. It is not, however, the only way of expressing the power which results from non-sinusoidal waveforms of voltage and current. If we seek the average power in a specified resistance R (or conductance G) then it is only necessary to known either the r.m.s. current I through that element or the r.m.s. voltage V across it; for in a purely resistive element the average power is correctly given by eqn 3.56, that is

$$P = I^2 R = V^2/R = V^2 G = I^2/G,$$

for *any* periodic waveform, provided only that V and I are r.m.s. values (as is easily confirmed from the definitions of these values and of instantaneous power). Given the choice, it is here even more important than before to choose the simplest of these expressions to suit a particular case. It would for example be folly to undertake a tedious Fourier calculation to find all the currents in a circuit in order to evaluate I in a resistance across which the r.m.s. voltage is already known. In any case, it is worth noting that in the case of pure resistance there is no need to calculate its current from its voltage, or vice versa, term by term because there is no frequency dependence and overall r.m.s. values, like individual terms at a given frequency, are simply related by Ohm's law in the form $V = IR$.

Example 9.4 Find, by accounting for components up to and including the fifth harmonic, the average power dissipated (i) in the circuit of Example 9.2, and (ii) if the same voltage waveform were applied to the same values of R and C connected in parallel.

(i) In this case there are only the fundamental and the third and fifth harmonics to be considered, since the d.c. current is zero and there are no even harmonics in the applied voltage. If we are to use eqn 9.12, we need first the appropriate r.m.s. voltage and current values. The required r.m.s. voltages, from Example 9.2, are

$$V_1 = \sqrt{2}/\pi, \ V_3 = \sqrt{2}/3\pi, \text{ and } V_5 = \sqrt{2}/5\pi \text{ V}$$

(V_0 is not required since there is no d.c. current), and the corresponding currents are

$$I_1 = 1/\pi, \ I_3 = 1/\pi\sqrt{5}, \text{ and } I_5 = 1/\pi\sqrt{13} \text{ A}.$$

The phase angles of Z_1, Z_3, and Z_5, from the same example, have tangents given by $1/n$ and hence their cosines are $n/\sqrt{1+n^2}$ or $1/\sqrt{2}$, $3/\sqrt{10}$, and $5/\sqrt{26}$ respectively. The power, by eqn 9.12, is then

$$P = P_1 + P_3 + P_5 = (1/\pi^2)(1 + 1/5 + 1/13) = \mathbf{0.129 \ W}.$$

A simpler way to calculate the power in this case is to use the fact that dissipation occurs only in the 1 Ω resistance, the r.m.s. current in which has already been found in Example 9.2 to be approximately 0.360 A on the basis of the same three components. Hence the power is simply

$$P = I^2 R = 0.360^2 = \mathbf{0.130 \ W}$$

in close agreement with the first result (the difference in the last digit is caused only by rounding errors; both calculations should give exactly the same result since both are based on the same truncation of the series).

(ii) Here the same voltage waveform as before is applied to a different impedance and the currents are therefore also different. There is, however, no need to calculate any current, not even that in the resistance, because the whole of the applied voltage now appears across the latter. The r.m.s. value of the voltage has already been calculated in Example 9.3, but as it happens neither of the figures found there is quite what is required here, namely an answer based on truncation after the fifth harmonic. Here the r.m.s. components to be included are

$$V_0 = 0.5, \ V_1 = \sqrt{2}/\pi, \ V_3 = \sqrt{2}/3\pi, \ \text{and} \ V_5 = \sqrt{2}/5\pi \ \text{V},$$

and these give an overall r.m.s. voltage

$$V = \sqrt{1/4 + 2/\pi^2 + 2/9\pi^2 + 2/25\pi^2} = 0.695 \ \text{V}$$

(an answer which, as should be expected, lies between the two previously obtained in Example 9.3). The power is then simply

$$V^2/R = 0.695^2 = \mathbf{0.483 \ W}.$$

The reactive volt-amperes which result from periodic non-sinusoidal waveforms can be found in the same way as power, according to the definitions of Section 3.9.2, but in practice such a calculation is rarely called for. It would also be possible to define the product of the resultant r.m.s. values of voltage and current as 'volt-amperes' in the sense used in Section 3.9.3, but the product would have little significance for non-sinusoidal waveforms since neither quantity in general corresponds to a single phasor, nor are they related by the unique phase angle which in the sinusoidal case allows the product to be resolved into power and reactive components.

Finally, we should note that non-sinusoidal waveforms arise from sinusoidal sources when these are applied to *non-linear* circuits: a familiar example is that shown in Fig. 9.4(a), which can represent the output voltage of a simple full-wave rectifier with a sinusoidal input.

9.1.4 Harmonics in polyphase circuits

Polyphase circuits are for the most part used for electrical power systems, and for that reason are designed to generate and transmit voltages and currents as nearly sinusoidal as possible. In practice there is some departure from this ideal, caused in large part by the ferromagnetic materials which are essential to all large electrical machines and in which the dependence of magnetic flux density on field strength is highly non-linear (Appendix B.3), especially at the relatively high densities required for economic design; a second contributing factor is the incidence of loads which are non-linear by nature, for example rectifying and frequency-changing equipment. As a result, nominally sinusoidal voltage and current waveforms at various points in a power system may have an appreciable content of harmonics. The arrangements used in polyphase systems, described in Chapter 6, turn out to be highly significant in respect of these harmonics.

Consider an n-phase set of balanced voltages. The fundamental components of the phase voltages are separated from each other in time by $1/n$ th part of a period, which is a whole period of the n th harmonic and m whole periods of the mn th harmonic where m is another integer; hence, provided that the harmonics bear identical relationships to the fundamental in all the phases (highly likely in a balanced system), all the

n th harmonics must be in exact time phase with each other, and likewise for every integer multiple of n.

At this point it is convenient to restrict our attention to the three-phase case, by far the most common in practice, and therefore to third, or more generally $3m$ th, harmonics. Suppose now that a balanced set of three-phase voltages is connected in star. Any one line voltage is given by the difference of two phase voltages (Section 6.2), a fact which follows only from Kirchhoff's laws and is therefore instantaneously true for any waveform. But the difference of two sinusoids of equal magnitudes in phase with each other is zero, so harmonics of order $3m$ which may exist in the phase voltages cannot appear in the line voltages. Since it is the actual phases of three-phase generators which form the voltage sources in a power system, it follows that the star connection removes from the line voltages all such generated harmonics.

Now suppose a similar balanced set of voltage sources is connected in delta. Each $3m$ harmonic now constitutes three sources, identical in magnitude and phase, connected in a closed loop or, in other words, short-circuited. A corresponding harmonic current must flow, in practice limited by the internal impedances of the sources; since the voltage drop in each phase must balance its contribution to the loop e.m.f. it follows once more that no $3m$ harmonic voltage can appear between lines. We may therefore conclude that whatever the form of the voltage sources in a three-phase system, $3m$ harmonics *cannot occur in the line voltages* so long as the phases in any set have identical waveforms with their fundamental components balanced.

Similar arguments apply to currents. The line currents to or from a delta-connected device are each given by the difference of two phase currents, so any $3m$ harmonics flowing in the phases of a balanced delta cannot appear in the lines. For a star connection with an unconnected star point the phase currents must sum to zero; but since $3m$ harmonics of a balanced set would all be in phase with each other, each must be zero. Hence, apart from unbalanced conditions which we have not considered, harmonic line currents of order $3m$ can flow *only in a four-wire system*, that is in a star with connected neutral (Section 6.4).

The behaviour of $3m$ harmonics in balanced three-phase systems should not be confused with the zero-sequence component (Section 6.6) which can arise in unbalanced conditions but which represents sinusoids at the fundamental frequency and has nothing to do with non-sinusoidal waveforms. Both represent component sets of currents and voltages which are cophasal, and have properties which are attendant on that; but they are otherwise quite different.

9.2 Non-periodic waveforms

In Chapter 7 we met examples of non-periodic functions (in particular impulse, step, and ramp functions) in connection with transients. Many non-periodic functions occur commonly, and not merely as a passing disturbance, in certain areas of engineering: for example the single rectangular pulse shown in Fig. 9.6 is an ideal representation of the pulses which occur widely in digital circuits of all kinds, not in general at regular intervals. The response of a circuit to such a pulse could evidently be found by the transient methods discussed in Chapter 7, by treating it as a superposition of steps or perhaps as an approximation to an impulse; and it appears at first sight that this may be the only possible approach. In fact, it happens that many non-periodic waveforms (exceptions are mentioned in Section 9.3.1) can be analysed in a way comparable to the use of Fourier series in the last section. As we shall see, and might expect, the method is closely related to the Laplace transformation which we have used for transients; but it has the convenience of a steady-state approach. It is based on the **Fourier transformation**, and it is most easily introduced by way of the Fourier series.

9.2.1 The Fourier transformation

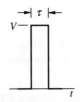

Fig. 9.6
A single rectangular pulse.

The pulse of Fig. 9.6 may in principle be derived from a periodic succession of such pulses by allowing the period T, between the start of one pulse and the start of the next, to become infinite. In that case, whatever the shape of the waveform, two things follow: first, because the harmonic frequencies of the Fourier series are separated by intervals equal to the fundamental frequency ω_1, which is defined as $2\pi/T$, they now become infinitesimally close, so that all frequencies are represented; secondly, it is clear from the definitions of the a, b, and c coefficients in eqns 9.2 and 9.8 that for any finite $f(t)$ the amplitudes of the harmonic, fundamental, and d.c. components alike must now tend to zero. The idea of dividing the waveform into an infinite number of infinitesimally close and vanishingly small components seems to offer little practical help, but it happens that we can extract much more useful information by concentrating on the integral defined for the complex c coefficient. Let us rewrite the definition of c_n in eqn 9.8 as

$$c_n T = \int_{-T/2}^{T/2} f(t)e^{-jn\omega_1 t}dt.$$

Now let $T \to \infty$; this also means that the fundamental frequency ω_1 is infinitesimally small, but $n\omega_1$ still represents any harmonic frequency since n can be as large as we please. So let us replace $n\omega_1$ by a general frequency ω, so that $c_n T$ becomes a function of this frequency; we write it as $F(j\omega)$, so that we have

$$F(j\omega) = \int_{-\infty}^{\infty} f(t)e^{-j\omega t}dt \tag{9.13}$$

which is called the **Fourier transform** of $f(t)$. It is also sometimes designated simply $F(\omega)$; and it may also be defined in terms of the cyclic frequency $f = \omega/2\pi$ (in which case, to avoid confusion, different symbols are used for the functions F and f). We shall have occasion to use both forms, according to convenience. It is worth noting immediately that whatever the physical dimensions to be attributed to $f(t)$, those of $F(j\omega)$ include in addition the dimension of time (or whatever other variable might replace t in different contexts).

Equation 9.13 can in turn be used to rewrite the definition of $f(t)$ given in eqn 9.8. The latter can be written

$$f(t) = \sum_{-\infty}^{\infty} T c_n e^{jn\omega_1 t} \frac{1}{T} \tag{9.14}$$

and in this, for $T \to \infty$, we may again set $n\omega_1$ to ω; also, because the frequency interval between harmonics, ω_1 or $2\pi/T$, is now infinitesimal we may write it as $d\omega$, so that

$$1/T = d\omega/2\pi.$$

In the limit of small $d\omega$ the summation in eqn. 9.14 becomes an integral which, with $F(j\omega)$ defined by eqn. 9.13, can now be written

$$f(t) = \frac{1}{2\pi} \int_{-\infty}^{\infty} F(j\omega) e^{j\omega t} \, d\omega \tag{9.15}$$

and is the **inverse Fourier transform** of $F(j\omega)$.

A function $f(t)$ and its Fourier transform $F(j\omega)$ are known as a Fourier **transform pair**. The function $F(j\omega)$ may be regarded as the equivalent of the Fourier series in the limit of infinite T, that is for a non-periodic waveform. The symmetry between the transform, eqn 9.13, and its inverse, eqn 9.15, is obvious. (It is possible to adjust the definitions so as to move the factor 2π from eqn 9.15 to eqn 9.13 or to improve the symmetry by writing both in terms of cyclic frequency f.) It is also clear that there is a resemblance between the Fourier transform and the Laplace transform, eqn 7.22. But the evident differences should also be noted: first, the variable in F is now $j\omega$, so there is no question of a complex frequency such as was represented by the variable s, although F itself is in general complex; second, the integration in eqn 9.15 is done over all t, negative as well as positive. We return in Section 9.2.3 to a comparison of the two transforms. As for the Laplace transform, there is seldom need in engineering practice to integrate eqns 9.13 or 9.15, for tables of common Fourier transform pairs are readily available; a selection is given in Appendix G.

The transform $F(j\omega)$ given by eqn 9.13 can be regarded as the **frequency spectrum** (or spectral density) of $f(t)$; it is in general a continuous function and for a non-periodic waveform it replaces the line spectrum represented by the Fourier series. The range of integration in eqn 9.15 shows that the spectrum must include negative frequency as well as positive, and it may therefore be most easily compared with the complex form of the series: the complex value of F at a given ω corresponds to the value of the complex coefficient c for a given harmonic (but with an essential difference, to which we return below). Just as even functions gave series containing only cosines (and real c_n) and odd functions gave only sines (and imaginary c_n), so it can be shown that an even $f(t)$ has real $F(j\omega)$, while odd $f(t)$ has imaginary $F(j\omega)$. It is also readily shown that the real part of $F(j\omega)$ is always an even function of ω and the imaginary part always an odd function; from this it follows that $F(-j\omega)$ is the complex conjugate of $F(j\omega)$, and that the magnitude of F is always even while the phase is always odd. Graphically, any spectrum $F(j\omega)$ can be completely represented by separate diagrams of magnitude and phase. In this respect it resembles Bode diagrams, but it must be remembered that the latter gives the properties of a circuit or system while a spectrum, strictly, gives those of a waveform. (There is an analogous distinction in the case of the Laplace transformation, between the transform of a function of time such as voltage, $V(s)$, and a generalized system function such as impedance, $Z(s)$.)

The rectangular pulse of Fig. 9.6 is a particularly important case. It is an even function of time if the origin is chosen as shown in Fig. 9.7(a), and its transform $F(j\omega)$ is then real and given by

$$F(j\omega) = V\tau\frac{\sin(\omega\tau/2)}{\omega\tau/2} \tag{9.16}$$

which is the spectrum shown in Fig. 9.7(b). It has the form $(\sin x)/x$, which we have already encountered, in Section 9.1.1, as the form of the envelope of the Fourier series for a periodic train of pulses like the single pulse now considered.

Equation 9.16 actually gives the envelope shown in Fig. 9.5(c) multiplied by the factor T. Unlike the lines of that spectrum, therefore, $F(j\omega)$ does *not* itself give the amplitude of a component of $f(t)$ at each frequency; indeed it was precisely because the component amplitudes tend to zero when $T \to \infty$ that $F(j\omega)$ was defined above as c_nT. This can be written in the non-periodic limit as $2\pi c_n/\mathrm{d}\omega$; hence, as that limit is approached, we may write

$$c_n = (1/2\pi)\,F(j\omega)\,\mathrm{d}\omega \tag{9.17}$$

which is, as we expect, a small quantity going to zero with $\mathrm{d}\omega$. The significance of eqn 9.17 is to show that $f(t)$ 'contains' a component c_n at the frequency ω, of amount proportional to the small range of frequency considered; thus in Fig. 9.7(b) the 'content' of the pulse $f(t)$ at ω can be represented by the small area $F(j\omega)\,\mathrm{d}\omega$ under the curve at that frequency. In this respect the spectrum $F(j\omega)$ can be seen to be a kind of distribution function: in statistics, the distribution of some attribute, h say, among the members of a population can be represented by a continuous function $F(h)$ such that the number of members for which the value of h lies between h and $h + \mathrm{d}h$ is $F(h)\,\mathrm{d}h$. In any situation where h has a continuous spread of values a curve showing $F(h)$ is often

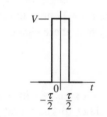

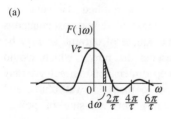

(a)

(b)

Fig. 9.7

A rectangular pulse and its spectrum.

preferable to a so-called bar chart which shows the actual numbers within chosen intervals of h. The latter is equivalent to giving the coefficients c of a line spectrum, the former to the function $F(j\omega)$. In our case the line spectrum is perfectly correct for periodic functions, in which only discrete frequencies are present, but the continuous spectrum is better for non-periodic functions.

The idea that a certain proportion of a voltage waveform, say, has frequencies lying within a range ω to $\omega + d\omega$ is perhaps not so easily appreciated as the distribution of some physical attribute among the members of a population, especially since in the former case $F(j\omega)$ has the apparently strange dimensions of voltage per unit frequency. Some physical justification can be given on the grounds that harmonics which are very close in frequency can be directly summed since they must also be close in phase; in a range $d\omega$ there will be $d\omega T/2\pi$ of them with a total amplitude $c_n d\omega T/2\pi$ which we may interpret as an amplitude $c_n T/2\pi$, that is $F(j\omega)/2\pi$, per unit interval of angular frequency (or, what is equivalent, $c_n T$ or $F(j\omega)$ per unit of cyclic frequency). However, as it happens, it is much more usual to consider the spectrum $F(j\omega)$ in terms of *energy* content, rather than voltage or current, and we proceed to this in the next section.

It may seem strange at first sight that the spectrum of a given pulse should depend on the choice of origin for time. In fact the *shape* of the spectrum is independent of this choice: only the angle of $F(j\omega)$ is affected. As we shall see later, the delay theorem for Fourier transforms means that a shift T in time corresponds simply to a phase change ωT in the spectrum.

9.2.2 Parseval's theorem

It is not strictly possible to talk of the power or energy in a single waveform of voltage or of current; we need either to know both or to know the circuit to which one or other is applied. However by convention it is usual in the present context to define instantaneous power as the square of voltage or as the square of current, as may be appropriate, which in each case represents the power which would be dissipated in unit resistance. (To keep dimensions correct we may regard the square of voltage as watts per siemens, or watt-ohms; and the square of current as watts per ohm.) This measure of power, which allows certain general results to be expressed rather simply, is also quite convenient in those areas of electrical engineering, such as communications, where matching is common practice (Section 3.11.4) and the same value of resistance may therefore appear at several points in a system. Energy, being the time integral of power, can be defined in an analogous way. Thus we may define the total energy in a non-periodic voltage $v(t)$, say, as

$$E = \int_{-\infty}^{\infty} v^2(t) \, dt. \qquad (9.18)$$

The integrand in this can be written as

$$v^2(t) = v(t)\frac{1}{2\pi}\int_{-\infty}^{\infty} V(j\omega)\,e^{j\omega t}d\omega.$$

according to the inverse transform, eqn 9.15, if $V(j\omega)$ signifies the Fourier transform of $v(t)$. It can then be shown, by changing the order of integration and using the Fourier transformation itself, eqn 9.13, that eqn 9.18 may also be written as

$$\int_{-\infty}^{\infty} v^2(t)\,dt = E = \frac{1}{2\pi}\int_{-\infty}^{\infty} \mid V(j\omega)\mid^2 d\omega. \tag{9.19}$$

Equation 9.19 is usually called **Parseval's theorem** for non-periodic $v(t)$. It asserts, in essence, that the energy of a non-periodic signal can be regarded not only as the sum of successive contributions in time but equally well as the sum of contributions for intervals of frequency, i.e. across the spectrum. The quantity $\mid V(j\omega)\mid^2$ is called the **energy density** per unit bandwidth; it can be shown graphically, like the spectrum $\mid V(j\omega)\mid$, as a symmetrical curve, that is an even function of ω. It may be noted from eqn 9.19 that, since $d\omega/2\pi$ is an increment of the cyclic frequency f, $\mid V\mid^2$ must have the dimensions joule-ohms (or V^2s) per hertz. To find E the density must be integrated over all f, negative as well as positive; alternatively we may simply take twice the integral over all positive f. The energy contained in any given bandwidth may also be obtained by integrating between the appropriate limits.

Parseval's theorem in the form of eqn 9.19 is sometimes attributed to Rayleigh, who deduced it from a corresponding (and earlier) form due to Parseval for periodic $v(t)$. In the latter case the theorem must be expressed in terms of average power rather than energy (which would of course be infinite for a periodic waveform); and it merely indicates that the average power is the sum of the contributions due to the various components of its (line) spectrum, i.e. of the powers due to the various terms in its Fourier series, a result already expressed by eqn 9.12 in Section 9.13.

Example 9.5 Find the Fourier transform of the voltage pulse described by

$$v(t) = 0 \qquad\qquad (t < 0)$$
$$v(t) = 2e^{-1000t}\ \text{V} \qquad (t > 0)$$

and the proportion of its energy contained in the frequency range 1–2 kHz when t is time in seconds.

The Fourier transform is, by eqn 9.13,

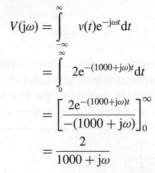

$$V(j\omega) = \int_{-\infty}^{\infty} v(t)e^{-j\omega t}dt$$

$$= \int_{0}^{\infty} 2e^{-(1000+j\omega)t}dt$$

$$= \left[\frac{2e^{-(1000+j\omega)t}}{-(1000+j\omega)} \right]_{0}^{\infty}$$

$$= \frac{2}{1000+j\omega}$$

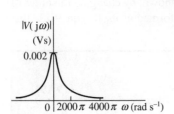

Fig. 9.8

The spectrum of an exponential pulse (Example 9.5).

and this gives the spectrum shown in Fig. 9.8. The energy density is

$$| V(j\omega) |^2 = \frac{4}{10^6 + \omega^2} \ \mathrm{V^2 s^2}$$

and the energy content between 1 and 2 kHz can be found from eqn 9.19 by changing the limits of ω to 2000π and 4000π and doubling the integral (to include the same range of negative frequencies); this gives an energy

$$E = 2 \times \frac{1}{2\pi} \int_{2000\pi}^{4000\pi} \frac{4}{10^6 + \omega^2} d\omega$$

$$= \frac{4}{1000\pi} \left[\tan^{-1} \frac{\omega}{1000} \right]_{2000\pi}^{4000\pi}$$

$$= \frac{4}{1000\pi} (\tan^{-1} 4\pi - \tan^{-1} 2\pi)$$

$$= 9.985 \times 10^{-5} \ \mathrm{V^2 s}.$$

The same answer would of course be found by evaluating the integral in terms of the cyclic frequency f instead of ω.

The total energy in the pulse may be found either by integrating the same energy density over all ω (or f) or by integrating $v^2(t)$ in time. The latter gives

$$E_t = \int_{0}^{\infty} 4e^{-2000t}dt$$

$$= -\frac{1}{500} \left[e^{-2000t} \right]_{0}^{\infty}$$

$$= 2 \times 10^{-3} \ \mathrm{V^2 s}.$$

The proportion of the total energy contained in the range 1–2 kHz is therefore

$$E/E_t = 9.985 \times 10^{-5}/2 \times 10^{-3} = \mathbf{0.0499},$$

or about **5%**.

9.3 Fourier transforms in circuit analysis

9.3.1 General

We saw in Section 9.1.2 that the response of a linear circuit to any periodic waveform could be readily found by applying steady-state principles to each component frequency: each term in the series for an input waveform may be multiplied by the value of a system function, say $H(j\omega)$ in general, at the corresponding frequency, and the results superimposed. In a logical extension to this idea, we may deduce that the response to a non-periodic waveform is obtained by multiplying $H(j\omega)$ by the spectrum of the input, say $X(j\omega)$, to obtain the spectrum of the output, say $Y(j\omega)$; taking the inverse Fourier transform of Y is then equivalent to superimposing components at different frequencies. Thus we may write the output transform as

$$Y(j\omega) = H(j\omega)X(j\omega) \tag{9.20}$$

and invert it to obtain the response $y(t)$ due to the input $x(t)$. Our deduction of this by analogy with the Fourier series for periodic $x(t)$ is by no means rigorous, if only because (as was pointed out earlier) $X(j\omega)$, unlike the terms of a Fourier series, does not represent an input amplitude. Nevertheless, eqn 9.20 is correct; its formal justification is based on the convolution integral (Section 7.4.6), as is likewise the corresponding relationship between Laplace transforms in terms of a generalized function $H(s)$ (Section 7.4.2).

It appears, then, that the similarity between the Laplace and Fourier transformations, pointed out in Section 9.2.1, extends to the use to which we may put them, and the question arises whether there is any need to consider both. The fact that the Laplace transformation was defined as an integral over positive t only, and the Fourier over all t, is not fundamentally important: we use the 'one-sided' Laplace transform simply because for engineering purposes it is nearly always applied to functions which are non-zero only for $t \geq 0$. The restriction to $j\omega$ in the Fourier case, as against the complex variable s in the Laplace, is more significant. Since it appears that the former can solve transient problems (because it can deal with most waveforms) in terms of the steady-state function $H(j\omega)$, it may seem surprising that we troubled even to introduce the Laplace transformation with its complex variable and the consequential implication of complex frequency. In fact however there are some advantages to the Laplace form. One is the mathematical fact that its transformation integral, with the complex variable s, exists for more functions than the Fourier: the latter does not converge for some functions of practical interest; for example the Fourier transform of a step function cannot be found directly, but only as a limiting case of the transform for an exponential decay (like that in Example 9.5). Another advantage of the Laplace form is the relative ease of dealing with initial conditions (Section 7.4).

Fourier transforms are most useful in dealing with input and output waveforms where there is advantage in spectral representation, including pulses and waveforms which arise from combinations of sinusoidal signals, and with systems whose bandlimited behaviour (Section 9.3.3) can be represented by functions $H(j\omega)$; all of these are common in communications engineering. In such contexts engineers are often concerned much more with frequency spectra of input and output than with waveforms as functions of time; and spectra, unlike functions of the variable s, are easily represented graphically. Also, although we introduced Fourier transforms in order to deal with non-periodic waveforms, it is also quite straightforward to use them for periodic functions if need be. As we know from Fourier series, such functions have discontinuous line spectra; but as transforms these are easily represented by delta (or impulse) functions. Here it should be recalled that these functions have infinite height and a magnitude defined as an area: strictly, therefore, a harmonic component of amplitude c say should be represented by the impulse $c\delta(t)$, not by a line of height c as in Fig. 9.5(a), although in practice we often take height to be a useful measure of magnitude (Section 7.4.3). The impulse corresponds to the elementary area $F(j\omega)d\omega$ for a continuous spectrum.

It should, finally, be pointed out here that as may be expected there are theorems for the Fourier transformation corresponding to those for the Laplace, which enable us to deal with delay, differentiation and integration when required. Some of them are given in Appendix G.

9.3.2 Duality and convolution

Equations 9.13 and 9.15 show that the Fourier transform and its inverse have **dual** forms; that is they have identical form, save for the sign of ω and the factor 2π. The latter can be removed by using the cyclic frequency f instead of ω and the consequence of this duality can then be most compactly expressed in the following statement: if $x(t)$ and $X(f)$ are a Fourier transform pair, so too are $X(t)$ and $x(f)$. If $X(f)$ happens not to be real, then neither is $X(t)$, which appears to cast doubt on the physical relevance of the statement; but in many cases of interest $x(t)$, which we always take to be real, can be chosen to be even so that X is then both real and even and $x(f) = x(-f)$. In plain terms, duality therefore means that if a waveform of shape A has a spectrum of shape B, then a waveform of shape B will have a spectrum of shape A.

One or two examples of this duality are particularly important. We have seen that a rectangular pulse as in Fig. 9.7(a) has the spectrum of form $(\sin x)/x$, shown in (b) and usually referred to as the **sinc function**.[†] It follows that a pulse having a sinc-function waveform has a perfect rectangular spectrum, as indicated in Fig. 9.9. Although such a pulse seems highly artificial, it is nonetheless a very useful idea (and for that matter a rectangular pulse is itself an idealized version of what is practically attainable). For example, the sinc pulse is the only waveform we have encountered, apart from a pure sinusoid, which does not contain an infinite range of frequency and which would therefore

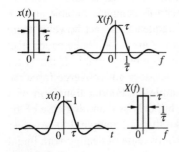

Fig. 9.9
Pulses and their spectra: an example of duality.

[†] The function sinc x is defined as $(\sin \pi x)/\pi x$. Sometimes $(\sin x)/x$ is called the **sampling function** and written Sa(x).

pass undistorted through an ideal filter of appropriate bandwidth (Section 5.4.1). We return to this point in the next section.

There is an important special case of this example of duality. By taking the rectangular pulse to the limit $\tau = 0$, it is easily confirmed that the Fourier transform of a unit impulse or delta function, at $t = 0$, is unity (as is also its Laplace transform); thus an impulse contains all frequencies equally, as shown in Fig. 9.10(a). It also follows, from duality, that the Fourier transform of a constant is an impulse at zero frequency, a result to be expected since a constant can be regarded as a waveform of zero frequency.

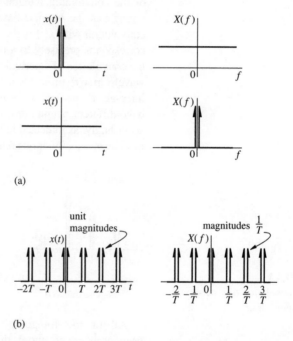

Fig. 9.10

Duality: (a) the spectra of an impulse and a constant; (b) impulse trains.

Another special case is a periodic sequence of equal impulses at equal intervals of time T, shown in Fig. 9.10(b). For this the transform can be shown to be an equispaced line spectrum of constant amplitude, that is another series of equal impulses, at frequency intervals $1/T$. (This can be confirmed by considering the spectrum of a train of rectangular pulses, Fig. 9.5(a), in the limit $\tau = 0$, $V \to \infty$.) In this case the function and its transform not only show the duality property but also have exactly the same form.

Duality can be seen not only in the forms of a given transform pair, but also in the basic properties of the transformation. For example the time-shift property (which is defined in Appendix G and corresponds to that for the Laplace transformation) means that a waveform delayed by time T, say, has its Fourier transform multiplied by the factor $e^{-j\omega T}$, representing a spectrum unchanged in amplitude but shifted in phase in proportion to frequency; duality then means that shifting a frequency spectrum by an amount Ω, say, corresponds to multiplication of the

waveform by the factor $e^{j\Omega t}$, a result to which we shall return below
when considering modulation.

Convolution has the same duality for Fourier transforms as for
Laplace: the transform of the product of two waveforms is the
convolution of their spectra, and the inverse of the product of two
spectra is the convolution of the corresponding waveforms. Certain
instances of convolution are particularly important in the application of
Fourier transforms. The most basic is the result that the convolution of a
single unit impulse at $t = T$ with a waveform $x(t)$ is the same waveform
delayed by T, that is $x(t - T)$. This may be confirmed from the definition
of the convolution integral (Section 7.4.6). It may also be deduced by
noting that the impulse has the transform $e^{-j\omega T}$, so the transform of its
convolution with $x(t)$ is the spectrum $X(f)$ multiplied by that factor; this
corresponds precisely to a time shift T, as we saw above. By extension,
the convolution of a periodic train of unit impulses, at intervals T, with the
waveform $x(t)$ produces a sequence of the same waveform repeated at
intervals T, that is $\Sigma x(t - nT)$ in which n is an integer. If the impulses
have different magnitudes, then the waveform amplitudes are multiplied
accordingly, as indicated in Fig. 9.11. These examples of convolution are,
of course, equally applicable in the frequency domain.

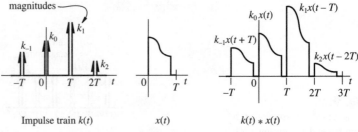

Fig. 9.11
The convolution of a pulse with
an impulse train.

Impulse train $k(t)$ $x(t)$ $k(t) * x(t)$

All of the foregoing indicates that in the context of Fourier
transforms, even more than Laplace, we may transfer to and fro
between and frequency according to mathematical and/or conceptual
convenience; while this offers great advantage, we need to remember
that the physical quantities with which we start and finish are in the main
observed as functions of time.

9.3.3 Bandwidth and bandlimiting

In earlier sections we met the idea of a bandwidth as a definition of a
range of frequency, usually that over which a frequency-dependent
quantity remains within certain limits. For many purposes this notion
can usefully be idealized, as in the case of the ideal filter described in
Section 5.4.1 which has a transfer function $H(f)$ with magnitude unity
within the specified range of frequency f and zero outside it, and phase
shift af where a is a real negative constant (or zero). These ideal
properties are illustrated in Fig. 9.12 for a band-pass filter with cut-off
frequencies f_1 and f_2; the bandwidth B is $f_2 - f_1$, and the only difference
from the response shown in Fig. 5.13(a) is that negative frequencies are

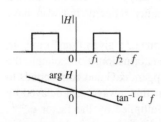

Fig. 9.12
The ideal band-pass filter characteristic.

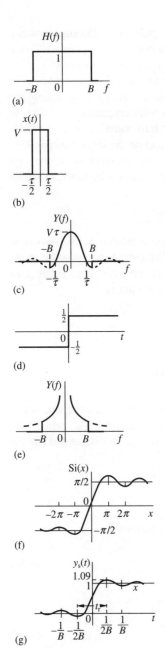

Fig. 9.13

Bandwidth limitation of a rectangular pulse: (a) low-pass filter characteristic; (b), (c) the pulse and its filtered spectrum; (d), (e), (f) a sign function, its filtered spectrum, and the function Si(x); (g) the filtered step function.

now included, since we are considering Fourier transforms. The significance of the phase property can now be seen from the result noted in the previous section, that a spectrum multiplied by $e^{-j\omega T}$ corresponds to a waveform shifted by time T but otherwise unchanged; in other words, an ideal filter is **distortionless** for those frequencies which lie within its passband. This can also be shown without reference to transforms, by noting that shifting a sinusoid at frequency ω by a phase angle ϕ is equivalent to shifting it in time by the interval ϕ/ω; it follows that to shift a range of sinusoids by the same time interval requires that ϕ/ω be the same for all.

The ideal filter is, not surprisingly, impossible to attain in practice but filters which more or less approach the ideal are extensively used, especially in communications engineering where they play an important part in signal processing, dealing not only with the frequencies of the actual signals but also with the much wider range of carrier frequencies used in their transmission. Communications apart, every engineering device has in practice a finite bandwidth, in that no device can handle an infinite range of frequency uniformly. A frequency response $H(f)$ can therefore represent either a bandwidth chosen to eliminate unwanted frequencies, as in a filter, or an unavoidable limitation imposed by a given device or system. In both cases the bandwidth defines a range of frequencies which are 'passed' (for a filter we have called it the pass band), as opposed to others which are 'stopped' (the stop band). There can be a third and different meaning: if a signal has a spectrum containing only frequencies in a certain range, that range is often said to be its bandwidth (the signal is said to be **bandlimited**). On this basis many of the signals we have met, such as the single rectangular pulse, have in theory infinite bandwidth; a pure sinusoid on the other hand has infinitesimal bandwidth. This use of the term to describe a signal must be distinguished according to context from its use to describe the properties of a system.

The effect of a system having an (ideal) bandwidth B on a signal $x(t)$ containing frequencies beyond the range of B is to truncate its spectrum accordingly; the resulting output signal differs from the input. Consider, for example, the effect of an ideal low-pass filter with a bandwidth defined by the magnitude $H(f)$ as shown in Fig. 9.13(a), on the rectangular pulse of width τ, shown at (b). The spectrum $X(f)$ of this pulse is the function given by eqn 9.16 and shown in Fig. 9.7(b), and it can be written $V\tau$ sinc $(f\tau)$. The filter removes the outer parts of the spectrum, beyond the frequencies $\pm B$: the form of $H(f)$ here causes it to act simply as a 'gating' function, giving in this case the output spectrum $Y(f)$, that is $X(f)H(f)$, shown in Fig. 9.13(c). To find the output $y(t)$ we can proceed in various ways. One is to invert $Y(f)$ directly by means of eqn 9.15. A second is to note that the inverse of $X(f)H(f)$ is the convolution of $x(t)$ with $h(t)$ and evaluate the convolution integral as described in Section 7.4.6. Here $x(t)$ is the input pulse, while $h(t)$ is the impulse response of the filter and in this case is (by duality) a sinc function. Either method is perfectly feasible but we shall instead use a third, taking the pulse to be the superposition of a positive step and, after an interval τ, a negative step; this will bring out more clearly one or two important features of the response.

Let us first therefore find the effect of the ideal low-pass filter on a unit step function. This in turn is made easier by considering the step to be made up of the odd function with amplitude $\frac{1}{2}$, shown in Fig. 9.13(d) and sometimes called a **sign function**[†], and the constant $\frac{1}{2}$. We know that a low-pass filter has no effect on the second component, which is a d.c. quantity, so we need only find its response to the sign function and add $\frac{1}{2}$ to that. The Fourier transform of the function shown at (d) is

$$X(\omega) = 1/j\omega$$

and multiplication by $H(f)$ for the filter truncates it, as shown in Fig. 9.13(e), at $f = \pm B$ or $\omega = \pm 2\pi B$. The output may now be found by applying the inversion integral, eqn. 9.15, between these limits; thus the contribution of the sign function to the output is

$$y(t) = \frac{1}{2\pi} \int_{-2\pi B}^{2\pi B} X(\omega) e^{j\omega t} d\omega$$

$$= \frac{1}{2\pi} \int_{-2\pi B}^{2\pi B} \left(\frac{\cos \omega t}{j\omega} + \frac{j \sin \omega t}{j\omega} \right) d\omega.$$

In this, because the cosine is an even function of ω while the sine and ω itself are odd, the first term is odd and the second even; hence with symmetrical limits of integration the first term can be discounted and the integral becomes

$$y(t) = \frac{1}{\pi} \int_{0}^{2\pi B} \frac{\sin \omega t}{\omega} d\omega.$$

By setting ωt to u, so that $d\omega = du/t$, this becomes

$$y(t) = \frac{1}{\pi} \int_{0}^{2\pi Bt} \frac{\sin u}{u} du = \frac{1}{\pi} \text{Si}(2\pi Bt) \qquad (9.21)$$

when we define

$$\text{Si}(x) = \int_{0}^{x} \frac{\sin u}{u} du. \qquad (9.22)$$

The function $\text{Si}(x)$ is the integral of the sinc function which we have already encountered; it cannot be expressed in a closed form but its values are tabulated and it has the form shown in Fig. 9.13(f).

If, finally, the output given by eqn 9.21 is added to the contribution $\frac{1}{2}$ due to the d.c. input component, the total response of the ideal filter to a unit step becomes

[†] Strictly, this name should be given only to the odd function whose amplitude is ± 1, and which is usually given the symbol $\epsilon(t)$; we are thus using $\frac{1}{2}\epsilon(t)$ here.

$$y_s(t) = \frac{1}{2} + \frac{1}{\pi} \text{Si}(2\pi B t) \qquad (9.23)$$

which is the curve shown in Fig. 9.13(g). It can alternatively be expressed as

$$y_s(t) = \frac{1}{\pi} \int_{-\infty}^{2\pi B t} \frac{\sin u}{u} \, du \qquad (9.24)$$

because

$$\int_{-\infty}^{0} \frac{\sin u}{u} \, du = \frac{\pi}{2}.$$

Certain features of the response in Fig. 9.13(g) deserve notice before we proceed to the complete pulse response. First is the remarkable fact that the response begins before the step is applied; indeed it apparently oscillates about zero for all negative time. This particular feature is plainly not physically possible: it appears in the result only because of our initial assumption that the filter characteristic is ideal, which is itself physically impossible. The oscillation, or ripple, in the output is known as the **Gibbs' phenomenon** or Gibbs' ripple and has the frequency B, but its amplitude is independent of the bandwidth. Most important of all, the figure shows that the time between the last negative peak before $t = 0$ and the first positive peak after that is

$$t_r = 1/B. \qquad (9.25)$$

This interval represents one of several ways to define the **rise time** of the response, somewhat artificial though it may seem; it could alternatively be defined, for example, as the time between the output last crossing the value zero and first reaching the value unity, which is about $t_r/2$ or $1/2B$, or as the time between $t = 0$ and the following peak, which is precisely $1/2B$. However it is defined, the rise time is clearly inversely proportional to bandwidth and is in the order of $1/B$. This is also true for the step response of an actual system when its bandwidth is defined in some appropriate way such as that given in Section 5.3.5. A simple example which can confirm the relationship is the first-order low-pass Butterworth filter defined by eqn 5.32 of Section 5.4.2 and realized by the circuit of Fig. 9.14(a). Its open-circuit voltage transfer function is

$$\frac{V_{\text{out}}}{V_{\text{in}}} = \frac{1}{1 + j\omega CR}$$

which has the Bode diagram shown in (b) and a bandwidth defined by the breakpoint as

$$B = \omega_0/2\pi = 1/2\pi CR.$$

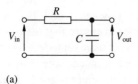

(a)

(b)

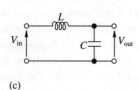

(c)

Fig. 9.14

(a) A first-order low-pass filter;
(b) its response; (c) an L–C filter.

The voltage output for a unit step voltage input is readily shown, by Laplace transform or otherwise, to be

$$v_{\text{out}}(t) = 1 - e^{-t/CR}.$$

The time constant of this response, defined as the time to reach the value $1 - 1/e$, or 0.63, is CR (Section 7.2.1) which is also $1/2\pi B$; if we choose a more arbitrary but better measure of rise time to be the time to reach 0.9 then we need

$$t/CR = \ln 10 = 2.30$$

so that

$$t_r = 2.30CR = 0.36/B.$$

Returning to the ideal filter, we can now easily find its response to the rectangular pulse of height V and width τ shown in Fig. 9.13(b). Since the pulse is a positive step of magnitude V at $t = -\tau/2$ followed by an equal negative step at $t = \tau/2$, eqn 9.23 gives the total response to be

$$y(t) = (V/\pi)[\text{Si}\{2\pi B(t - \tau/2)\} - \text{Si}\{2\pi B(t + \tau/2)\}], \qquad (9.26)$$

a function which might appear somewhat as shown in Fig. 9.15(a). Evidently its appearance must depend on the relative values of B and τ, and one or two other possibilities are shown in (b). Generally, we may deduce that the output pulse is reasonably like the input pulse when $\tau \gg 1/B$; that it is recognizable as a pulse, more or less reaching the amplitude V before decaying again, when $\tau = 1/B$; and that it is vestigial, with an amplitude much less than V, when $\tau \ll 1/B$. These results are quite important in practice: in order to ensure that the output pulse from a system of bandwidth B is a fairly accurate reproduction of a rectangular input pulse we should clearly need B much more than $1/\tau$; but if a roughly triangular pulse of about the right height is good enough (as it would be if we merely needed to count pulses, for example), then a bandwidth of about $1/\tau$ will do; a bandwidth much less than this will not produce a reliably recognizable output. A bandwidth $1/\tau$, according to the above discussion on the step response, is equivalent to having the rise time t_r in the same order as the pulse width τ; and in that case Fig. 9.15(b) indicates that two successive output pulses could be separately distinguished if the second began at a time not appreciably less than τ after the first. Recognizable pulses of a given height can therefore be transmitted at rates up to about $1/\tau$ per second; or, in other words, the maximum pulse rate which can be satisfactorily transmitted through a system of bandwidth B hertz is in the order of B pulses per second. (The **information capacity** of the system depends on the maximum possible pulse rate, which is often taken to be $2B$.) The result is particularly important for the digital transmission of information, and information theory can provide formal confirmation of it.

It may be observed here that a continuous train of rectangular pulses, each of width τ, transmitted at the rate of $1/\tau$ per second would

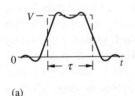

(a)

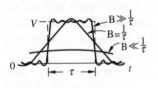

(b)

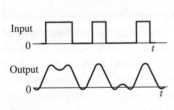

(c)

Fig. 9.15

Bandwidth limitation of a rectangular pulse; (a) a typical output; (b) the effect of varying bandwidth; (c) a bandwidth-limited pulse train.

constitute merely a continuous d.c. level, and the output, which according to the analysis above is the superposition of a train of responses like that in Fig. 9.15(b), must likewise reduce to a constant value for any non-zero bandwidth. However, in transmitting information digitally it is the presence *or absence* of a pulse which is important: it is irregularity which conveys information; a periodic or d.c. waveform, being predictable, does not. Hence the idea of a maximum pulse rate is significant, in our present context, only when some are missing, as indicated in Fig. 9.15(c).

If, on the basis of Fig. 9.15, the optimum bandwidth of a filter to transmit rectangular voltage pulses acceptably is taken to be $B = 1/\tau$, this implies that the spectrum of the pulse, shown for the general case in Fig. 9.13(c), is cut off at its first zeros, $f = \pm 1/\tau$. The energy transmitted by the filter for each pulse in this case can be calculated from the energy density of the pulse spectrum which is, from eqn 9.16,

$$| V(j\omega) |^2 = V^2\tau^2 \frac{\sin^2(\omega\tau/2)}{(\omega\tau/2)^2}$$

for a pulse of height V, and has the form shown in Fig. 9.16. From eqn 9.19 the energy between the limits $\pm 1/\tau$ is therefore

$$E = V^2\tau^2 \frac{2}{2\pi} \int_0^{2\pi/\tau} \frac{\sin^2(\omega\tau/2)}{(\omega\tau/2)^2} \, d\omega$$

which gives the computed value $0.92V^2\tau$. Since the energy of the input pulse is simply $V^2\tau$, it follows that the filter transmits 92% of the input energy. (Because our definition of energy is slightly artificial, it does not follow that the actual energy loss is 8%; the actual energy levels at input and output depend on the effective resistance there, and on its frequency dependence. Any filter made up of only L and C, like that of Fig. 9.14(c) for example, must in theory be lossless; on the other hand, it has by no means the ideal characteristic on which the foregoing calculation is based.)

Finally, because the impulse response of a system with transfer function $H(f)$ is simply its inverse $h(t)$, it follows that the impulse response of an ideal filter with rectangular $H(f)$ and bandwidth B is the sinc function with zero crossings at $\pm 1/2B$ as shown in Fig. 9.17; the width of this output pulse, however defined, can clearly be said to be in the region of $1/B$. Hence we may conclude that a system of bandwidth B has a rise time for a step input, an output pulse width for an impulse input, and an optimum input pulse width for fastest transmission which are all in the order of $1/B$. Calling each of these times τ, we may write in general

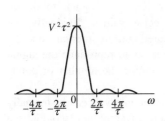

Fig. 9.16
The energy density in a rectangular pulse.

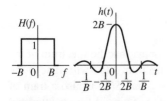

Fig. 9.17
Impulse response of the ideal low-pass filter.

$$B\tau \approx 1 \qquad\qquad (9.27)$$

a result which has something in common with the uncertainty principle of physics. Even for an ideal filter, it has been derived here on the basis of arbitrary definitions; it can be given a more formal basis, but it represents an important and useful working rule in practice. The balance between bandwidth and time which it represents for a system has a counterpart for the signals themselves, which can be appreciated immediately from the spectrum of a rectangular pulse shown in Fig. 9.7: the narrower the pulse the more spread out is its spectrum. The two extreme cases of this interdependence have already been mentioned in Section 9.3.2, namely the uniform spectrum of the impulse $\delta(t)$ and the single impulse which is the line spectrum of a continuous signal at a given frequency. It can be shown that no signal can in theory occupy a limited range in both time and frequency; so a signal which is non-zero for a finite time has an infinitely extended spectrum, and a bandlimited signal has an infinitely extended waveform.

9.3.4 Sampling

A common requirement in communicating data is to convert a continuous function of time, a so-called analogue signal, to digital form by giving it a succession of precise values. Evidently the magnitude of the signal can only be measured at finite intervals, and the question is how often this must be done so as to reproduce the signal exactly; it is a problem comparable with finding the number of points needed to specify a curve.

First the idea of measuring the magnitude at intervals can be formally expressed. A nearly instantaneous measure of amplitude would be a narrow pulse having the same amplitude as the signal over a fixed duration; the time integral of such a pulse is proportional to the average signal value over its duration, and because that is a small interval the pulse is equivalent to an impulse whose magnitude is a measure of the instantaneous signal amplitude. Hence a sequence of samples at regular intervals τ, say, can be formally expressed as the product of the original signal $x(t)$ and a periodic train of unit impulses which we may denote as $s(t)$. Thus the sequence of samples forms a new waveform given by the product

$$y(t) = s(t)\,x(t).$$

It follows from this, by the convolution theorem, that the spectrum of the samples is the convolution of the signal spectrum $X(f)$ with that of the impulse train $S(f)$. We saw in Section 9.3.2 that $S(f)$ is also a train of unit impulses, with spacing $1/\tau$; in the same section it was pointed out that the convolution of a unit impulse train with any function (in the same domain) simply repeated the function at the impulse intervals. Hence we find that the spectrum $Y(f)$ of the sample is periodic, as shown in Fig. 9.18(a). From this follows, finally, the vital result that the the spectrum of the original signal $x(t)$ is an identifiable portion of that of the samples, and that $x(t)$ can therefore be recovered by filtering, provided that the repeated spectra do not overlap. This last requirement

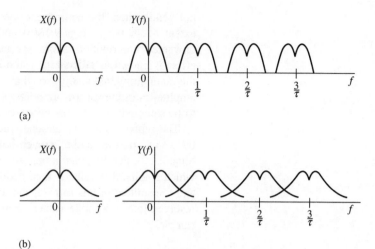

Fig. 9.18
The spectra of a sampled signal
(a) with, and (b) without, bandwidth limitation.

means that the original spectrum must be zero outside the range $\pm 1/2\tau$, or one half of the sampling frequency, and is the basis of the **sampling theorem** which may be stated as follows:

> The minimum sampling rate for accurate reproduction of a waveform containing a finite range of frequencies is twice the maximum frequency.

The minimum sampling rate is known as the **Nyquist rate**. The theorem actually corresponds to an interpolation formula which states that a function $x(t)$ can be written as

$$x(t) = \sum_{n=-\infty}^{\infty} x(n\tau)\frac{\sin\{\pi(t-n\tau)/\tau\}}{\pi(t-n\tau)/\tau}$$

$$= \sum_{n=-\infty}^{\infty} x(n\tau)\,\text{sinc}\frac{t-n\tau}{\tau}$$

(9.28)

in which τ is a constant interval of t and n an integer, provided that the Fourier transform $X(f)$ is zero for $|f| > 1/\tau$. Equation 9.28 asserts, essentially, that the series of values $x(n\tau)$ can be interpolated by means of the sinc function to yield the values of such an $x(t)$ everywhere; and it is in fact the response of an ideal filter to the train of samples $y(t)$. Because of its significance in eqn 9.28, $\text{sinc}\{(t-n\tau)/\tau\}$ is sometimes called the **interpolation function**.

Although in practice neither the sampling impulses nor the filter can be ideal, and the signal $x(t)$ may not be bandlimited, the theorem remains useful. Sampling pulses of small but finite width, as opposed to impulses, affect only the relative amplitudes of the repeated spectra in $Y(f)$ and not the shape of each. The effect of the somewhat extended frequency response of an imperfect filter can be minimized by making the sampling frequency appreciably more than twice the maximum signal frequency, giving a gap between adjacent spectra. If the signal is not bandlimited, the component spectra of $Y(f)$ will overlap to some extent as shown in Fig. 9.18(b) and even an ideal filter would pass unwanted contributions from spectra adjacent to that selected. This effect is known as **aliasing**; it too can be counteracted by increasing the sampling frequency, so that overlap occurs only where the amplitudes of the spectra are relatively small, and by filtering the signal to be sampled so that is as nearly bandlimited as possible.

The representation of an analogue signal by sampled values is the basis of **pulse-code modulation** (PCM). Each sample is 'quantized', that is, given a precise value according to the range within which it falls; this value is then codified as a sequence of binary digits and transmitted in this form. If the maximum signal amplitude is to be resolved into n possible values, then $\log_2 n$ digits are needed for each sample.

9.3.5 Modulation, detection, and multiplexing

The basic signals of communications, for example the audio signals of speech and hearing, are not generally suitable for direct transmission by broadcast or otherwise. For one thing their frequencies are in many cases too low for economic transmission, and for another a number of different signals containing more or less the same frequencies have somehow to be kept separable. To overcome the difficulty, **carrier** frequencies are used: the carrier is a sinusoidal or pulsed periodic waveform, at a frequency suitable for the kind of transmission to be used, which is combined with the actual signal in such a way as to carry all the information of the latter at frequencies near to that of the former. The process is known in general as **modulation**; after transmission the signal itself can be recovered by a process known as demodulation or **detection**.

In **amplitude modulation** a sinusoidal carrier waveform, $A \cos \omega_c t$ say[†], has its amplitude varied according to the instantaneous value of a signal $x(t)$ to give a resulting waveform

$$y(t) = \{a + bx(t)\} \cos \omega_c t \tag{9.29}$$

in which a and b are positive constants. Depending on the method of detection to be used, either $a = 0$, in which case the modulation is simply multiplication, or a and b are chosen such that $| bx(t) | < a$ for all t. The resulting waveform in the latter case, which we take as more general, would appear typically as shown in Fig. 9.19; eqn 9.29 shows that it contains the original carrier frequency, unless $a = 0$, together with a component

$$bx(t) \cos \omega_c t = \tfrac{1}{2} bx(t)(e^{j\omega_c t} + e^{-j\omega_c t}). \tag{9.30}$$

But we saw in Section 9.3.2 the frequency-shift property of the Fourier transform by which multiplication of a waveform by $e^{j\Omega t}$ merely shifts

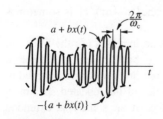

Fig. 9.19
Amplitude modulation of a carrier.

[†] The use of the symbol ω_c here is unconnected with its use to designate the cut-off frequency of a filter, as in Section 5.4.

its spectrum by the frequency interval Ω. It follows that the spectrum of the modulated carrier $y(t)$ contains, apart from a line component at f_c, the shifted signal spectra $\frac{1}{2}bX(f \pm f_c)$ as indicated in Fig. 9.20(a). The halves of each spectrum, centred on $\pm f_c$, are called **sidebands**. Since each sideband contains all the information in the original waveform, the signal can be sent with a saving in power by transmitting only one of them: the carrier can be removed by appropriate modulation, for example with $a = 0$, and the other sideband by filtering. This is known as **single-sideband suppressed-carrier** transmission. If the signal contains frequencies down to d.c., the sidebands are contiguous at the carrier frequency and the precise selection of one is difficult; in that case the low-frequency end of the second sideband may be transmitted as a **vestigial sideband**. In practice, despite reduced efficiency, both sidebands are sometimes retained to give **double-sideband**, suppressed-carrier transmission. Sometimes the carrier too is retained.

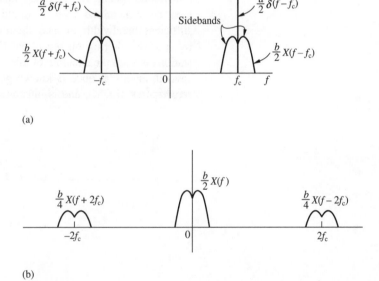

(a)

(b)

Fig. 9.20

The spectra of amplitude modulation: (a) after modulation; (b) in synchronous detection.

Since amplitude modulation involves the multiplication of functions, it cannot be done with purely linear components. Various methods are used in practice, all depending in some way on non-linear behaviour; the use of a diode for this purpose is outlined in Section 11.4.1. Detection too is done by various means. That known as **synchronous detection** again depends on the multiplication of waveforms: here one or both sidebands are themselves multiplied by the same carrier (possibly transmitted for the purpose, otherwise reconstructed). By the same argument as before, the sidebands are now shifted in frequency by $\pm f_c$, which produces two replicas of their spectra, one pair centred on the origin, corresponding precisely to the original signal, the other pair centred on the frequency $2f_c$. The result is shown in Fig. 9.20(b). The high-frequency pair of shifted sidebands is easily removed by filtering to leave the signal only. A simple alternative to the synchronous detector is the **envelope detector**, which uses the rectifying properties of a diode (Section 11.4.2).

Amplitude modulation is only one of several schemes of modulation used in practice, both for analogue and for digital signals, which are based on the idea of a carrier frequency. A sinusoidal carrier can be modulated not only in amplitude but also in frequency and in phase, by either analogue signals or by digital signals in the form of pulses. Similarly, a train of carrier pulses can be modulated in amplitude, duration, and position by an analogue signal; pulse-amplitude modulation (PAM) is equivalent in waveform to sampling the signal at the carrier frequency, and is often converted into pulse-code modulation, as described in the last section. The bandwidth needed for satisfactory transmission of a given signal varies somewhat from one method to another, but in general it increases according to the accuracy required.

By using different carrier frequencies for a number of signals, the transmitted spectra can be made adjacent but not overlapping; they can then be transmitted over a single system or 'channel' which has sufficient bandwidth to pass them all, and can be readily separated by suitable filtering on reception. (The tuning of a radio to different stations is a simple example of this process.) This means of transmitting several signals at once is known generally as **frequency division multiplex** (FDM), and is illustrated in Fig. 9.21(a).

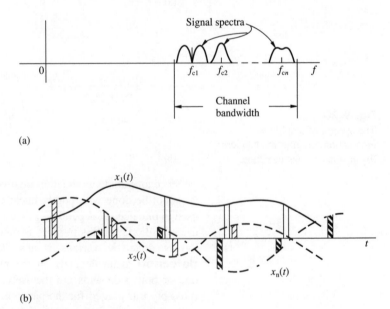

(a)

Fig. 9.21
Multiple signals: (a) in frequency division multiplex; (b) in time division multiplex.

(b)

Example 9.6 A telephone conversation is estimated to need frequencies up to 3 kHz for satisfactory reproduction. If a transmission link is available for frequencies between 400 and 600 kHz, find how many conversations can be transmitted simultaneously, and the necessary carrier frequencies, on the assumptions that a gap of 1 kHz between adjacent spectra is provided and that only the lower sidebands are used.

If n conversations are possible, they require $3n$ kHz plus an extra $(n-1)$ kHz for gaps, so the total bandwidth needed is

$$B = (4n - 1) \text{ kHz.}$$

Since the available bandwidth is 200 kHz we have

$$4n - 1 = 200$$

so $n = 201/4$, which gives, to the nearest lower integer, **50** conversations.

The lowest carrier frequency will occur at the upper limit of the first sideband, that is at 403 kHz, and the others at 4 kHz intervals above that. Hence the carrier frequencies are required to be at **403, 407, 411, ..., 599 kHz**.

A signal based on pulses, whether in the form of modulated pulses or in the binary code of PCM, may be transmitted by a system of a bandwidth large enough to accommodate several other pulses during the interval between the separate pulses or coded groups needed to send one signal. In that case the system can be devoted in turn to the pulses which represent several signals as shown in Fig. 9.21(b). This interleaving of different signals in sequence is known as **time division multiplex** (TDM). The transmitted pulses, on reception, must be correctly allocated in sequence to different signal channels and the signals then reconstructed, a process that corresponds to the filtering and detection required for FDM.

Example 9.7 How many telephone conversations might be transmitted by the channel in Example 9.6 if TDM were used instead of FDM?

The minimum sampling rate for a telephone conversation containing frequencies up to 3 kHz is, by the sampling theorem, 6000 per second. The rate at which pulses can be transmitted by a channel of bandwidth 200 kHz may be taken to be 200,000 per second (Section 9.3.3). Hence the number of conversations which may be accommodated is in the order of

$$n = 200,000/6000 = \mathbf{33}.$$

It is noticeable that the answers to Examples 9.6 and 9.7 are not very different. Indeed, the two calculations are almost identical, although the latter hinged on ideas of sampling and channel capacity which were not considered in the former, nor in the preceding discussion of FDM. (The answers differ only because of the arbitrary allowance for gaps in FDM, and the choice of channel capacity to be precisely its bandwidth in the TDM case.) The similarity illustrates the close correspondence between time and frequency domains. However, it should also be pointed out that if the conversations were transmitted by PCM instead of amplitude-modulated pulses then each would require an increased pulse rate and an increased bandwidth, so reducing the number of conversations accommodated by the channel. Despite this, PCM is often used in practice because in other respects it provides a superior performance.

9.4 Noise

9.4.1 The nature of electrical noise

All of the waveforms so far encountered, whether periodic or not, have been precisely definable and therefore predictable. However, there arise in practice waveforms which are best described as random and whose instantaneous values are therefore unpredictable. Voltages and currents with such waveforms are generally known as electrical **noise**. They arise not so much from defective design or operation (although these can have comparable effects) as from the fundamental statistics of physics. We exclude signals which convey information, such as a speech waveform or an irregular train of coded pulses, and which are therefore random in the sense that they are unpredictable by the recipient.

In the sense in which we use the term, noise is nearly always an undesirable intrusion in any circuit or system. In some cases it is on a relatively large scale and caused for example by disturbances in atmospheric conditions, including lightning, which interfere with radiation and hence with broadcast signals; or by relatively gross imperfections of physical contact, as in the more obvious kinds of telephone noise and in faulty electrical connections generally. But the most important forms of electrical noise are unavoidable, and irreducible below a natural minimum for given circumstances. The importance of this basic noise becomes acute in data-handling and communications equipment for which its minimum level may compare in intensity with the signals to be handled.

There are two such fundamental forms. One is **thermal noise** (or **Johnson noise**) and arises from the random thermal motion of charged particles in any material. It can be shown by a statistical argument that in any resistance R this noise constitutes a voltage source which over a frequency interval or bandwidth B has a mean square voltage

$$\overline{v^2} = 4kTBR \qquad\qquad (9.31)$$

in which k is the Boltzmann constant and T the absolute temperature of the material.

The other fundamental source is **shot noise** (so called from an analogy between shot, in the sense of small pellets, and discrete charge carriers) which arises, whenever a resultant current flows, from the fact that charge must then cross any boundary in its path in discrete amounts; so in fine detail the resultant current is made up of a superposition of random pulses. By considering the spectra of these pulses it can be shown that over a frequency interval or bandwidth B they represent a noise current whose mean square value is proportional to the average current flow I and is given by

$$\overline{i^2} = 2eIB \tag{9.32}$$

where e is the magnitude of the electronic charge.

Shot noise is important in active electronic devices. However, the question arises as to whether eqn 9.32 applies to *any* current-carrying element, including resistance (which we have already identified as the source of thermal noise). In fact eqns 9.31 and 9.32 are both always valid in theory, in that there is always thermal noise in resistance and always shot noise in the current crossing a given surface; but shot noise becomes significant only when the average drift velocity of charge carriers needed to maintain the current I is comparable with their random thermal velocities. As it turns out, this is not the case in conducting materials under most conditions, but happens in, for example, vacuum devices and in the junction regions of semiconductor devices. Consequently, in practice eqn 9.32 is appropriate for such devices but in any resistance we need only consider thermal noise, eqn 9.31. Of course, noise current generated in a device may flow in a resistance and there cause a corresponding voltage drop.

There are two further points to be made here about thermal noise, both arising from its basis in particle statistics: one, that it applies to *any* resistance, even if it is only an effective or equivalent value, because it does not depend on the properties of a material; second, eqn 9.31 fails at very high frequencies (around 10^{13} Hz) because it is based on classical statistics and ignores quantum effects, which can be included only at the cost of simplicity and, particularly, of the very convenient linear dependence on B. Finally, it is worth noting that thermal noise, expressed as a noise voltage in eqn 9.31, is easily converted to a noise current, for comparison with eqn 9.32 or for other reasons. The straightforward conversion between voltage source and current source (Section 1.6.3) shows that a mean square voltage $\overline{v^2}$ generated in a resistance R is equivalent to a current source of mean square value $\overline{v^2}/R^2$ in parallel with R, as shown in Fig. 9.22. (These sources do not carry the usual direction signs, since they represent random processes.) It follows that eqn 9.31 corresponds to a noise source of mean square current

$$\overline{i^2} = \frac{\overline{v^2}}{R^2} = \frac{4kTB}{R} \qquad (9.33)$$

and of output impedance R. Equation 9.32, on the other hand, is not associated with resistance and represents an ideal current source (having infinite output impedance, or zero output admittance).

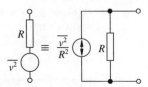

Fig. 9.22

Equivalent sources of thermal noise.

All the above equations have in common their linear dependence on the bandwidth B. Because they express the square of voltage or current, they represent power in the sense used in connection with spectra (Section 9.2.2). The proportionality to B therefore seems to suggest that noise from either kind of source has a constant spectral density of *power*. We have not yet met this idea, because on the one hand periodic waveforms have line spectra with discrete power contributions at each frequency, and on the other hand the non-periodic waveforms discussed earlier were appropriately described by an *energy* density, as defined in Section 9.2.2. Before considering power density further, we may observe that the apparently constant intensity of noise power density with frequency, implying that all frequencies must be equally represented, leads by optical analogy to the description **white noise**; it also leads to the conclusion that there must be some natural limit (such as was mentioned above for thermal noise) to this linear dependence on B, since otherwise every noise source would generate an infinite mean square voltage or current although the bandwidth of measuring equipment would in practice impose a limitation on what can be observed.

There is clearly an important difference between a continuous but random signal and any we have as yet encountered: such a function, unlike the non-periodic waveforms so far considered, must have infinite total energy and energy density; but equally it evidently cannot be periodic and so contain only certain frequencies. To introduce power in these circumstances a new definition is needed, namely that of the **power spectral density**, which we denote by $W(f)$. If $X_\Theta(f)$ is the Fourier transform of that part of a function $x(t)$ which lies in the time interval $-\Theta/2 < t < \Theta/2$ then the power spectral density is defined as

$$W(f) = \lim_{\Theta \to \infty} \frac{1}{\Theta} \mid X_\Theta(f) \mid^2 \qquad \textbf{(9.34)}$$

and it can be shown from Parseval's theorem that the power obtained by integrating $W(f)$ over all f equals, as it must, the mean square value of

$x(t)$. We may now reasonably suppose that what we have been claiming for white noise, on account of its proportionality to B, is that it has constant $W(f)$: thus, if $v(t)$ is the noise voltage whose mean square value is given by eqn 9.31, it appears that its power spectral density must be given by

$$W(f) = 4kTR$$

and measured in (volts)^2 per hertz. However, we are no further forward in showing that constant $W(f)$ is indeed a property of a random function, and it is not encouraging to recollect from Section 9.3.2 that we have already met a waveform which contained all frequencies equally: that was the impulse, which seems very far removed from the continuous and random nature of noise. The difficulty lies in the fact that $x(t)$ cannot itself be precisely defined in such a way as to allow $X(f)$ to be evaluated. What is required is some time-dependent property of $x(t)$ which *is* precise; as it turns out we find this in the **autocorrelation function**, which for a waveform $x(t)$ is defined as

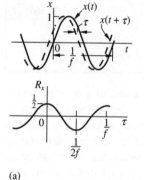

(a)

$$R_x(\tau) = \lim_{\Theta \to \infty} \frac{1}{\Theta} \int_{-\Theta/2}^{\Theta/2} x(t)x(t + \tau)dt \qquad (9.35)$$

and is a measure of the correlation between the value of x at any instant and its value after an interval τ. It is the long-term average of the product of such pairs of values, the averaging time being infinite (as in other averaging processes) so that R_x will be a unique characteristic of $x(t)$. The function R_x can of course be found for any $x(t)$ whether random or not; and because the substitution of $-\tau$ for τ would clearly not affect the principle, it must be an even function. At $\tau = 0$, R_x becomes simply the mean square value of $x(t)$, which can also be shown to be the maximum of R_x. If $x(t)$ is periodic, so too is R_x periodic in τ.

Example 9.8 Find the autocorrelation function of (a) $\sin 2\pi ft$ and (b) a square wave of unit amplitude and period $1/f$.

(a) From eqn 9.35 we have

$$
\begin{aligned}
x(t)x(t + \tau) &= \sin 2\pi ft \sin\{2\pi f(t + \tau)\} \\
&= \tfrac{1}{2}[\cos\{2\pi ft - 2\pi f(t + \tau)\} - \cos\{2\pi ft + 2\pi f(t + \tau)\}] \\
&= \tfrac{1}{2}[\cos 2\pi f\tau - \cos\{2\pi f(2t + \tau)\}]
\end{aligned}
$$

in which the second term has a zero time-average; so the long-term time average of the whole is

$$R_x = \tfrac{1}{2}\cos 2\pi f\tau,$$

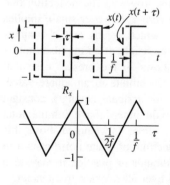

(b)

Fig. 9.23

Waveforms and their autocorrelation functions for Example 9.8.

which is the function, periodic in τ at the frequency f, shown in Fig. 9.23(a). The plausibility of the result can be seen by noting its agreement with the facts that $\frac{1}{2}$ is the mean square value of a unit sinusoid; that when τ is one half-period of $x(t)$, i.e. $1/2f$, the product $x(t)x(t+\tau)$ is $-x^2(t)$, so that $R_x(1/2f) = -R_x(0)$; and that when τ is a whole period $1/f$, the product is $x^2(t)$ and so $R_x(1/f) = R_x(0)$.

(b) For the square wave R_x is most easily found by inspection. From Fig. 9.23(b) it is clear that R_x is $+1$ when $\tau = 0$, is -1 when $\tau = 1/2f$, is again $+1$ when $\tau = 1/f$, and so on; between these extremes it must vary in a linear fashion, so R_x is the function shown.

For random waveforms $x(t)$, the usefulness of the autocorrelation function is that it can be precisely defined as a statistical property which is precisely related to the power spectral density: from eqns 9.34 and 9.35 it can be shown that they are in fact a Fourier transform pair; thus

$$R_x(\tau) = \int_{-\infty}^{\infty} W(f)e^{j2\pi f\tau}\,df$$

$$W(f) = \int_{-\infty}^{\infty} R_x(\tau)e^{-j2\pi f\tau}\,d\tau. \tag{9.36}$$

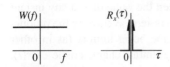

Fig. 9.24

White noise and its autocorrelation function.

Now we are in a position to deduce the significance of a constant power spectrum $W(f)$, for we know that a constant is the Fourier transform of an impulse function. Hence if white noise does indeed have constant $W(f)$, as we suspected, then its autocorrelation function must be an impulse at $\tau = 0$, as shown in Fig. 9.24; this means that there can be no correlation between the values of $x(t)$ at *any* two different instants. Although we have not invoked any formal definition of randomness, we should certainly expect it to include this property. The deduction that there can be no correlation for any non-zero τ, however small, implies that $x(t)$ can change infinitely quickly, which means that it contains infinitely high frequencies. Also, since the height of an impulse (not its magnitude) is in theory infinite, it follows in the present case that $R_x(0) \to \infty$ and therefore that $x(t)$ has an infinite mean square value. This is not so absurd as it may seem; for a signal having a constant power density over all frequencies would indeed have infinite total power and hence an infinite mean square value. As we noted above, this simply means that the ideal constant spectrum of white noise cannot in reality extend over all frequencies, although in practice it may to a sufficiently good approximation extend over all relevant frequencies.

There is a statistical property of white noise which is important in estimating error rates in digital communication: the probability of its instantaneous amplitude lying within a certain range follows the normal or Gaussian distribution curve with mean zero, so the probability that it has amplitude between x and $x + dx$ is $p(x)dx$ where

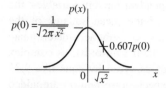

Fig. 9.25

The Gaussian distribution for the amplitude of white noise.

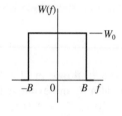

(a)

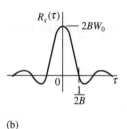

(b)

Fig. 9.26

Bandlimited noise and its autocorrelation function.

$$p(x) = \frac{1}{\sqrt{2\pi x^2}} e^{-x^2/2\overline{x^2}} \tag{9.37}$$

which is the curve shown in Fig. 9.25. The fact that it indicates a zero mean value for a signal whose spectrum includes zero frequency, that is d.c., can be explained by noting that the power at any frequency f is proportional to the range df, and the power content of a continuous spectrum at *precisely* zero frequency is zero. (Alternatively, the same conclusion follows from the fact that a unit impulse, which has the same spectrum as white noise, has no d.c. component, because its average height over all time is zero.) The mean square value $\overline{x^2}$ in 9.37, which in statistical terminology is the square of the standard deviation and is called the **variance**, is of course taken to be finite, which implies that the spectrum is limited in bandwidth.

Although the foregoing ideas appear to be realistic only over a limited bandwidth, both because classical predictions fail at high frequencies and because infinite power cannot be admissible, this is not necessarily an obstacle to their use in practice, for any circuit or system itself limits the bandwidth over which noise can be measured. If a system has a frequency response which approximates to the ideal filter characteristic of bandwidth B, lying within the range of validity of eqns 9.31 and 9.32, then the power spectrum of its output noise, at least that part arising from white noise in its own components (external input may have a spectrum limited by something else) will be the rectangular spectrum shown in Fig. 9.26(a). The inverse Fourier transform of this, which is also its autocorrelation function, will be the sinc function shown in (b). It can be seen that there is little correlation between values more than $1/2B$ apart in time, but values closer than that are more or less correlated: in other words the noise amplitude cannot be expected to change much in times appreciably less then $1/2B$, which is another example of the relationship between response time and bandwidth discussed in Section 9.3.3.

9.4.2 Noise in circuits

Although noise has a random waveform it must obey the fundamental laws of circuit behaviour. We have already seen how a source of noise voltage can be converted to a current source in the usual way (and vice versa), for example. Any noise source can be inserted in a representation of a linear circuit as required, with or without other sources, and by the principle of superposition its effect can be found independently. However the question arises as to how noise currents and voltages can be added according to Kirchhoff's laws when they are not known as determinate functions of time: that is, how can we combine two noise voltages in series, say, which are known only as mean square values, to give a resultant? The answer is that what was found appropriate for

Fourier series is also appropriate for independent random variables: the mean square of the whole is the sum of the mean squares of the components. Thus where we would add (or subtract) determinate voltages or currents as functions of time or, if sinusoidal, as complex phasors, we now add mean square values as given by eqns like 9.31 and 9.32; in other words, as we might expect, noise *powers* are added and there is no question of a noise source having a sign or direction.

To evaluate each noise source in a circuit the bandwidth B must be known. This is likely to be determined by the circuit itself, or possibly by some external source, and will generally be defined by some frequency response function rather than the precise characteristic of an ideal filter. For practical purposes any conventional measure of bandwidth can be used to estimate noise power, such as those based on half-power points (Section 5.3.5) or on cut-off frequencies (Section 5.4). If different parts of a system have different bandwidths, then the effective bandwidth for noise may depend on their sequence: clearly any one stage can act both as a filter for the noise due to earlier stages and as a source of noise over its own bandwidth.

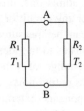

(a)

(b)

Fig. 9.27
The circuit for Example 9.9 and its noise equivalent.

Example 9.9 Find the noise current which flows in the two resistors shown in Fig. 9.27(a) within a bandwidth B, and evaluate this current for $R_1 = 100$ kΩ, $R_2 = 1$ MΩ, $T_1 = 300$ K, $T_2 = 400$ K, and $B = 1$ MHz. Find also the noise voltage which appears between the terminals A and B.

So far as noise is concerned, the circuit becomes that shown in (b). The total mean square voltage in the loop is

$$\overline{v^2} = \overline{v_1^2} + \overline{v_2^2} = 4kB(R_1T_1 + R_2T_2)$$

and the mean square current which flows is therefore

$$\overline{i^2} = \frac{\overline{v^2}}{(R_1 + R_2)^2} = \frac{4kB(R_1T_1 + R_2T_2)}{(R_1 + R_2)^2}.$$

With the figures given and with Boltzmann's constant taken to be 1.38×10^{-23} JK^{-1} this gives

$$\overline{i^2} = 1.96 \times 10^{-20} \text{ A}^2.$$

The noise voltage between A and B can be found either by converting circuit (b) into two current sources which can be added or by superposition of the voltage contributions obtained by potential division. Thus, due to v_1 there is a contribution

$$\overline{v_{AB}^2}' = \overline{v_1^2}\frac{R_2^2}{(R_1 + R_2)^2}$$

and due to v_2,

$$\overline{v_{AB}^2}'' = \overline{v_2^2} \frac{R_1^2}{(R_1 + R_2)^2}$$

giving a total

$$\overline{v_{AB}^2} = 4kB\left\{\frac{T_1 R_1 R_2^2 + T_2 R_2 R_1^2}{(R_1 + R_2)^2}\right\} = \frac{4kBR_1R_2}{(R_1 + R_2)^2}(T_1 R_2 + T_2 R_1)$$

and inserting values as before gives

$$\overline{v_{AB}^2} = \mathbf{1.55 \times 10^{-9} \ V^2}.$$

9.4.3 Available power, noise temperature, and noise factor

Because electrical noise is of particular importance in communications engineering, where signal sources and circuits generally are often matched to their loads in order to obtain maximum power transfer, the maximum available noise power from any source (meaning now actual power as opposed to the mean square of voltage or current) is an important quantity; since matching depends only on impedance levels, any circuit which is matched for signals is also matched for noise.

It was shown in Section 3.11.5 that an a.c. source of r.m.s. voltage V and resistance r can provide a maximum average power $V^2/4r$, and this remains true for any waveform of mean square voltage V^2. It therefore follows from eqn 9.31 that the maximum thermal noise power *in watts* available from any resistance is given by

$$N_{max} = kTB. \tag{9.38}$$

The maximum power directly obtainable from a shot-noise source is in theory unlimited, for eqn 9.32 implies an ideal current source; but the actual maximum shot noise available to a given load is of course determined by the remainder of the circuit.

Because the available thermal noise power given by eqn 9.38 is independent of a value of resistance, it can be usefully applied to almost any kind of noise source, such as the radiated noise received by an aerial, and leads to the idea of a **noise temperature**. For any source we may define an effective temperature T_e such that the maximum power actually received from it within a bandwidth B is kT_eB. A source will have a unique value of T_e only if it produces white noise, so for a frequency-dependent source of noise the value of T_e depends on the region of frequency in which the chosen bandwidth (e.g. that of a receiver) lies.

In a similar way an equivalent bandwidth for noise can be defined, irrespective of other definitions of bandwidth previously mentioned, for a system having any variation of frequency response. If its response is defined by a transfer function $H(f)$, say, then white noise of power density $W(f)$ at its input will produce at the output noise of density $W(f) \mid H(f) \mid^2$, giving a total power output

$$N = \int_{-\infty}^{\infty} W(f) \mid H(f) \mid^2 \, df$$

(in units of voltage squared or current squared, according to the units of W and H; the question of actual power need not arise here). Now suppose the same noise input to a fictitious system having an ideal low-pass response of bandwidth B with a value of H equal to the d.c. value $H(0)$ of the original, so that

$$H(f) = H(0) \qquad -B < f < B$$
$$H(f) = 0 \qquad \mid f \mid > B;$$

the total power output would now be

$$N = 2W(f)H^2(0)B.$$

Thus to produce the same total noise as the actual system the fictitious system would need to have bandwidth

$$B = \frac{\int_{-\infty}^{\infty} \mid H(f) \mid^2 \, df}{2H^2(0)} = \frac{\int_{-0}^{\infty} \mid H(f) \mid^2 \, df}{H^2(0)} \tag{9.39}$$

and this is known as the **equivalent noise bandwidth**. It can in principle be evaluated for any given $H(f)$.

An important quantity in communications or in any form of signal processing where noise is likely to be a significant factor is the ratio of the signal power to the noise power at any point, usually written S/N. Because every part of an electrical system inevitably contributes noise from its own components, the ratio must decrease at every stage between input and output. For any four-terminal device or circuit, this is taken into account by defining its **noise factor** F as the quotient of the ratio S/N at the input and that at the output; that is

$$F = \frac{(S/N)_{\text{in}}}{(S/N)_{\text{out}}} \tag{9.40}$$

which is always a positive quantity greater than unity (but would of course be unity for a perfectly noise-free circuit). Equation 9.40 applies as it stands when F and the S/N ratios are actual numerical values. For the ratio S/N it is common to use a logarithmic measure, usually decibels, so that each power ratio is expressed as $10(\log_{10} S - \log_{10} N)$ in decibels; the difference of the input and output values then gives F in like measure.

It would be possible to evaluate the noise factor of a circuit by calculating all the contributions to noise power from its components, but to do so would in practice be prohibitively laborious and it is more usual to deduce a value for F from measurements at input and output. For this purpose the above definition of F can be cast into alternative forms, and a number of other relationships deduced; these are often based on the assumption that the various parts of the system are matched, which would be likely in circuits handling the low power levels at which noise becomes important. If we suppose the circuit to be an amplifier with power gain A, say, then

$$S_{out}/S_{in} = A$$

and eqn 9.40 can be written

$$F = N_{out}/AN_{in}. \tag{9.41}$$

Whatever the bandwidth B of the amplifier, so long as the gain can be assumed constant over the range B the input noise power must also be amplified by A, and AN_{in} is therefore its contribution to the output noise N_{out}. The contribution due to the amplifier's own noise is thus $N_{out} - AN_{in}$ which by eqn 9.41 can be written

$$N'_{out} = N_{out} - AN_{in} = N_{out}(F - 1)/F. \tag{9.42}$$

(In this and in what follows F must be taken to represent an actual number rather than a logarithmic measure.) Alternatively this contribution can be expressed in terms of the input noise by writing N_{out} as FAN_{in}, from eqn 9.41, to give

$$N'_{out} = FAN_{in}(F - 1)/F = AN_{in}(F - 1). \tag{9.43}$$

In this, the input noise power N_{in} may arise from the resistance of the input source, or it may be amplified noise from a previous stage; in any case, according to the above definitions its value will affect the apparent value of F for a given amplifier. For the purpose of applying these definitions N_{in} is therefore taken to be the (unamplified) available thermal noise at the input terminals, implying a matched input and enabling N_{in} to be written as kTB. The noise factor can now be written, from eqn 9.41,

$$F = N_{out}/kTBA \tag{9.44}$$

and the output contribution from the amplifier is, from eqn 9.43,

$$N'_{out} = kTBA(F - 1).$$

It is usually convenient to convert this into an equivalent contribution at the *input*, by simply dividing it by the gain A. Thus the amplifier's own noise is equivalent to the extra input

$$N'_{in} = kTB(F - 1). \tag{9.45}$$

Example 9.10 Find the noise factor of two amplifiers in cascade, having respectively power gains and noise factors A_1, A_2, F_1, F_2, an overall bandwidth B and a common temperature T, on the assumption that the first has a matched input with thermal noise only and that the second is matched to the first.

The noise flow is shown in Fig. 9.28. The specification of the input means that the external input noise is kTB, to which, by eqn 9.45, the first amplifier in effect adds $kTB(F_1-1)$. This total input is amplified by A_1A_2; in addition the second amplifier similarly contributes an amount $kTB(F_2 - 1)$ amplified by A_2. Hence the total output noise is

$$N_{\text{out}} = \{kTB + kTB(F_1 - 1)\}A_1A_2 + kTB(F_2 - 1)A_2$$
$$= kTBA_2(F_1A_1 + F_2 - 1).$$

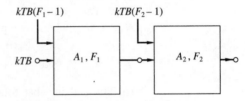

Fig. 9.28
Noise flow in the amplifiers of Example 9.10.

Since by eqn 9.44 the overall noise factor is given by $N_{\text{out}}/kTBA_1A_2$, we have finally

$$F = F_1 + (F_2 - 1)/A_1.$$

A similar calculation could of course be made for a circuit with more than two stages.

Problems

9.1 From the Fourier series for a triangular waveform given in Appendix F, write the first three terms for the output voltage of the integrator in Problem 8.11 and find the percentage error when these terms are used to estimate the r.m.s. value.

9.2 Estimate the r.m.s. current which flows when the triangular voltage waveform referred to in Problem 9.1 is applied to a capacitance of 1 μF.

9.3 Find the r.m.s. current which flows when the same voltage as in Problem 9.2 is applied to a resistance of 100 Ω.

9.4 Find what fraction of the power dissipated in the resistance of Problem 9.3 is due to the fundamental component of the voltage.

9.5 Find the angular frequency at which the Fourier transform of a Gaussian pulse (Appendix G) falls to half of its peak value.

9.6 Find the energy contained in the frequency range $0 < f < 1$ kHz of the voltage waveform in Example 9.5, and from the results of that example deduce the proportion of energy contained in frequencies above 2 kHz.

9.7 A signal contains frequencies up to 10 kHz. It is to be transmitted in digital form by sampling, with 20 levels available to specify a sampled value. Estimate the pulse rate and the minimum bandwidth needed for transmission.

9.8 A sinusoidal carrier of frequency 1 MHz and amplitude 10 V is modulated by a square-wave of period 100 μs and amplitude ± 1 V from which components above 60 kHz have been filtered out. What frequencies appear in the spectrum of the modulated carrier?

9.9 Shot noise due to a d.c. current of 1 mA flows in a 10 kΩ resistance. Estimate the mean square voltages across the resistance, per unit bandwidth, (a) due to shot noise and (b) due to thermal noise at 300 K.

9.10 Find the equivalent noise temperature of a source which radiates 10^{-15} W of noise over a bandwidth of 1 MHz.

9.11 Show that the noise bandwidth of a first-order low-pass Butterworth filter with cut-off frequency f_c is $\pi f_c / 2$.

9.12 Find the equivalent input noise, at 300 K, of an amplifier having a noise figure of 1 dB and 1 MHz bandwidth.

10 Distributed circuits

10.1 Ladder networks

10.1.1 General

Section 2.7 described circuits, belonging to a class known as ladder networks, which were made up of a supposedly infinite series of identical combinations of resistance connected in cascade, and it was shown that the input resistance and the variation of voltage or current from section to section could be readily found. In a resistive ladder both voltage and current must decay (to vanishing point if it is infinite), and the input power must be dissipated; although a few sections of such a circuit can be used in practice as an **attenuator**, it is hard to imagine many applications for it which could not be as well served by a potential divider or even a single resistance. It is, however, a different matter if each section of the network is made up of a combination of reactances, and we shall see that this can lead to interesting and useful properties which arise, in essence, from the fact that reactance produces phase difference. Ladder networks are often known also as **periodic networks**.

A few sections of a general ladder network are shown in Fig. 10.1(a). What may be called the series elements are shown as impedances Z, and the parallel, or shunt, elements as admittances Y; this choice turns out to be convenient as well as natural, and we revert to bold notation to indicate that we shall be considering mainly the steady state with voltages and currents represented by phasors. Before going further, we should observe that the network can be made up from four-terminal sections of several different configurations. Two simple possibilities are shown in (b) and (c); but if we wish to regard the circuit as made from symmetrical sections the choice is between the T section of (d) and Π section of (e). Clearly the form of the *first* section, at the input, would determine which of the two is literally correct (the output end is of no concern while we are considering an indefinite number of sections), and we should need that information in order to find, for example, the input impedance; but the essential behaviour of the circuit can be studied

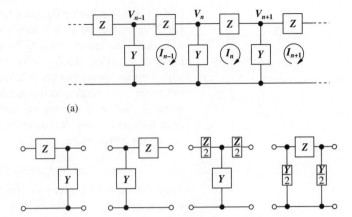

Fig. 10.1

Sections of a uniform ladder network.

(a)

(b)   (c)   (d)   (e)

without this knowledge, by considering the nth section and its neighbours as shown in (a). With the current phasors indicated Kirchhoff's second law around the I_n loop gives

$$(I_n - I_{n-1})/Y + I_nZ + (I_n - I_{n+1})/Y = 0$$

or

$$2I_n - I_{n-1} - I_{n+1} = -I_nZY. \tag{10.1}$$

This is a difference equation, and like comparable differential equations it is most easily solved by postulating a form of solution. Suppose, at a guess, that the current I_n in the n th section (counting from the input end) is given by $I_0e^{-\gamma n}$ where I_0 is some arbitrary current phasor, presumably depending on the input, and γ a quantity depending on the circuit. It immediately follows that

$$I_{n-1} = I_ne^{\gamma}$$
$$I_{n+1} = I_ne^{-\gamma}.$$

Substituting the proposed solution in eqn 10.1 therefore gives

$$I_n(2 - e^{\gamma} - e^{-\gamma}) = -I_nZY$$

which can be written

$$2(1 - \cosh \gamma) = -ZY$$

or

$$\cosh \gamma = 1 + ZY/2. \tag{10.2}$$

Exactly the same equation would have resulted had we written Kirchhoff's first law for the node with voltage V_n, for that yields a voltage equation having the same form as eqn 10.1. It is also evident that $I_0 e^{-\gamma n}$ is another possible solution to eqn 10.1 which would again yield eqn 10.2. Thus either sign may be attached to γ to give a valid solution for I_n or V_n: if eqn 10.2 is satisfied by a value γ it is also satisfied by $-\gamma$. We may therefore conclude that current and voltage behave in the same way from section to section; eqn 10.2 tells us what γ must be for given Z and Y and once its sign can be decided it specifies the relationship between successive currents which applies also to voltages. We return below to the question of sign.

10.1.2   Attenuation and phase shift

Since the value of γ given by eqn 10.2 may in general be complex we write it as

$$\gamma = \alpha + j\beta$$

and consider next the implication of this, taking only the solution $I_0 e^{-\gamma n}$ meantime. If γ is real, then $\beta = 0$ and successive currents are related by the factor $e^{-\alpha}$; this is a real quantity and means that their magnitude but not their phase changes from section to section. On the other hand if γ is purely imaginary then successive currents are in the ratio $e^{-j\beta}$ which has magnitude unity and means that only phase would change between sections, by an angle β. It follows that in general a complex value for γ implies that currents (and voltages) change in both magnitude and phase between sections. The quantity γ is known as the **propagation constant**, for reasons which will become clear in later sections; its real part α is the **attenuation constant** and its imaginary part β the **phase constant**. (An alternative notation for the same quantities is referred to in Section 10.2.1.) Although these are referred to as 'constants', their dependence on the product ZY means that in general they vary with frequency. Eqn 10.2 is actually an example of what is known as a **dispersion relation** between the propagation constant and frequency for a given system.

The attenuation constant α, being the natural logarithm of the ratio of currents or voltages at the input and output of a section, gives the attenuation in units of nepers per section. In previous chapters the decibel was used as the logarithmic measure for such ratios, and attenuation can easily be changed to this form by the conversion given in Section 5.2.1: the common logarithm of a voltage or current ratio $e^{-\alpha}$ is -0.4343α, and the ratio in decibels is therefore 20 times that, namely -8.686α; hence the value of 8.686α may be quoted as the attenuation in decibels. In this chapter we shall nevertheless refer to attenuation simply in terms of α.

We are now in a position to investigate how current and voltage will vary along the ladder for various combinations of Z and Y, as described by the following headings.

(i) *Similar elements*   If Z and Y represent pure resistance and conductance then the product ZY is real and so too, by eqn 10.2, is cosh γ and therefore γ itself; so $\beta = 0$ and only magnitude changes, by the factor $e^{-\alpha}$ per section. The same is true if Z and Y both represent inductances, say L_1 and L_2 respectively, for then ZY is L_1/L_2 at any frequency; similarly if they are both capacitive, ZY has the form C_2/C_1 and is again real. This makes perfect sense, for in a ladder composed entirely of elements of a kind we should expect the input amplitude to decay but all the currents to retain the same phase and all the voltages likewise; in a purely reactive ladder, unlike the resistive case, the currents will be in quadrature with the voltages and there will be no dissipation, but the attenuation of current and voltage is common to all three kinds. (Ladder networks used as attenuators in practice may comprise either resistances or capacitances.)

This case also gives us a clue as to the sign to be allocated to γ. If the ladder is infinite, the amplitude of voltage or current can only decay with increasing n, i.e. away from the input, so we must choose the solution which ensures that; thus, if γ has a positive real part α the currents and voltages must vary as $e^{-\gamma n}$ rather than $e^{\gamma n}$. If the ladder is finite, the actual solutions turns out in general to be a combination of both possible solutions (as would be expected mathematically). However we have already seen in Section 4.7 that a finite ladder can be made to behave as if it were infinite by terminating it in its iterative (or characteristic) impedance, in which case it is said to be matched; in that case too, only the solution which gives amplitude decreasing with n is required.

(ii) *Series inductance, shunt capacitance* If Z is an inductive reactance $j\omega L$ and Y a capacitive susceptance $j\omega C$, then ZY is the negative real quantity $-\omega^2 LC$; eqn 10.2 then shows that $\cosh \gamma < 1$. According to the properties of the cosh function this condition means that γ cannot be purely real and that there are the following two possibilities.

(a) If $-1 \leq \cosh \gamma < 1$, then γ must be purely imaginary, so $\alpha = 0$ and $\cosh \gamma = \cos \beta$. Eqn 10.2 shows that this applies when

$$-4 \leq ZY < 0, \tag{10.3}$$

which means that in this case

$$\omega^2 LC \leq 4.$$

This gives a critical frequency

$$\omega_c = \frac{2}{\sqrt{LC}} \tag{10.4}$$

below which there is no attenuation, so that currents and voltages have unchanged amplitude along the ladder but suffer a phase shift β per section given, according to the dispersion relation eqn 10.2, by

$$\cos \beta = 1 + \frac{ZY}{2} = 1 - \frac{\omega^2 LC}{2}$$

or

$$\sin^2 \frac{\beta}{2} = \frac{\omega^2 LC}{4}. \qquad (10.5)$$

This result does not itself reveal the sign of the phase shift, but confirms that either choice will satisfy the equations; nor can we in this case decide by a physical argument based on attenuation, since there is none. However, by considering just one section terminated in its iterative impedance (Section 10.1.3) so as to simulate an infinite ladder, it can be shown that for series L and shunt C the phase shift per section is a negative angle between 0 and π; so each current and voltage lags that of the previous section, and in the factor $e^{-\gamma n}$ we therefore take β to be positive.

(b) The other possibility is that $\cosh \gamma < -1$, in which case γ is a complex quantity of the form $\alpha \pm j\pi$. Equation 10.2 shows that this occurs for $ZY < -4$, which in this case requires that ω is above the critical frequency given by eqn 10.4, and that the attenuation constant is then given by

$$\cosh \alpha = -\left(1 + \frac{ZY}{2}\right) = \frac{\omega^2 LC}{2} - 1. \qquad (10.6)$$

Here again the sign of α is not defined, but for an infinite or matched network must be such that amplitudes decay for increasing n, away from the input; so if we choose the solution with the form $e^{-\gamma n}$ as before then α must be positive.

These results for the two possibilities (a) and (b) can be conveniently summarized in a diagram showing the variation of α and β with frequency, as in Fig. 10.2. Evidently the matched (or infinite) L–C ladder acts as a low-pass filter (section 5.4) for which the critical frequency of eqn 10.4 represents cut-off. We return to this aspect of ladder networks in Section 10.1.4.

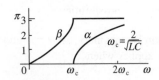

Fig. 10.2
Attenuation and phase constants for a ladder with series L, shunt C.

(iii) *Series capacitance, shunt inductance*   In this case ZY becomes the negative real quantity $-1/\omega^2 LC$. Here again, therefore, eqn 10.2 shows that $\cosh \gamma < 1$ and γ accordingly cannot be purely real. There are again the following two possibilities.

(a) If $-1 \leq \cosh \gamma < 1$, γ is purely imaginary and as before ZY must be in the range given by eqn 10.3. Now, however, this implies that

$$1/\omega^2 LC \leq 4$$

and that there is therefore a critical frequency which in this case is

$$\omega_c = \frac{1}{2\sqrt{LC}}. \tag{10.7}$$

Above this frequency the C–L ladder shows no attenuation and a phase constant β given by

$$\cos\beta = 1 + \frac{ZY}{2} = 1 - \frac{1}{2\omega^2 LC}$$

or

$$\sin^2\frac{\beta}{2} = \frac{1}{4\omega^2 LC}. \tag{10.8}$$

In this case it can be shown that the phase *advances* from section to section along a matched ladder, so for a solution in the form $e^{-\gamma n}$ as before we now need β to be negative. (Alternatively we could retain positive β and choose the solution $e^{\gamma n}$ for such a case.)

(b) If $\cosh\gamma < -1$ then as for the previous network γ takes the form $\alpha \pm j\pi$ and this now happens below the critical frequency given by eqn 10.7. The attenuation constant α is in this case given by

$$\cosh\alpha = -\left(1 + \frac{ZY}{2}\right) = \frac{1}{2\omega^2 LC} - 1 \tag{10.9}$$

and as before α is to be taken as positive for a solution in the form $e^{-\gamma n}$.

The overall variation of α and β with frequency for this ladder is as shown in Fig. 10.3; it clearly behaves as a high-pass filter (Section 5.4) when matched, with a cut-off frequency given by eqn 10.7. Below this the attenuation rises with decreasing frequency, reflecting the behaviour of the low-pass network of (ii) above.

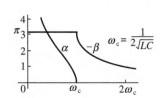

Fig. 10.3

Attenuation and phase constants for a ladder with series C, shunt L.

(iv) *Mixed reactances* If either Z or Y or both contain a combination of L and C, without resistance, each is purely imaginary but may carry a sign dependent on frequency. The product ZY is therefore again real but there will be in general a range of frequency for which it is positive, so that the network attenuates without phase shift as in case (i), except that the attenuation is now frequency dependent. There will be other frequencies for which ZY is negative and the network then behaves as in case (ii) or case (iii), with no attenuation over a certain range. Evidently a ladder of this kind can form a band-pass or band-stop filter (Section 5.4).

(v) *Reactance with resistance* If Z and Y between them include both resistance and reactance, then ZY is in general complex and so too, by eqn 10.2, is $\cosh\gamma$ and therefore γ itself. The ladder network will then show both attenuation and phase shift, both of them dependent on frequency. In the special case that either Z or Y is a pure reactance and the other a pure resistance, the product is imaginary but cosh γ and γ are still complex. If Z and Y happened to have equal and opposite angles (implying that both are inductive or both capacitive), the product would be positive real and as in case (i) the network would then show attenuation without phase shift.

It is, of course, only to be expected that any ladder network which includes resistance will show attenuation, since it must dissipate at least some of its input power. But attenuation need not imply resistance and dissipation: we saw in the last section that a ladder entirely composed of one kind of reactance showed attenuation, although it can dissipate no power.

Example 10.1 In a certain ladder network the series impedance comprises a 1 mH inductance and the shunt admittance a 0.1 μF capacitance. Find the change in phase per section at the frequencies (a) 50 Hz and (b) 30 kHz, and (c) the attenuation per section at 100 kHz.

(a) The series impedance at 50 Hz is given by

$$Z = 2\pi j \times 50 \times 10^{-3} = j\pi/10 \ \Omega$$

and the shunt admittance by

$$Y = 2\pi j \times 50 \times 10^{-7} = j\pi \times 10^{-5} \ \text{S}$$

from which

$$ZY = -\pi^2 \times 10^{-6} = -9.870 \times 10^{-6}.$$

Equation 10.2 then gives

$$\cosh \gamma = 1 + ZY/2 = 0.999\ 995\ 065$$

from which the phase shift per section is

$$\beta = \cos^{-1} 0.999\ 995\ 065 = 0.003\ 142 \ \text{rad}, \ \text{or } \mathbf{0.18°}.$$

(b) At 30 kHz the required values are now

$$Z = 2\pi j(3 \times 10^4)10^{-3} = j188.5 \ \Omega$$

and

$$Y = 2\pi j(3 \times 10^4)10^{-7} = j0.018\ 85 \ \text{S},$$

so

$$ZY = -3.553.$$

Equation 10.2 now gives

$$\cosh \gamma = 1 + (ZY/2) = -0.7765$$

and the phase shift per section is therefore

$$\beta = \cos^{-1}(-0.7765) = 2.46 \text{ rad, or } \mathbf{141°}.$$

(c) At 100 kHz the required values are

$$Z = 2\pi \text{j} \times 10^5 \times 10^{-3} = \text{j}628.3 \ \Omega$$

and

$$Y = 2\pi \text{j} \times 10^5 \times 10^{-7} = \text{j}0.062 \ 83 \ \text{S}$$

which give

$$ZY = -39.48$$

and therefore

$$\cosh \gamma = 1 + ZY/2 = -18.74.$$

Equation 10.6 now gives the attenuation constant as

$$\alpha = \cosh^{-1} 18.74 = \mathbf{3.62 \ nepers \ per \ section}.$$

The attenuation can also be expressed in decibels as 8.686α, which is **31.5 dB per section** or as the amplitude ratio of successive voltages or currents, which is $\text{e}^{-3.623}$ or **0.0267**.

Example 10.2 In each section of a ladder network the series impedance comprises a 1 mH inductance only and the shunt admittance a 1 mH inductance in parallel with a 0.1 μF capacitance. Find the ranges of frequency in which from section to section there are (a) loss of amplitude but no phase change, (b) phase change but no loss of amplitude, and (c) loss of amplitude and phase reversal.

(a) Loss of amplitude without phase change means that both Z and Y are the same type of reactance, as in case (i) above, which in this example means that Y must be inductive. This will occur when the admittance of 1 mH exceeds (in magnitude) that of 0.1 μF, that is when

$$1/(10^{-3}\omega) > 10^{-7}\omega$$

or

$$\omega < \mathbf{10^5 \ rad \ s^{-1}}.$$

This upper value of ω is also, as expected, the resonant frequency of the combination Y.

(b) Phase change with no loss of amplitude occurs when the product ZY satisfies the condition 10.3. With the components given we have

$$\mathbf{ZY} = 10^{-3}j\omega(1/(10^{-3}j\omega) + 10^{-7}j\omega)$$
$$= 1 - 10^{-10}\omega^2$$

which is zero for $\omega = 10^5$ and has the value -4 when

$$10^{-10}\omega^2 = 5,$$

that is when $\omega = 10^5\sqrt{5}$. The range of angular frequency is therefore given by

$$10^5 < \omega < 10^5\sqrt{5} \text{ rad s}^{-1}.$$

(c) Loss of amplitude with phase reversal, being the only other possibility for a purely reactive network, must arise for frequencies outside the ranges found in (a) and (b), that is for

$$\omega > 10^5\sqrt{5} \text{ rad s}^{-1}.$$

This is also, of course, the range which results from the condition that $\mathbf{ZY} < -4$, in violation of 10.3.

10.1.3 Input impedance

In the previous section we have seen how current and voltage vary along infinite or matched ladders of different kinds, but so far no attempt has been made to relate these two quantities. Because they behave in identical ways, their relationship at any point on such a ladder can be established by considering the input to the first section, where the ratio of voltage to current must be the input impedance $\mathbf{Z}_{in}$ which must also be the iterative impedance of each section. It is a straightforward matter to find $\mathbf{Z}_{in}$ by the method used in Section 2.7 for the case of an infinite resistance ladder, and again in Section 4.7 to find the iterative impedance of a four-terminal network.

Before deriving an expression for $\mathbf{Z}_{in}$ we must first decide which of the possible section configurations shown in Fig. 10.1 is to be used. It turns out that the simplest results are obtained for the symmetrical sections shown in Fig. 10.1(d) and (e) (in which case, according to the definitions in Section 4.7, $\mathbf{Z}_{in}$ is also the characteristic impedance $\mathbf{Z}_0$ of each section). The first of these, the T section, leads to the circuit shown in Fig. 10.4 and to the result

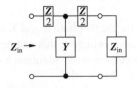

Fig. 10.4
The input impedance of an infinite ladder network.

$$\mathbf{Z}_{in}^2 = \frac{Z}{Y}\left(1 + \frac{ZY}{4}\right). \tag{10.10}$$

From this point we shall mainly consider circuits in which $\mathbf{Z}$ and $\mathbf{Y}$ represent opposite types of reactive elements, one inductance and the other capacitance. If we assume meantime that these are pure, with no associated resistance, the ratio $\mathbf{Z}/\mathbf{Y}$ is positive real and the product $\mathbf{ZY}$ is negative real as in cases (ii) and (iii) of the previous section. Now consider the pass band in either of these cases, that is to say frequencies for which eqn 10.3 is satisfied: this means that $\mathbf{ZY}/4$ must have a value between -1 and zero. It follows that at those frequencies Z_{in}, or Z_0, is a positive real quantity, in other words a pure resistance, although it varies with frequency because of the product ZY. It is evident from eqn 10.10 that at the cut-off frequency in each case (eqns 10.4 and 10.7) $ZY = -4$ and Z_{in} becomes zero: the ladder appears to be a short circuit because at the cut-off frequency its reactances, like those of the series resonant circuit (Section 5.3.3), exactly cancel each other. (This can easily be seen by considering an actual T section with its output short-circuited.) Beyond the cut-off frequency, in which case there is attenuation given by eqn 10.6, $ZY < -4$ and eqn 10.10 shows that Z_{in} is then imaginary, that is a pure reactance. The equation does not indicate the sign, but it is not difficult to show (by considering an actual section again) that the imaginary Z_{in} has the same sign as Z, the series reactance.

The other symmetrical choice, the Π section of Fig. 10.1(e), leads in the same way to the result

$$Z_{in}^2 = \frac{Z}{Y}\left(1 + \frac{ZY}{4}\right)^{-1} \tag{10.11}$$

which can also, as might be expected, be written as the dual of eqn 10.10, namely

$$Y_{in}^2 = \frac{Y}{Z}\left(1 + \frac{ZY}{4}\right). \tag{10.12}$$

Although the input impedance now differs from that of a ladder made up of T sections, here too it is clearly a pure resistance in the pass band and a pure reactance in the stop band. However, eqn 10.12 shows that for the ladder of Π sections it is the admittance Y_{in} which becomes zero at the cut-off frequency, when $ZY = -4$, and the ladder then behaves like a parallel resonant circuit. Beyond cut-off, when Y_{in} is imaginary, it can be shown that it takes the same sign as Y.

For the remaining possible arrangements, namely the asymmetrical sections shown in Fig. 10.1(b) and (c), the input impedance can be quickly deduced from the foregoing results. If the first section is that shown in (b), the input impedance must be that of a ladder beginning with the T section of (d) with an additional $Z/2$; from eqn 10.10 it is therefore given by

$$Z_{in} = \frac{Z}{2} + \sqrt{\frac{Z}{Y}\left(1 + \frac{ZY}{4}\right)}. \tag{10.13}$$

Similarly, the input admittance of a ladder comprising sections like Fig. 10.1(c) is that of the symmetrical section (e) with the addition of an extra $Y/2$; from eqn 10.12 this gives

$$Y_{in} = \frac{Y}{2} + \sqrt{\frac{Y}{Z}\left(1 + \frac{ZY}{4}\right)} \tag{10.14}$$

Equations 10.13 and 10.14 show that the input impedance of an infinite ladder of asymmetrical sections is not, as for the T and Π sections, a pure resistance but has in each case a reactive component. However, an important feature can be seen in all of the eqns 10.10 to 10.14: in every case the input impedance becomes $\sqrt{Z/Y}$ when the product ZY is small. We shall hear more of this in later sections, when considering transmission lines, but it can be noted here that for both of the network types now considered, L–C and C–L, this limiting value $\sqrt{Z/Y}$ is given by $\sqrt{L/C}$.

Example 10.3 If the ladder in Example 10.1, with Z comprising an inductance of 1 mH and Y a capacitance of 0.1 μF, is made up from T sections, find its input impedance (a) at 50 Hz, (b) at 30 kHz, and (c) at 100 kHz.

The impedance required is that given by eqn 10.10, and we need both the ratio Z/Y and the product ZY. For the former, we have at any frequency

$$Z/Y = L/C = 10^{-3}/10^{-7} = 10^4 \ \Omega^2.$$

The product ZY is real, given by $-\omega^2 LC$ and takes the following values:

at 50 Hz: $\quad ZY = -4\pi^2 \times 50^2 \times 10^{-10} = -9.870 \times 10^{-6}$;

at 30 kHz: $\quad ZY = -4\pi^2 \times 30^2 \times 10^6 \times 10^{-10} = -3.553$;

at 100 kHz: $\quad ZY = -4\pi^2 \times 10^{10} \times 10^{-10} = -39.48$.

Eqn 10.10 then gives the results

(a) $Z_{\text{in}} = \{10^4(1 - 2.468 \times 10^{-6})\}^{1/2} \approx \textbf{100 }\boldsymbol{\Omega}$;

(b) $Z_{\text{in}} = \{10^4(1 - 0.8883)\}^{1/2} = \textbf{33.4 }\boldsymbol{\Omega}$;

(c) $Z_{\text{in}} = \{10^4(1 - 9.87)\}^{1/2} = \pm\textbf{j298 }\boldsymbol{\Omega}$.

The input impedance is a pure reactance in case (c) because, as eqn 10.4 confirms, 100 kHz is above the cut-off value of (cyclic) frequency for the network; although eqn 10.10 does not determine the sign in this event, it was mentioned earlier that Z_{in} has the same sign as Z for a ladder of T sections and it is therefore inductive in this example.

10.1.4   The ladder network as a filter

It was shown in Section 10.1.2 that matched ladder networks comprising sections with both L and C could behave as filters, showing a pass band with zero attenuation and a stop band in which the attenuation α increased beyond cut-off. This should not be very surprising, since we saw in Section 5.4 that the functions discussed there arose from combinations of L and C.

Cases (ii) and (iii) of Section 10.1.2 showed that low-pass and high-pass behaviour arose when Z and Y comprised opposite kinds of reactance; in both cases the ratio Z/Y is therefore given by L/C, which is a constant for a given section, independent of frequency. A constant ratio can also result from a suitable choice of mixed reactances for Z and Y (in general impedances Z and $1/Y$ which have this property are known as **inverse impedances**), so it can therefore apply to a ladder acting as a band-pass filter, as in case (iv) of Section 10.1.2. At one time matched ladder networks of this kind were in common use as filters; the ratio Z/Y was designated k^2, and the networks known as **constant-k** filters.

Although it would be possible in principle to convert the expression for α in eqn 10.6 into a voltage transfer function for any number of matched sections, we should not find that the network yielded the relatively simple rational functions met in Section 5.4: this too is hardly surprising, since its currents and voltages are those of a network having an infinite number of reactive elements. However it is evident that the more sections there are before the termination the better the filter, since the overall attenuation in the stop band is αn. Even so, constant-k filters are not very good: the roll-off rate due to attenuation beyond cut-off is not as rapid as is desirable, and there is difficulty in matching them. Equations 10.10 to 10.12 show that their iterative impedance in the pass band, for symmetrical sections, is a pure resistance which depends strongly on frequency; it ranges in value from $\sqrt{L/C}$, at d.c. in the low-pass case or at very high frequencies in the high-pass case, to zero (for T sections) or infinity (for Π sections) at cut-off. It follows that any choice of terminating resistance will inevitably result in mismatch over most of the pass band; this, we shall see, results in a loss of transmitted power (Section 10.2.3). Usually constant-k filters were terminated in the resistance $\sqrt{L/C}$, so that the worst mismatch occurred near cut-off.

Filters based on ladder networks were considerably improved, as to both attenuation and matching, by making certain systematic adjustments to both Z and Y of a constant-k section taken as the prototype; these adjustments were expressed in terms of a factor m, between zero and unity, and the resulting filters were known as **m-derived**. The sections used were not necessarily identical or symmetrical, and designs were therefore based on image impedances (Section 4.7). These m-derived filters and their constant-k prototypes (which correspond to $m = 1$) together formed the basis of what is often called classical filter design; even before the widespread use of active filters they were overtaken by the kind of approach to analogue filters outlined in Section 5.4.

10.2 Waves

10.2.1 Travelling waves

In previous sections it was found that a ladder network of L and C elements could produce a phase shift β per section, and βn over

n sections, without change of amplitude. Because voltages and currents were supposed to be steady-state a.c. quantities, they have been represented by phasors $V_n = V_0 e^{-\gamma n}$ and $I_n = I_0 e^{-\gamma n}$ at the nth section, where V_0 and I_0 are reference phasors at the input; we now consider the case of no attenuation, so γ is simply jβ and these become $V_0 e^{-j\beta n}$ and $I_0 e^{-j\beta n}$. However further insight can be had by writing V_n and I_n as functions of time. Suppose that the phasor V_0 represents the actual voltage $v(t) = V_m \cos \omega t$; this can be written in exponential form as $V_m e^{j\omega t}$ if it is understood as usual that the real part is to be taken (Section 3.3.2). In this notation it then follows that

$$V_n = V_m e^{j\omega t} e^{-j\beta n} = V_m e^{j(\omega t - \beta n)} \tag{10.15}$$

of which the real part gives the actual time function

$$v(t) = V_m \cos(\omega t - \beta n). \tag{10.16}$$

Equation 10.16 has a very important form: it describes a **travelling wave**. It can equally well be expressed as a sine, or given an additional arbitrary phase angle, or left in exponential form; that makes no difference to its basic attributes, which arise from the angle $(\omega t - \beta n)$. A wave of voltage and current on a network is no different in its basic character from the many other kinds of wave which arise in physical systems and in nature, but since these are mostly encountered in a continuous medium it is more usual to see expressions like eqn 10.16 with a distance variable instead of the integer n; that affects the dimensions of β but not the principle, and it is not difficult to think of n as a distance along the network measured in discrete intervals from the input. (Sometimes an entirely artificial quantity is introduced to represent the 'length' of a section so that β can be a phase shift per unit length, as in other contexts.) Later in this chapter we shall consider continuous waves of voltage and current which really do depend upon a physical distance rather than a number of sections.

Equation 10.16 shows that at any instant of time the voltage v varies in a sinusoidal pattern along the network. After a small interval τ, say, the changes in value at each section have the net effect of moving the pattern to a different position, as shown in Fig. 10.5, in a direction dependent on the sign of β; with positive β (which with our solution in the form $e^{-\gamma n}$ means that current and voltage in a section lag those of the previous section) the pattern moves away from the input, in the direction of increasing n. This movement is said to constitute a **forward wave**. Negative β, such as we found for the C–L ladder in case (iii) of Section 10.1.2, results in a **backward wave**, in which the pattern moves towards the input, in the direction of decreasing n. We shall find later that it is often preferable to describe a backward wave by the alternative solution with $e^{\gamma n}$, retaining β as a positive quantity; the angle $(\omega t - \beta n)$ in the above expressions then becomes $(\omega t + \beta n)$. Because

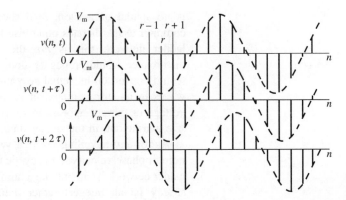

Fig. 10.5
A travelling wave of voltage on a
ladder network.

voltage and current vary with n in the same way, a forward wave of voltage is accompanied by a forward wave of current and likewise for backward waves. Whether the quantity concerned is expressed as a cosine, as in eqn 10.16, an exponential (as in eqn 10.15), or a sine, it is the relationship of sign between ωt and βn which governs the movement of the wave pattern: thus, for positive β, $\sin(\beta n - \omega t)$ and $\sin(\omega t - \beta n)$ would both correctly represent a forward wave, travelling in the direction of increasing n, although it is conventional to write the argument in the latter form.

At this point an alternative notation should be mentioned which is often used for waves in general. The phase constant which we have designation β, following the more common practice in electrical engineering, is often instead given the symbol k, and it may be called either a wave number or a wave **vector** since it can be related to the direction of propagation of waves in a continuous medium. Attenuation is then accounted for, when necessary, not by a separate constant like α but by regarding k itself as the complex quantity $k' - jk''$. It then follows that

$$\mathrm{e}^{-jkx} = \mathrm{e}^{-jk'x}\mathrm{e}^{-k''x}$$

in which k' evidently corresponds to β and k'' to α. It may seem slightly perverse to have attenuation defined by the imaginary part of a phase constant in order to yield a real exponent; for practical purposes the α, β notation is usually preferred, and we retain that in what follows.

10.2.2 Wave velocities

From eqn 10.16 it is a simple matter to deduce the speed with which the voltage pattern moves along a reactive network. The cosine variation means that in the forward wave the voltage reaches a peak value when $\omega t - \beta n = 0$, so at node n a peak is reached when $t = \beta n / w$, and at node $n + 1$ when $t = \beta(n + 1)/\omega$, a difference in time of β/ω for one section; hence the velocity of the wave can be expressed as ω/β sections per second. For reasons which will become evident later, this quantity is called the **phase velocity** and given the symbol v_p. (Since velocity is formally a vector quantity and in the context of ladder networks there is no spatial dimension, ω/β should strictly be called a speed; but common usage dictates otherwise.) It is important to realize that for a ladder network this is *not* the velocity with which an electrical disturbance would propagate along an initially quiescent network (that would depend on its actual size and on the velocity of electromagnetic waves in its surroundings); it is simply the rate at which a pattern of voltage or current appears to move along the sections, whatever their size and layout, in the steady state.

As for any travelling wave, a **wavelength** can be deduced as the ratio of phase velocity to the cyclic frequency f, although in the case of a ladder network it can only be a number of sections (and should therefore strictly be an integer) rather than an actual length. In general the wavelength is the distance between peaks of the same sign on a travelling wave like that shown in Fig. 10.5; it is given the symbol λ and can be expressed as v_p/f or $2\pi v_p/\omega$ or $2\pi/\beta$; if this represented an actual distance rather than, as here, a number of sections, then $\beta/2\pi$ would represent the number of wavelengths in unit distance. Although this has little significance for ladder networks, it explains why β is sometimes referred to as a **wave number**.

In principle it is possible to evaluate the phase velocity ω/β in terms of the network parameters from dispersion relations such as eqns 10.5 and 10.8 of Section 10.1.2, illustrated in Figs 10.2 and 10.3. One thing is immediately clear from these diagrams: even in the pass band the velocity ω/β is not constant but depends on the frequency applied to the network; it is the analogy between this fact and optical dispersion which givens eqns like 10.2 their name. A consequence of a frequency-dependent velocity is that an input waveform which includes different frequencies will in general give rise to different waveforms at subsequent sections, and this clearly has adverse implications for the accurate transmission of signals. For a constant velocity, and no dispersion, β must be proportional to ω; in other words phase shift must be a linear function of frequency. This is exactly the condition specified for an ideal filter (Sections 5.4.1 and 9.3.3).

Because, as we saw in Section 10.1.2, the phase constant β can take different signs in different networks, so too can the phase velocity. Thus, the ladder with series L and shunt C gives a positive phase velocity, as we expect for a forward wave; while that with series C and shunt L, which we know to produce a backward wave, has a negative phase velocity.

In the above discussion we have been considering only the pass band of each ladder, that is the range of frequency for which the relation 10.3 is satisfied and for which there is no attenuation. This is precisely the range over which current and voltage propagate as waves in each case, and which the dispersion diagrams of Figs 10.2 and 10.3 immediately indicate. The stop band is not merely that range of frequency which gives attenuation; it is also the range over which waves cannot propagate. (Although in general it is progressive phase shift which gives rise to a travelling wave, the limiting value of $\pm\pi$ which in

Section 10.1.2 was found for β in the stop band does not correspond to such a wave.)

If two forward waves, both like that of eqn 10.16 except that their frequencies ω differ by a small amount $d\omega$, are superimposed it is simply a matter of trigonometry to show that their resultant can be written in the form

$$2V_m \cos(\omega t - \beta n) \cos(\tfrac{1}{2}t\,d\omega - \tfrac{1}{2}n\,d\beta)$$

where $d\beta$ is the small change in β which corresponds to the increment $d\omega$ according to the dispersion relation. This evidently represents a wave travelling at the phase velocity ω/β (approximately the same for both component waves) but modulated by the factor $\cos(\tfrac{1}{2}t\,d\omega - \tfrac{1}{2}n\,d\beta)$. By the same argument as we used before, the **envelope** of the modulation must be travelling at velocity $d\omega/d\beta$; this quantity, for any system which can support travelling waves, is known as the **group velocity** v_g. Its significance here lies in the fact that a signal in the form of amplitude modulation, which makes it the envelope of a carrier (Section 9.3.5), will traverse a dispersive network at the velocity v_g with its waveform intact provided that v_g is constant over the frequency range of the signal, in other words if the $\omega - \beta$ dispersion relation is linear over the relevant bandwidth, even though the phase velocity v_p of the carrier may be quite different. Figure 10.6 shows the variation with frequency of the two velocities, for both the $L-C$ ladder (from eqn 10.5) and the $C-L$ (from eqn 10.8). There are two main points to be noted: one, for each ladder the group velocity falls to zero at the cut-off frequency; secondly, although the $C-L$ network supports a backward wave, so that v_p is negative, its group velocity is positive. (We return to this point in the next section.) A further point about this ladder is the unlimited increase in both velocities with increasing frequency: in theory this is quite possible because velocity here involves a number of sections rather than a physical length. In practice, of course, the physical dimensions of a network would impose limits. We can see why this must be so from the fact that the actual connections between sections must have some series inductance and shunt capacitance, which at high enough frequencies would tend to cancel the intended reactances of each section and eventually cause the ladder to behave like the $L-C$ version. Expressed more simply, at high enough frequencies the connections behave like transmission lines, which, as we shall see later in this chapter, have natural limits to the velocity of waves which they can support.

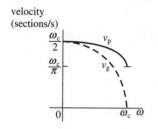

(a) series L, shunt C

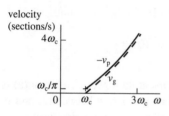

(b) series C, shunt L

Fig. 10.6

Phase and group velocities on $L-C$ and $C-L$ ladders.

Example 10.4 Find the phase and group velocities of waves on the network of Example 10.1, at (a) 50 Hz and (b) 30 kHz.

The phase velocity is immediately found from the previous results as ω/β. At 50 Hz β was found to be 0.003142 rad, which gives

(a) $v_p = 2\pi \times 50/0.003142 = \mathbf{10^5}$ sections per second;

and at 30 kHz β was 2.46 rad, giving in the same units

(b) $v_p = 2\pi \times 3 \times 10^4/2.46 = \mathbf{7.66 \times 10^4}$.

To find the group velocity we first need to find an expression for $d\omega/d\beta$. The appropriate dispersion equation for the series-L, shunt-C ladder is 10.5; differentiating that gives

$$2\sin(\beta/2)\cos(\beta/2)d\beta/2 = 2\omega LCd\omega/4$$

so the group velocity is

$$v_g = d\omega/d\beta = \sin\beta/\omega LC.$$

Substituting the appropriate values gives, at 50 Hz,

(a) $v_g = 0.003142/(2\pi \times 50 \times 10^{-3} \times 10^{-7}) = \mathbf{10^5}$,

again in sections per second; and at 30 kHz in the same units

(b) $v_g = 0.63/(2\pi \times 3 \times 10^4 \times 10^{-3} \times 10^{-7})$

$= \mathbf{3.34 \times 10^4}$.

The apparent identity of the two velocities at the lower frequency (a) is due to the fact that for small β the dispersion relation between ω and β, given by eqn 10.10 and shown in Fig. 10.4, is very nearly linear.

10.2.3   Reflection and standing waves

In previous sections it has been shown that certain ladder networks of indefinite extent can support waves of voltage and current in both directions within a certain range of frequency, and we have deduced the velocities of these waves. Mathematically, forward and backward waves are equally acceptable solutions to the circuit equations for the general section of a ladder network as illustrated in Fig. 10.1(a): the dispersion relation 10.2 is not affected by the sign allotted to γ. However, the possibility of a given kind of wave is no guarantee of its existence; in physical fact waves need to be excited by some source of energy, and for ladder networks we have assumed this to be an input to the first section. In that case we have found that an infinite or matched ladder with series L and shunt C sustains a forward wave while that with series C and shunt L sustains a backward wave. In the latter case it may seem unnatural that there there should be a backward wave, moving *towards* the input. However, it is still true that *energy* must flow in the forward direction, away from the input, for with a resistive Z_{in} there must be an input of power of which none can be dissipated in a purely reactive network and all of which must therefore be delivered to the remote end, whether at infinity or a terminating resistance. We can shed more light on this by recalling that although the C–L network has a negative phase velocity, its group velocity is positive. Hence, even although the wave is backward the group velocity is nevertheless in the forward direction, which is also the direction of energy flow.

The above discussion has assumed that Z and Y consist of only one reactive element each. If one or other, or both, comprise mixed reactances (case (iv) in Section 10.1.2) it can be shown that the phase velocity is then in the forward direction at frequencies for which Z is inductive and Y capacitive, and conversely. This remains true even when Z or Y includes some resistance, in which case propagation is accompanied by attenuation.

Now that we know what kind of wave an input source will produce on each type of ladder when it is infinite or matched, we should ask what will happen when the ladder is finite but unmatched, i.e. not terminated in its iterative impedance. Of course, one can imagine another source connected to the far end, which would presumably produce both backward and forward waves, by superposition; but a more important case in practice is that of some arbitrary passive termination.

Consider, for example, a network with its output terminals open-circuit. Clearly the relations assumed for a general section, which lead to eqns 10.1 and 10.2, simply cannot apply now, at least not to the last section: for one thing, its current output, far from having the same amplitude as that of the penultimate section, is zero; and we may suspect that the whole approach breaks down. On the other hand, the mathematical argument which showed that forward and backward waves were possible did not actually depend on setting n to infinity. In fact this difficulty with an end section, which has so far restricted us to infinite or matched networks, is one of satisfying boundary conditions, such as zero output current, or some ratio of voltage to current dictated by a load impedance. Termination by the iterative impedance simply ensures a boundary condition which can be met by a single wave. But any boundary conditions can be satisfied by the superposition of forward and backward waves of appropriate relative amplitude; so we are led to conclude that in general there will be a second travelling wave. The second wave will, of course, carry energy back towards the source; although this seems impossible on an infinite or matched network, it is precisely what happens with any other termination: the phenomenon is called **reflection**, because the second wave carries energy from the termination back towards the input. Like the original wave, this reflected wave may have its phase velocity v_p, or ω/β, in either direction, depending on the constituents of the network, but its group velocity, like the energy flow, is in the backward direction. The current in each series impedance Z due to the reflected wave will also flow in the opposite direction for the same node voltage; consequently, if we use the same reference direction for current in the two waves, the net current I_n in any Z is the difference of the forward and backward contributions, whereas the total node voltage V_n across any Y is by superposition the sum of the two. (This does not of course tell us about the *actual* directions of reflected voltage and current.) We shall return to the question of reference directions.

A finite ladder network which is terminated in its iterative or characteristic impedance carries no reflected wave and it follows that this termination will receive all of the power transmitted from the input source. Whether this is an absolute maximum depends, of course, on whether the source itself is matched to the network; but the analogy with power transfer in circuits generally (Sections 2.9 and 3.11.5) explains why a network terminated in this way is said to be matched.

The relationship between the incident and reflected waves is governed by the terminating impedance, since that dictates the boundary condition to be satisfied, and we shall see in a later section how **reflection coefficients** can be deduced. But an important point arises immediately: for a reflected wave the *input* end represents a termination which may not be matched, so it in its turn gets reflected there and presumably produces a third wave superimposed on the first. This seems to lead to an infinite series of waves: in a transient state of affairs that would indeed happen, but we are here considering only the steady state; that in general consists of two waves only, one in each direction, but they must satisfy boundary conditions at *both* ends, dictated respectively by the terminating impedance and by the properties of the input source.

The input impedance of a finite network also depends on the termination, and in general differs from those given by the expressions 10.10 to 10.14, which are correct only if the network is terminated in its iterative impedance and is therefore matched. The question of input impedance in other cases is considered further in those later sections which deal with transmission lines.

We know from previous sections that one of the properties of a travelling wave in a purely reactive network is its constant amplitude: only the phase changes from section to section. However, it is not immediately obvious how the resultant voltage and current will vary along a network which carries both forward and backward waves. To find out, it is a simple matter to combine two expressions in the form used for a forward wave in eqn 10.16. Thus a network carrying a forward wave of voltage, say, with amplitude A and a backward wave with amplitude B has a resultant voltage

$$v = A\,\cos(\omega t - \beta n) + B\,\cos(\omega t + \beta n + \delta) \tag{10.17}$$

where the phase constant β is taken to be a positive quantity, and the arbitrary phase angle δ is introduced because we have as yet no means of knowing the phase relationship of the two voltages at a given value of n. Expanding this expression gives

$$v = \cos \omega t \{A \cos \beta n + B \cos (\beta n + \delta)\}$$
$$+ \sin \omega t \{A \sin \beta n - B \sin (\beta n + \delta)\}$$

which can be written

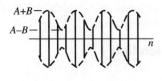

(a) $A \neq B$

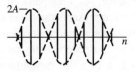

Fig. 10.7
Standing waves on a ladder
network.

$$v = X \cos \omega t + Y \sin \omega t$$
$$= \sqrt{X^2 + Y^2} \, \sin(\omega t + \phi) \qquad (10.18)$$

in which ϕ is $\tan^{-1}(X/Y)$; the coefficients X and Y are evidently periodic functions of n, i.e. of 'distance' along the network, but they are not time dependent. Hence v is an a.c. voltage whose amplitude varies periodically with position, forming a stationary envelope; it is said to constitute a **standing wave**, and an example is shown in Fig. 10.7(a). Clearly the maximum amplitude occurs where the forward and backward waves of eqn 10.17 are in phase, i.e. at the sections for which $\omega t - \beta n$ is equal to $\omega t + \beta n + \delta$ (or as nearly so as an integer n allows) and has there the value $A + B$; similarly the minima occur where the two waves are as nearly as possible in antiphase, i.e. where $\omega t - \beta n$ and $\omega t + \beta n + \delta$ differ by π or as near to that as possible, and have the value $A - B$. In optical terminology, the voltage shows the interference pattern formed by the superposition of the two waves. For the special case $A = B$ the above expressions give

$$v = 2A \, \cos(\beta n + \tfrac{1}{2}\delta) \, \cos(\omega t + \tfrac{1}{2}\delta) \qquad (10.19)$$

which is a standing wave whose peak amplitude can vary with n between $2A$ and zero and whose phase would be constant but for the change of sign at every zero which means that the phase changes by π. This case is illustrated in Fig. 10.7(b). Both positive and negative peak amplitudes are shown in the figure, but it is usually sufficient to show only the positive envelope and we change to this convention later.

It is evident from the above expressions that the change in n between consecutive peaks of amplitude (irrespective of phase) must be such that βn changes by π, and n therefore by π/β. From the definitions of Section 10.2.2 it follows that the distance between the peaks of a standing wave is *one half* of a wavelength, in contrast to the separation of the peaks of a travelling wave, which is one wavelength (the reason lies in the fact that both forward and backward waves are changing phase by 2π in a wavelength, so their phase difference changes by that much in a half wavelength). It should also be emphasized here that the standing-wave patterns of current and voltage, when they exist, are not coincident (that is, they do not have their maxima at the same sections); the reason lies in the difference in the effects of an unmatched termination on current and on voltage: it was pointed out earlier that the currents in the two waves oppose each other when the voltages reinforce each other.

The ideas of reflection and standing waves, or for that matter any waves, on a ladder network containing only a small number of sections n become rather academic, although none of the relationships so far encountered needs any restriction on n beyond the fact that it is an integer. These ideas are, however, important in continuous systems and we shall consider them further in the following sections, in which we deal with the limiting form of the reactive ladder network with series L

and shunt C, namely the transmission line. We shall also return to the question of how the inevitable presence of resistance, which makes Z and Y complex in general, can be taken into account.

10.3   The continuous transmission line

10.3.1   The transmission-line equations

In Section 10.1.3 the possibility of a ladder having very small Z and Y was mentioned in connection with input impedance. As it happens, this limiting case is of great practical importance, for it represents the general **transmission line** such as is very commonly used to carry electrical energy and/or information over a distance. For our purpose we may take a transmission line to be any arrangement of two or more continuous conductors having the required length and a uniform cross-section. It can take many forms in practice: they include the three-phase lines, whether underground or suspended on pylons, which carry over quite large distances most of the electrical energy we use; the twin wires used in many everyday applications; the coaxial cables which are familiar in communication equipment generally; and even the thin copper strips used for connections on printed circuit boards, sometimes known as **strip lines** or **microstrip** lines.

Like any device which carries current and sustains a voltage the transmission line must inevitably have some inductance and capacitance, resistance in its conductors and, at least in theory, conductance between them due to the material which separates them. In a continuous line these parameters are inextricably mingled, in contrast to the circuits so far considered which are made up of separate or **lumped** components each of which is usually designed to have one parameter dominant. It is nevertheless clear that the values of all four parameters for the whole line must increase with its length, and if follows that all can be given a value per unit length for an axially uniform line; it also follows that inductance and resistance must act like series elements, since they each are additive when in series, while capacitance and conductance, which are each additive when in parallel, are shunt elements. The intermingling of the parameters in a physically continuous line of any length can therefore be represented by dividing the line into small elements of length dx, say, each of which is like a section of a ladder network having infinitesimally small components dZ and dY. If L, R, C, and G now designate values *per unit length* then we may write

$$\begin{aligned}
dZ &= (R + j\omega L)\,dx = Z dx \\
dY &= (G + j\omega C)\,dx = Y dx
\end{aligned} \tag{10.20}$$

where Z and Y now likewise represent values per unit length. The elementary section can be represented by any of the arrangements shown in Fig. 10.1(b) to (e): the section corresponding to (b) is shown in Fig. 10.8. At this stage it would be possible to investigate the behaviour of a transmission line by using eqns 10.20 directly in the relations found

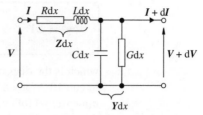

Fig. 10.8

An elementary section of transmission line.

earlier for ladder networks; however, it is informative to redo some of the analysis. Since the changes in steady-state voltage and current from section to section can now be written as dV and dI, the application of Kirchhoff's laws to Fig. 10.8 now yields the equations

$$dV = -ZdxI$$
$$dI = -YdxV$$

which combine to give

$$\frac{d^2I}{dx^2} = ZYI \qquad (10.21)$$

and

$$\frac{d^2V}{dx^2} = ZYV. \qquad (10.22)$$

These two differential equations are versions of what are known as the **transmission-line equations**; the current eqn 10.21 corresponds to the difference eqn 10.1, and 10.22 to its counterpart for voltage. Their solutions turn out to be travelling waves, as for ladder networks; now, however, the continuous nature of the line and the consequently infinitesimal nature of each section lead to some differences, and indeed to simpler results which allow us to pursue further in the following sections some of the ideas encountered for discrete networks. Ladder networks are sometimes known as **artificial transmission lines**.

10.3.2 Propagation, dispersion, and attenuation

Following the procedure for a ladder network in Section 10.1, the transmission-line equations in phasor form can be readily solved by postulating a solution with current and voltage depending not on a number of sections n but on physical distance x; thus, if γ is now the propagation constant per unit distance, representing as before a complex quantity $\alpha + j\beta$, we postulate a solution to eqn 10.21 of the form

$$I(x) = I(0)e^{-\gamma x}.$$

Substitution of this in the equation gives

$$\gamma^2 = ZY \tag{10.23}$$

which is the dispersion equation of the transmission line. It corresponds to eqn 10.2 for a network, and indeed its form (although not its distance dimensions) follows immediately from that equation when the former γ, the propagation constant *per section*, is taken to be very small. As before, the equations can be satisfied by both forward and backward waves but, compared with the former result, eqn 10.23 has some interesting consequences.

First there is now no apparent restriction on the magnitude of ZY which will allow propagation, and therefore we may expect no cut-off frequency. Secondly the nature of a transmission line means that Z is bound to be inductive and Y capacitive, as eqns 10.20 indicate, so according to our earlier results for networks we should expect an input source to produce propagation with both phase and group velocities in the forward direction. If R and G are taken to be zero for the time being (implying a **lossless** line, in which conductors and insulation are both perfect), so that $Z = j\omega L$ and $Y = j\omega C$, then the dispersion eqn 10.23 can be written

$$\gamma^2 = -\omega^2 LC. \tag{10.24}$$

It follows that, as we might expect, there is no attenuation and that the phase constant is

$$\beta = \omega\sqrt{LC}, \tag{10.25}$$

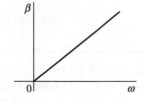

Fig. 10.9
The dispersion relation for a lossless line.

which is the linear relationship shown in Fig. 10.9. There is *no* dispersion: the phase velocity is independent of frequency and is given by

$$v_p = \frac{\omega}{\beta} = \frac{1}{\sqrt{LC}} \tag{10.26}$$

which is also the group velocity v_g at all frequencies. It follows that the lossless line transmits any signal perfectly: not only is there no loss of amplitude, but all frequency components take the same time to traverse the same distance (or, what is equivalent, suffer a total phase shift proportional to frequency), so preserving the original waveform. The lossless transmission line shares this particular property with the ideal filter (Section 9.3.3).

It is worth a digression here to note that the product LC for a uniform transmission line in a homogeneous medium, whatever its cross-section, is given to a good approximation by $\mu\epsilon$, the product of the magnetic permeability and the electrical permittivity of the medium (Appendix B); the velocity of propagation given by eqn 10.26 is therefore close to $1/\sqrt{\mu\epsilon}$, which is the velocity of light in the medium. Waves of current and voltage on the line therefore travel at the speed with which their fields propagate in the surroundings as electromagnetic waves.

Thus far we have found that a lossless transmission line will propagate forward waves (ignoring reflections for the time being) at the speed of light, at all frequencies, without dispersion. Now let us consider losses in the form of finite resistance R in the conductors, and conductance G in the dielectric or insulation. Setting Z to $(R + j\omega L)$ and Y to $(G + j\omega C)$ in eqn 10.23 leads to the more general dispersion equation

$$\gamma^2 = (R + j\omega L)(G + j\omega C) \qquad\qquad (10.27)$$

and this clearly gives complex values of γ indicating, as expected, that there is both attenuation and propagation.

From eqn 10.27 we can deduce the properties of a travelling wave, whether of voltage or current, on a line with losses represented by R and G. It seems obvious that propagation, which simply requires γ to have an imaginary part, is still possible at all frequencies, but it appears unlikely that we shall still find the constant velocities and dispersionless behaviour which followed from the simple form of eqn 10.24. An example will quickly illustrate the point.

Example 10.5 Find the phase velocity and the attenuation of waves on a transmission line with the parameters $L = 2$ mH km^{-1}, $C = 50$ nF km^{-1}, $R = 5\ \Omega$ km^{-1} and $G = 0$, at frequencies (a) 50 Hz and (b) 100 kHz.

From eqn 10.27 the propagation constant is here given by

$$\gamma^2 = (5 + j\omega \times 2 \times 10^{-3})\, j\omega \times 50 \times 10^{-9}.$$

(a) At 50 Hz this gives, if kilometre units are retained for convenience,

$$\gamma^2 = (5 + j200\pi \times 10^{-3})\, j5000\pi \times 10^{-9}$$
$$= (-9.870 + j78.54) \times 10^{-6}$$
$$= 79.16\underline{|97.16°} \times 10^{-6}\ \text{km}^{-2}$$

from which

$$\gamma = \alpha + j\beta = 8.897\underline{|48.58°} \times 10^{-3}$$
$$= (5.886 + j6.672) \times 10^{-3}\ \text{km}^{-1}.$$

The phase velocity at 50 Hz is therefore

$$v_p = \omega/\beta = (100\pi/6.672) \times 10^3$$
$$= \mathbf{4.71 \times 10^4}\ \textbf{km s}^{-1}$$

and the attenuation can be expressed as

$$\alpha = 5.89 \times 10^{-3} \text{ nepers km}^{-1}$$

or as

$$8.686\alpha = 0.0511 \text{ dB km}^{-1}.$$

(b) At 100 kHz a similar calculation gives

$$\gamma^2 = -39.48 + \text{j}0.1571 = 39.48\,\underline{|179.77}\ \text{km}^{-2}$$

and so

$$\gamma = 6.283\,\underline{|89.89°}\ \approx 0.0125 + \text{j}6.283 \text{ km}^{-1}.$$

This now gives a phase velocity

$$v_\text{p} = \omega/\beta = 2\pi \times 10^5/6.283 = 10^5 \text{ km s}^{-1}$$

and an attenuation

$$\alpha = 0.0125 \text{ nepers km}^{-1}$$

or

$$8.686\alpha = 0.109 \text{ dB km}^{-1}.$$

It will be observed that the phase velocity at the higher frequency is (to the accuracy of the calculation) the same as the value for a lossless line given by $1/\sqrt{LC}$ but at the lower frequency it is reduced by the effect of the resistance R, which means that the line shows dispersion. The attenuation too is greater at the higher frequency despite the fact that the propagation constant has an angle very close to 90°.

Equation 10.27 does not in general yield simple expressions for α and β, but one or two special cases (apart from the lossless line already considered) are summarized in what follows.

(i) *The distortionless line*

In the event that the line parameters are in the relationship

$$R/L = G/C = a, \tag{10.28}$$

where a is a constant, then eqn 10.27 can be written

$$\gamma^2 = LC(a + \text{j}\omega)^2$$

and therefore

$$\gamma = (a + \text{j}\omega)\sqrt{LC}$$

from which

$$\alpha = a\sqrt{LC} = R\sqrt{\frac{C}{L}} = G\sqrt{\frac{L}{C}} = \sqrt{RG};$$

(10.29)

$$\beta = \omega\sqrt{LC}.$$

Thus the line has exactly the same phase constant as the lossless line (eqn 10.25) and the same constant propagation velocity (eqn 10.26); moreover, its attenuation is independent of frequency. It follows that the shape of any original waveform is preserved; the only difference from the lossless line is that the amplitude decays with distance. Such a line is said to be **distortionless** (like the ideal filter mentioned in Section 9.3.3). It is of course unlikely that a normal transmission line will have parameters that happen to satisfy the condition of eqn 10.28. Usually R/L is greater than G/C, because with a good dielectric G is likely to be very small whereas there are practical limits to the reduction of the conductor resistance R. To satisfy eqn 10.28 it is possible artificially to adjust the parameter values, and the most effective way of doing this is to increase L by **loading** the line, either by using magnetic material as well as dielectric, or by adding lumped inductive elements at intervals.

(ii) *Low frequencies*

Because, as pointed out above, R/L is likely to exceed G/C there may exist a range of frequency for which $R \gg \omega L$ but $G \ll \omega C$. In that case eqn 10.27 reduces to the approximation

$$\gamma^2 = j\omega CR$$

and it follows that

$$\gamma = \alpha + j\beta = \sqrt{j\omega CR} = (1 + j)\sqrt{\omega CR/2}$$

from which

$$\alpha = \beta = \sqrt{\omega CR/2}.$$

(10.30)

Under these conditions the line is evidently dispersive, since β is not proportional to ω.

By setting ω to zero in eqn 10.27, we find that in the d.c. limit

$$\alpha = \sqrt{RG}$$
$$\beta = 0.$$

Thus the amplitude of voltage and current on the line is attenuated but, as we might expect, no wave is propagated. We have met this situation before, in non-propagating ranges of frequency for ladder networks. It can arise in different physical contexts, and is often said to constitute an **evanescent** wave; but on a normal transmission line it is less significant since it arises only for d.c. and does not occur at any non-zero frequency. It is interesting to note that if G were zero there would be no d.c. attenuation, however large R may be. This apparently odd result makes sense if we reflect that for d.c. no current would flow to an infinite line with $G = 0$ because the shunt admittance Y is then zero and the input impedance therefore infinite.

(iii) *The high-frequency or low-loss line*

If the frequency is so high, or the losses so low, that $R \ll \omega L$ and $G \ll \omega C$ then it can be shown from the binomial expansion of eqn 10.27 that, to a first approximation,

$$\alpha \approx \frac{1}{2}\left(R\sqrt{\frac{C}{L}} + G\sqrt{\frac{L}{C}}\right)$$

(10.31)

$$\beta \approx \omega\sqrt{LC}.$$

Here again we have approximately the same dispersionless phase constant as for the lossless and distortionless lines, and an attenuation which is independent of frequency; so it appears that if the frequency is high enough any line will become distortionless and will tend to a limiting value of attenuation. This is not quite true in practice, for as the frequency increases the parameters of the line change: for example, the skin effect increases R and dielectric losses increase the effective value of G (Appendix B). It should also be noted that, while the propagation velocity ω/β according to eqns 10.31 has to a first approximation the same constant value $1/\sqrt{LC}$ as for the lossless and distortionless lines, a more precise analysis shows that the effect of losses generally is somewhat to reduce the phase velocity. It can be shown from eqn 10.27 that the fractional decrease in v_p from the lossless value $1/\sqrt{LC}$ is closely given by

$$\frac{\delta v_p}{v_p} = \frac{1}{8}\left(\frac{R}{\omega L} - \frac{G}{\omega C}\right)^2.$$

(10.32)

Example 10.6 Compare the results for the transmission line of Example 10.5 at the frequencies (a) and (b) with those obtained from the low- and high-frequency approximations respectively.

(a) Since the given line has $G = 0$, eqn 10.30 is the appropriate low-frequency approximation. The phase velocity becomes

$$v_p = \omega/\beta = \sqrt{2\omega/CR}$$

$$= \sqrt{200\pi/(5 \times 10^{-8} \times 5)} = \mathbf{5.01 \times 10^4 \ km \ s^{-1}}$$

which is some 6% higher than the more precise answer obtained in Example 10.5. Equation 10.30 also gives for the attenuation at 50 Hz

$$\alpha = \sqrt{(100\pi \times 250 \times 10^{-9}/2)} = \mathbf{6.27 \times 10^{-3} \ nepers \ km^{-1}}$$

which again is about 6% greater than the previous figure. Since at 50 Hz $\omega = 314.2$ and R/L is $5/0.002 = 2500$, the low-frequency approximation $\omega \ll R/L$ is not particularly good in this case; nevertheless the results it leads to are not dramatically inaccurate.

(b) At 100 kHz the phase velocity according to the high-frequency approximation, eqns 10.31, is given by the expression for the lossless line and is therefore identical to the figure obtained in Example 10.5. The high-frequency attenuation is given by eqn 10.31 as

$$\alpha = \tfrac{1}{2}R\sqrt{\frac{C}{L}} = 2.5\sqrt{\frac{5 \times 10^{-8}}{2 \times 10^{-3}}} = \mathbf{0.0125 \ nepers \ km^{-1}}.$$

which is again indistinguishable from the result found in Example 10.5. In this case it therefore appears that the high-frequency approximation gives results easily accurate enough for most practical purposes.

Equation 10.32 can be used to estimate the error in phase velocity which arises from the approximation and was too small to appear in the previous example. In this case we have

$$R/\omega L = 5/(2\pi \times 10^{5} \times 2 \times 10^{-3}) = 3.979 \times 10^{-3}$$

and the error is therefore

$$\delta v_p/v_p = (R/\omega L)^2/8 = 15.83 \times 10^{-6}/8 \approx 2 \times 10^{-6}$$

which confirms the closeness of the approximation.

10.3.3 Characteristic impedance and reflection

In Section 10.1.3 it was seen that for small Z and Y per section the input impedance of an infinite ladder network, and the characteristic impedance of a symmetrical section, becomes $\sqrt{(Z/Y)}$. It follows that the characteristic impedance of the infinitesimal sections of a transmission line is given by

$$Z_0 = \sqrt{\frac{Z}{Y}} \qquad\qquad (10.33)$$

where Z and Y are values per unit length. Since the ratio Z/Y is independent of length, Z_0 for a continuous line is a property which depends on the geometry and materials of its cross-section but not on its length; as for a network, it is measured in ohms. By analogy with ladder networks, it represents not only the input impedance Z_{in} of an infinite length of line but also that of a finite line terminated in an impedance of value Z_0, in other words of a matched line which can carry no reflected wave. (Although a continuous line of finite length can be infinitely subdivided into sections, it is nevertheless analogous to a finite network in that reflections must occur to satisfy boundary conditions at an unmatched termination.) The presence of a reflected wave means that the input impedance is no longer given by Z_0.

In general a line has $Z = R + j\omega L$ and $Y = G + j\omega C$, so eqn 10.33 can be written

$$Z_0 = \sqrt{\frac{R + j\omega L}{G + j\omega C}}. \tag{10.34}$$

This is in general a complex quantity, but for a lossless line, or in the high-frequency limit, it becomes

$$Z_0 = \sqrt{L/C} \tag{10.35}$$

which is a real quantity, representing a pure resistance. At sufficiently low frequencies it approaches the d.c. limit

$$Z_0 = \sqrt{R/G}$$

which is again, of course, real.

Example 10.7 Find the components needed to match the line of Example 10.5 (a) at 50 Hz and (b) at 100 kHz.

With the parameters given, the characteristic impedance is

$$Z_0 = \sqrt{Z/Y} = \sqrt{\{(5 + j2\omega \times 10^{-3})/(j50\omega \times 10^{-9})\}}$$
$$= \sqrt{\{-(j/10\omega)10^9 + 4 \times 10^4\}}.$$

At 50 Hz, $\omega = 314.2$ and this becomes

$$Z_0 = \sqrt{(4 \times 10^4 - j3.183 \times 10^5)} = \sqrt{(3.208 \times 10^5 \underline{|-82.84°})}$$
$$= 566.4\underline{|-41.42°}$$
$$= 424.7 - j374.7 \ \Omega.$$

This is the value of the impedance required to match the line. The negative imaginary term implies capacitance C of such value that

$$1/314.2C = 374.7$$

so the necessary impedance can be provided by a capacitance

$$C = 1/(314.2 \times 374.7) = \mathbf{8.5} \ \mu\mathbf{F}$$

in series with a resistance of **425 Ω**. A parallel combination having different values (found from the real and imaginary parts of the admittance $1/Z_0$) would, of course, serve equally well.

At 100 kHz only the imaginary part of Z_0^2 is different (in this example) and since it is reduced in inverse proportion to frequency the required impedance now becomes

$$\mathbf{Z}_0 = \sqrt{(4 \times 10^4 - \mathrm{j}159.2)} = \sqrt{(4 \times 10^4 \underline{|-0.228°})}$$
$$= 200\underline{|-0.114°}$$
$$= 200 - \mathrm{j}0.398 \; \Omega.$$

Hence the required impedance is now a resistance of **200 Ω** in series with a capacitance of value

$$C = 1/2\pi \times 10^5 \times 0.398 = 4 \; \mu F.$$

In this case, however, with a very small negative reactive component, it is much more economical to use parallel components (because the small reactance requires a large capacitance whereas the corresponding small susceptance requires a small capacitance). The required susceptance is the imaginary part of the admittance $1/\mathbf{Z}_0$ and is very nearly $0.398/200^2 = 9.95 \times 10^{-6}$ S; since this is the value of $\omega C'$ for a parallel capacitance C', we find

$$C' = 9.95 \times 10^{-6}/2\pi \times 10^5 = \mathbf{15.8 \; pF},$$

a reduction in the order of 10^6 compared with the series value.

Because travelling waves of voltage and current vary in the same way along the line, namely as $\mathrm{e}^{-\gamma x}$, it follows that $\mathbf{Z}_0$ is not only the ratio of voltage to current at the input to a matched (or infinite) line, it is also their ratio at any point on the line; it is in fact the ratio of the two everywhere *on any one travelling wave*, whether forward or reflected, since their variation has the same form for both directions. There may, however, be a change of sign, depending on the reference directions for the two currents. It can be confirmed directly from the elementary section of Fig. 10.8 that voltage and current are related by the impedance $\sqrt{\mathbf{Z/Y}}$, and that (as we noted for ladder networks) the series currents due to forward and backward waves are in opposition at a given point on the line.

If a line of finite length is terminated by some load impedance $\mathbf{Z}_L$ there will be a reflected wave unless $\mathbf{Z}_L$ happens to have the value $\mathbf{Z}_0$. Let us now address the question which was postponed when reflections on ladder networks were discussed in Section 10.2.3, namely the relation between the forward and reflected waves and its dependence on $\mathbf{Z}_L$. A forward wave of voltage in the steady state may be written as the phasor $V_1\mathrm{e}^{-\gamma x}$ where V_1 is its value at the input[†]; likewise a backward wave of voltage can be written as $V_2\mathrm{e}^{\gamma x}$ where V_2 is its value at the *input* end and γ is the same as before, with both parts positive.

[†] Sometimes the output end is taken to be the origin of x, which has some advantages.

(An attenuated backward wave *increases* in amplitude with x, so includes the factor $e^{\alpha x}$.) The total voltage phasor at any x is then

$$V(x) = V_1 e^{-\gamma x} + V_2 e^{\gamma x}. \tag{10.36}$$

It is worth noting that in this V_1 is *not* the actual input voltage to the line, which must be $V_1 + V_2$: in the steady state this is a boundary condition which the two contributions must satisfy; at present, however, we are more interested in the boundary condition at the output termination.

For each of the two voltage waves there is a corresponding current wave, and the ratio of voltage to current in each direction is given by the characteristic impedance Z_0. However, the forward- and backward-wave currents flow in opposite directions for a given node voltage; so, if we keep the same (forward) reference direction for both the backward-wave current can be written

$$I_2 = -V_2/Z_0$$

(which is sometimes expressed as the equivalent statement that the backward-wave impedance is $-Z_0$.)

Thus the total current can be written

$$I(x) = I_1 e^{-\gamma x} + I_2 e^{\gamma x} = (V_1 e^{-\gamma x} - V_2 e^{\gamma x})/Z_0. \tag{10.37}$$

Now at the termination the ratio of total voltage to total current must be equal to the load impedance Z_L; if the line has length l then, keeping the input as origin, we must have

$$Z_L = \frac{V(l)}{I(l)} = Z_0 \frac{V_1 e^{-\gamma l} + V_2 e^{\gamma l}}{V_1 e^{-\gamma l} - V_2 e^{\gamma l}}.$$

Dividing this by $V_1 e^{-\gamma l}$ and rearranging it gives the ratio of the reflected, i.e. backward, voltage to the incident, i.e. forward, voltage at the termination as

$$\rho_v = \frac{V_2 e^{\gamma l}}{V_1 e^{-\gamma l}} = \frac{Z_L - Z_0}{Z_L + Z_0} \tag{10.38}$$

and this dimensionless quantity, complex in general, is the **reflection coefficient** for voltage. For a lossless line terminated by a pure resistance, both Z_L and Z_0 are real, and so too therefore is ρ_v; and as we should expect ρ_v is zero when $Z_L = Z_0$. Because we have defined I_2 as $-V_2/Z_0$ it follows immediately that the reflection coefficient for current is

$$\rho_i = -\rho_v \tag{10.39}$$

provided that we retain the same reference direction for both currents. (It is however not uncommon to use different reference directions, and there is then only a single reflection coefficient for both voltage and current.) The plausibility of eqn 10.39 is evident on considering a line terminated by an open circuit, in which case the two currents must cancel but the total voltage might be expected to be maximum; or a short-circuit termination, when the reverse is true. It is this difference between the reflection of currents and voltages, whatever sign convention is used, which means that the standing waves formed by their resultants (on an unmatched line) are not coincident. This was pointed out for ladder networks in Section 10.2.4; we return to standing waves in a later section.

Because a travelling wave of voltage $V(x)$ is always accompanied by a corresponding current wave $I(x) = \pm V(x)/Z_0$, the power flow due to the wave at any x depends on the angle of Z_0 according to the usual calculation of a.c. power (Section 3.9). If the line is lossless, Z_0 is real; the power flow in a single wave is then simply V^2/Z_0 or I^2Z_0, where V and I are r.m.s. values, and does not vary with x. If there is attenuation, Z_0 has a phase angle and the power flow decreases with distance by dissipation in R and G. At the termination of the line, unless it is matched, the reflected wave carries part of the incident power back to the source and only the difference can be transferred to the impedance Z_L (if Z_L happens to be purely reactive, it follows from energy conservation that all the incident power must be reflected). Since V and I have the same angle between them in each wave, and eqn 10.39 shows that on reflection they suffer the same phase change after allowing for the change in direction of I, we may deduce that the reflection coefficient for power is $|\rho_v|^2$ (or $|\rho_i|^2$, which is the same thing).

Example 10.8 Find the voltage reflection coefficient for the line in Example 10.7 at the frequencies given if, to match it on the assumption that it is lossless, it is terminated by a pure resistance of value $\sqrt{L/C} = 200\ \Omega$.

(a) Taking the result of Example 10.7 for 50 Hz and inserting values in eqn 10.38 gives

$$\rho_v = (-224.7 + \mathrm{j}374.7)/(624.7 - \mathrm{j}374.7)$$
$$= (-2.247 + \mathrm{j}3.747)(6.247 + \mathrm{j}3.747)/(6.247^2 + 3.747^2)$$
$$= (-28.08 + \mathrm{j}14.99)/53.07$$
$$= -0.529 + \mathbf{j}0.282.$$

In this case the mismatch is **therefore considerable.**

(b) At 100 kHz the same process gives

$$\rho_v = \mathrm{j}0.398/(400 - \mathrm{j}0.398)$$
$$= (\mathrm{j}159.2 - 0.158)/400^2$$
$$= \mathbf{j}9.95 \times \mathbf{10^{-4}} - 9.88 \times \mathbf{10^{-7}}$$

which evidently represents a good approximation to the matched condition.

Example 10.9 Find the reduction, in comparison with the matched case, in the power delivered from a lossless transmission line of characteristic impedance 100 Ω to a resistive load of (a) 99 Ω and (b) 50 Ω.

From eqn 10.38 the voltage reflection coefficient here is

$$\rho_v = (Z_L - 100)/(Z_L + 100).$$

(a) For $Z_L = 99$ Ω this gives

$$\rho_v = -1/199$$

and the power reduction is therefore $1/199^2$, or 2.53×10^{-5}.

(b) For $Z_L = 50$ Ω we have

$$\rho_v = -50/150 = -1/3$$

and the power reduction is then $1/9$ or 0.111.

It may be deduced from this last example that the reduction in power delivered due to reflection is serious only for a gross mismatch.

10.3.4 Input impedance

Reflected waves, whether on a ladder network or a continuous line, must affect the input impedance so that it is no longer that for the matched condition. We can find the input impedance of a line in the general case by using the expressions given in the previous section for resultant voltage and current, eqns 10.36 and 10.37, and the definition of reflection coefficient, eqn 10.38.

From eqn 10.38, the reflected voltage which reaches the input can be written

$$V_2 = \rho_v V_1 e^{-2\gamma l}.$$

Inserting this in eqns 10.36 and 10.37 for $x = 0$ enables us to write the ratio of total input voltages V_S to total input current I_S as

$$Z_{in} = \frac{V_S}{I_S} = Z_0 \frac{V_1 + V_2}{V_1 - V_2} = Z_0 \frac{1 + \rho_v e^{-2\gamma l}}{1 - \rho_v e^{-2\gamma l}} \tag{10.40}$$

which of course reduces to Z_0 for $\rho_v = 0$. This result can also be expressed in terms of the terminating or load impedance Z_L by using eqn 10.38; after some rearrangement this gives

$$Z_{in} = Z_0 \frac{Z_L \cosh \gamma l + Z_0 \sinh \gamma l}{Z_L \sinh \gamma l + Z_0 \cosh \gamma l}. \tag{10.41}$$

Example 10.10 Find the input impedance of 1 km of the line of Example 10.9, with load impedances (a) and (b), at a frequency such that the phase constant is 0.01 m^{-1}.

Since the line is lossless γ is simply $j\beta$ and has the value here of j0.01, so γl is j0.01 $\times 10^3$ or j10 rad; hence

$$\cosh \gamma l = \cos 10 = -0.8391,$$
$$\sinh \gamma l = j \sin 10 = -j0.544.$$

(a) The input impedance for $Z_0 = 100\ \Omega$ and $Z_L = 99\ \Omega$ is therefore, by eqn 10.41,

$$Z_{in} = 100(-99 \times 0.8391 - j54.4)/(-j99 \times 0.544 - 83.91)$$
$$= 100(83.07 + j54.4)/(83.91 + j53.86)$$
$$= \mathbf{99.6 + j0.911\ \Omega}.$$

As might be expected, because the line is nearly matched, the input impedance is close to the value of Z_0.

(b) With $Z_L = 50\ \Omega$, we now have

$$Z_{in} = 100(41.95 + j54.4)/(83.91 + j27.2)$$
$$= \mathbf{64.3 + j44.0\ \Omega}.$$

The input impedance now differs considerably from Z_0, and has a large reactive component despite the fact that the line is lossless and the load resistive.

Apart from the matched condition, for which $\rho_v = \rho_i = 0$ and $Z_{in} = Z_L = Z_0$, the following two special cases of the terminating impedance are important.

(i) *The short-circuited line: $Z_L = 0$*
For this case eqns 10.38 and 10.41 give

$$\rho_v = -1; \quad \rho_i = 1; \quad Z_{in} = Z_0 \tanh \gamma l. \tag{10.42}$$

The values of ρ here are clearly necessary to fit the boundary conditions, since at a short circuit the two voltages must be in antiphase and the currents therefore in phase.

(ii) *The open-circuited line: $Z_L \to \infty$*
The same equations now give

$$\rho_v = 1; \quad \rho_i = -1; \quad Z_{in} = Z_0 \coth \gamma l. \tag{10.43}$$

Again the values of ρ agree with the boundary conditions, which at an open circuit required currents to be in antiphase and the voltages therefore in phase.

Although it may seem unlikely that a healthy transmission line should have either of these terminations, the two results are of practical importance (apart from representing possible faults). If we denote the two input impedances by Z_{SC} and Z_{OC} respectively, we evidently have

$$Z_{SC}Z_{OC} = Z_0^2$$

and a convenient practical measure of characteristic impedance is therefore

$$Z_0 = \sqrt{Z_{SC}Z_{OC}}. \qquad (10.44)$$

Equations 10.42 or 10.43 also yield the result

$$\gamma l = \tanh^{-1}\sqrt{Z_{SC}/Z_{OC}}.$$

This relation appears to afford a method of measuring the propagation constant for a line of known length, but since it can be satisfied by many values of γl it does not immediately yield a unique answer. The same difficulty arises in using eqns 10.42 and 10.43 to deduce a value for the distance to an open or short circuit when the line parameters are known; it can be resolved by finding the variation of Z_{in} with frequency.

Since Z_0 is real and γ is purely imaginary for a lossless line, and tanh γl is then also imaginary, it follows from eqns 10.42 and 10.43 that the input impedance of such a line terminated in an open circuit or short circuit is a pure reactance which may have either sign, and therefore behave like an inductance or like a capacitance, depending on its length and on the value of γ, i.e. on frequency. We deal with an application of this in the next section.

Other special cases of input impedance arise from particular values of γl, that is from particular combinations of length and frequency. These we consider in the next section.

Finally, it should be observed that all the statements made above about input impedance apply equally well to the impedance at any other point on the line, that is the impedance of the length of line between any point and the termination. It is therefore possible, for example, to use expressions for input impedance to sketch the variation of impedance along an unmatched line; at any point this would simply indicate the ratio of voltage to current at that point. Similarly the term 'reflection coefficient' is sometimes used for the ratio of reflected to forward voltage or current at any point, even when, as in a lossless line, only the phase of that ratio varies with distance.

10.3.5   Standing waves

As was seen for ladder networks in Section 10.2.3, the effect of a backward wave superimposed on a forward is to produce a standing wave; and eqns 10.18 and 10.19 are equally valid for continuous lines when β is a phase shift per unit distance rather than per section and βn therefore becomes βx. The wave itself becomes a smooth variation of amplitude along the line, rather than the series of discrete changes which occur between the sections on a ladder. As before, the standing wave due to a forward wave of amplitude A and a backward wave of amplitude B has a maximum amplitude of $A + B$ and a minimum of $A - B$. Now, however, we can relate B to A by means of a reflection coefficient: retaining voltage as an example, and again assuming the line to be lossless so that neither A nor B depend on distance, we may write B as $|\,\boldsymbol{\rho}_v\,|\,A$; the maximum voltage amplitude in the standing wave is then $A(1+ |\,\boldsymbol{\rho}_v\,|)$ and the minimum is $A(1- |\,\boldsymbol{\rho}_v\,|)$. The ratio of these is the **voltage standing-wave ratio**, or VSWR, and it is evidently given by

$$s = \frac{1+ |\,\boldsymbol{\rho}_v\,|}{1- |\,\boldsymbol{\rho}_v\,|}. \qquad\qquad (10.45)$$

Because $|\,\boldsymbol{\rho}_i\,| = |\,\boldsymbol{\rho}_v\,|$, the current standing-wave ratio is exactly the same and it is therefore usual to refer to the voltage ratio only.

In the case of a lossless line terminated in a short circuit or an open circuit, for which the reflection coefficients are unity (eqns 10.42 and 10.43), the VSWR is infinite, and the standing wave has the form illustrated for a ladder network in Fig. 10.7(b). The equation of the standing wave in these cases corresponds to eqn 10.19 for the ladder, and the resultant voltage has the same phase (or is in antiphase) at every point on the line; now, however, for a short-circuit termination we may set its amplitude to zero at $x = l$ so if we take the arbitrary angle δ to be zero the voltage can be written

$$v = 2A \sin \beta(l - x) \cos \omega t. \qquad\qquad (10.46)$$

For an open-circuit termination the voltage at $x = l$ has its maximum amplitude $2A$, since $\boldsymbol{\rho}_v = 1$, and the wave can be written

$$v = 2A \cos \beta(l - x) \cos \omega t. \qquad\qquad (10.47)$$

For a matched line there is no reflection and no standing wave, and the ratio given by eqn 10.45 is unity.

One or two further deductions about the standing wave can be made. Equation 10.38 shows that for a lossless line terminated in a pure resistance R_L the voltage reflection coefficient is real, and positive if $R_L > Z_0$; it follows that at such a termination the forward and backward voltages are in phase, so they add directly and the standing wave is therefore at a maximum. For $R_L < Z_0$ the reflection coefficient is negative real, the two voltages are in antiphase and there is a voltage minimum at the terminating load. Exactly the reverse argument applies to the currents, because $\boldsymbol{\rho}_i = -\boldsymbol{\rho}_v$. Hence we may state that a resistive termination on a lossless line is a point of voltage maximum and current minimum for $R_L > Z_0$, and vice versa for $R_L < Z_0$. The same cannot of course be said of the input end of the line, even if the source has a resistive output impedance, because there the ratio of total voltage $V_1 + V_2$ to total current $I_1 + I_2$ is the input impedance of the line, not the output impedance of the source.

Because the phase of voltage or current in a travelling wave changes with distance in opposite senses for forward and backward waves, it follows that the distance between maxima in a standing wave (that is, the distance between successive points where its forward and backward components are in phase) is that over which each travelling wave changes in phase by π, i.e. that distance x for which $\beta x = \pi$. The separation of maxima, and therefore also of minima, is thus π/β or precisely one half wavelength by the arguments outlined for networks in Section 10.2.2. Hence for both voltage and current the maxima occur at half-wave intervals, but by the argument of the previous paragraph the maxima of one coincide with the minima of the other.

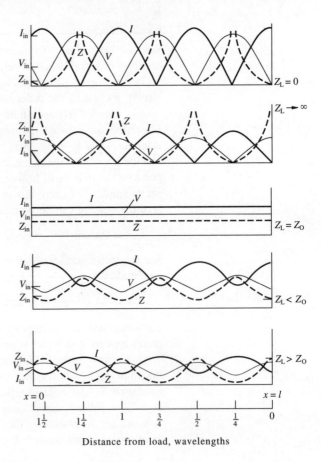

Fig. 10.10

Standing waves of voltage and current, and the variation of impedance, on a lossless line with different resistive terminations and fixed input voltage.

Distance from load, wavelengths

Attenuation causes the standing-wave ratio s to vary with distance along the line, being larger near the load termination: nearer the input the reflected wave is a smaller fraction of the forward wave and the maxima and minima, of both current and voltage, becomes less pronounced.

Figure 10.10 shows some of the standing wave patterns for a lossless line and resistive load discussed in this section; for clarity, and in contrast to Fig. 10.7, only the positive envelope is shown here. It is clear from the discussion above that in general the input conditions must correspond to a point on the standing wave which depends on the length of the line and the frequency, or, more simply, on the number of wavelengths on the line. This is no more than another way of stating that the input impedance depends on this length, as we already know from eqn 10.41, but it can be seen at a glance from the diagrams of Fig. 10.10, in which the impedance shown as Z is simply the ratio of total voltage to total current at any point and is the value which the input impedance would take if that point were the input. For example the input impedance will be a maximum or minimum when the line is a whole number of quarter wavelengths long: if that number is even then the input conditions are a replica of those at the output for a lossless line (if the output is a point of maximum voltage, for example, so too is the input); and if the number of quarter wavelengths is odd then the input conditions are the inverse of those at the output, in the sense that one is a voltage maximum and current minimum (i.e. an impedance maximum) while the other is a voltage minimum and a current maximum (an impedance minimum).

These deductions are easily confirmed from the expression for input impedance, equation 10.41: an even number of quarter wavelengths means that $\gamma l = j\beta l = j\pi, j2\pi, j3\pi, \ldots$ so $\sinh \gamma l = 0$ and the equation gives $Z_{in} = Z_L$; the input impedance of such a line is thus simply the load impedance, as it would be for a matched line. Now, however, the result does not depend on Z_0 and in general there is a standing wave. On the other hand an odd number of quarter wavelengths means that $\gamma l = j\beta l = j\pi/2$ or $j3\pi/2, j5\pi/2, \ldots$ so $\cosh \gamma l = 0$ and the same equation gives

$$Z_{in} = Z_0^2/Z_L, \tag{10.48}$$

an inverse dependence on Z_L. Thus, for example, in the extreme case of a short-circuited line, $Z_L = 0$, the input impedance of an odd number of quarter wavelengths is infinite; but if the same line is terminated by an open circuit it presents a short circuit at the input.

10.3.6 Matching

Equation 10.48 also shows that a lossless line whose length is an odd number of quarter wavelengths acts as an impedance transformer, effectively changing the value of the impedance presented to a source or another line by a load Z_L. This leads to a useful application in matching a load to a line. A load Z_L can evidently be matched to any line of

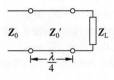

Fig. 10.11

The quarter-wave transformer.

impedance $\mathbf{Z}_0$ by interposing between the two a second line of such properties that its input impedance is $\mathbf{Z}_0$ when it is terminated by $\mathbf{Z}_L$. If the (real) characteristic impedance of the interposed line is $\mathbf{Z}_0{}'$ and its length is an odd number of quarter wavelengths, as shown in Fig. 10.11, then by eqn 10.48 we require

$$\mathbf{Z}_0 = \mathbf{Z}_0{}'^2/\mathbf{Z}_L$$

and the interposed line should therefore have characteristic impedance

$$\mathbf{Z}_0{}' = \sqrt{\mathbf{Z}_0 \mathbf{Z}_L}. \qquad (10.49)$$

A line which satisfies this conditions and is one quarter wavelength long (there being no point in using a line longer than necessary) is known as a **quarter-wave transformer**. Usually it is assumed that the line to be matched is also lossless, so $\mathbf{Z}_0$ too is real and eqn 10.49 can be satisfied only for resistive loads. Since the arrangement provides a matched load for the original line, there can be no standing wave in that line; on the other hand both junctions have a non-zero reflection coefficient since $\mathbf{Z}_0 \neq \mathbf{Z}_0{}'$ and $\mathbf{Z}_0{}' \neq \mathbf{Z}_L$. The apparent anomaly is due to the fact that a backward wave in the original line could be regarded as the sum of two components, one due to reflection of the forward wave as it arrives at the transformer and the other to transmission of the backward wave from the load as it arrives back at the line from the transformer: the conditions are such that these components precisely cancel.

There are other methods of matching a load to a transmission line. One is to use a normal coupled-circuit transformer, which was shown in Section 4.9 to transform impedance according to the square of its turns ratio. This is a preferred option at those relatively low frequencies for which a quarter-wavelength would be inconveniently long. If the frequency is low enough (not more than a few kilohertz or so) an iron-cored transformer is convenient but at rather higher frequencies the air-cored version is better; that too becomes unsatisfactory with increasing frequency, so that for higher radio frequencies, with wavelengths in the order of metres, the quarter-wave transformer is preferred.

Neither kind of transformer, however, can readily match a load with appreciable reactance to a virtually lossless line: a coupled circuit or a quarter-wave transformer can produce the correct effective resistance but cannot render the transformed load any less reactive (for eqn 10.48 shows that a quarter wavelength of lossless line can change the angle of $\mathbf{Z}_L$ only in sign). Fortunately, any reactance in the load can be cancelled by using an additional length of line to provide an equal and opposite reactance. In Section 10.3.4 it was seen that a length of lossless line terminated in either an open or a short circuit has a purely reactive input impedance, of value depending on length. A **matching stub** is a short length of line, up to one half-wavelength, which is terminated by a short circuit or an open circuit (the former is often preferred since the effective length may be more easily varied) and which may be connected at its input end in parallel with the load; if its susceptance cancels that of the load then the latter is reduced to a pure resistance. If the load resistance itself is not matched to the line, then a stub can be used to transform the complete load impedance as required, by connecting it across the line at some distance from the load: that distance and the length of the stub constitute two variables which enable the line as a whole to be correctly matched in respect of both resistance and reactance. In practice it may be more convenient to use instead two or three stubs of variable lengths but fixed spacings, connected to different points on the line. Where a stub is connected across the line its input admittance is added to that of the remainder of the line and load to give the admittance which in effect terminates the line at that point.

10.3.7 The Smith chart

Some of the relationships which govern the behaviour of transmission lines can be conveniently expressed graphically for practical purposes. Several kinds of diagram can be used; one of the most common is a polar form known as the Smith chart, of which a practical version is shown in Fig. 10.12. This circular diagram in essence relates the complex ratio, in polar coordinates, of the backward and forward voltage phasors at any point on the line to the real and imaginary parts of the line impedance or admittance at that point, when the value of the load impedance is specified. The chart is made universal by referring all impedances to the characteristic value Z_0, to give **normalized** values, and by expressing distances along the line in units of wavelength. It can be shown that the real and imaginary parts r and x of the normalized

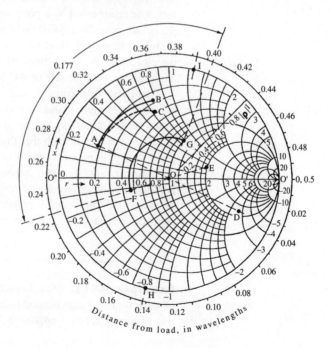

Fig. 10.12
The Smith chart.

impedance of the line (with its load) are defined by the two orthogonal families of circles given on the chart, each circle being a locus of constant r or x. The angular coordinate of any point, with the centre of the whole diagram as origin O, represents both the phase change of the voltage ratio and distance along the line; the radial coordinate represents not only the ratio of voltage magnitudes (as on the radial scale shown) but also, by a calibration based on eqn 10.45, the voltage standing-wave ratio. On the chart, the complex ratio of the two voltages at a point is usually referred to as the reflection coefficient ρ for that point; in other contexts it is more usual to apply that term only at the load (as in previous sections of this chapter) or at a discontinuity.

The Smith chart can be used in the following way, for example. If the normalized load impedance is $r + jx$, then the intersection of the circles for the values r and x specifies the point A, say, whose polar coordinates give the reflection coefficient at the load. The ratio of voltages at another point, a fraction m of a wavelength from the load, say, will differ from this only by the phase angle $4\pi m$ (equivalent to $2\beta d$ for an actual distance d) if the line is lossless; the new ratio is therefore specified by a point on the chart such as B and the impedance of the line at that point can then be read from the appropriate circles for r and x. The angle $4\pi m$ from A to B is easily marked off by means of the wavelength scale around the periphery. (This scale is arbitrarily chosen to be zero at the angle which corresponds to $r > 1$, $x = 0$; it therefore conveniently represents distance from a resistive load exceeding Z_0 on a lossless line, but in general it can be used only to give relative distances.) If the line is not lossless then the ratio of voltages will also decrease in magnitude with distance from the load, by the factor $2\alpha d$ or $4\pi m\alpha/\beta$, and is therefore represented by a point such as C having the same angle as B (if we ignore any effect on β) but a smaller radius; the impedance readings change accordingly. Starting at the point representing the load we may therefore find the conditions elsewhere on the line simply by moving along an arc such as AB or AC in a clockwise direction.

The chart can equally well be used with normalized values of admittance, which would be natural in dealing with stub matching, for example, as described in the previous section. The geometry of the impedance circles is such that the reciprocal of a complex value, that at point A say, is given by the diametrically opposite point, indicated on the figure by D; thus the impedance at point A is equivalent to the admittance whose normalized real and imaginary parts g and b are those of point D. One useful property of the chart, not necessarily in connection with transmission lines, is the inversion of complex numbers in this way.

Example 10.11 Find from the Smith chart the length and position of a short-circuit stub needed to match a load $140 + j35 \ \Omega$ to a lossless line of characteristic impedance $70 \ \Omega$.

The normalized load impedance is

$$z = r + jx = \mathbf{Z}_L/\mathbf{Z}_0 = (140 + j35)/70 = 2 + j0.5$$

which is the point E on Fig. 10.12. Since the stub is purely reactive, we need to find a point on the line where the normalized admittance has real part unity and then specify the length of the stub so as to cancel the imaginary part. The load admittance is represented by the point F, opposite to E; the conditions elsewhere on the line are represented by points on an arc through F centred on the origin O. To find an admittance with real part unity we must therefore trace such an arc as far as point G, a movement which by the peripheral scale represents about 0.177 of a wavelength λ; the stub should therefore be attached **0.177** λ from the load. The admittance at G has an imaginary part of about 0.8, so the stub needs to provide a normalized susceptance of –0.8. It short-circuited end can be represented by the infinite admittance point O′; from there a clockwise arc around the periphery must reach the point H in order to give a susceptance of –0.8. The distance represented by the arc O′H is approximately **0.14** λ, and this is the required length of the stub. It may be noted that the same result is obtained if the stub is specified in terms of impedance rather than admittance: its required input impedance is the reciprocal of –j0.8, namely j1.25; starting now from the zero point O″ to represent the short circuit a clockwise arc reaches the value j1.25 at the point I, and the arc O″I again gives a distance of about 0.14 of a wavelength.

10.3.8 Equivalent networks and the short line

Although in general its elements must be regarded as distributed rather than lumped, a transmission line of fixed length is still a passive four-terminal network whose input and output voltages and currents are in a frequency-dependent relationship. In this respect it is like any other four-terminal network, so we may expect that it can be described by the sets of parameters discussed in Section 4.2 and that it can be represented by T and Π equivalent circuits as outlined in Section 4.3. We must also expect, of course, that such representations of a line will involve its length as well as its other constants.

The essential relations between the input and output quantities of a transmission line are given by eqns 10.36 and 10.37 with $x = 0$ and $x = l$ respectively. Of the parameters described in Section 4.2 the transmission (a) and inverse transmission (b) sets are the most likely to be useful for a line. By rearranging the above equations and using subscripts S and L to denote the input (source) and output (termination, or load) terminals respectively, the input quantities can be readily expressed as

$$V_S = V_L \cosh \gamma l + I_L Z_0 \sinh \gamma l$$
$$I_S = V_L (\sinh \gamma l)/Z_0 + I_L \cosh \gamma l. \tag{10.50}$$

It should be recalled here that our transmission-line convention for current means that I_L flows out of the positive output terminal, and the coefficients in eqn 10.50 are therefore the A, B, C, D transmission parameters as opposed to the a parameters, for which the opposite convention is used (Section 4.2). The a parameters for the line are given by the same coefficients expect that those of I_L are changed in sign. Thus

$$a_{11} = \cosh \gamma l, \; a_{12} = -Z_0 \sinh \gamma l,$$
$$a_{21} = (\sinh \gamma l)/Z_0, \; a_{22} = -\cosh \gamma l$$

and we may note that these satisfy the reciprocity condition

$$a_{11}a_{22} - a_{12}a_{21} = -1$$

for a linear passive network, given in Section 4.2.3. The inverse transmission parameters can be obtained either by inverting the a matrix or by rearranging eqns 10.50 to give

$$V_L = V_S \cosh \gamma l - I_S Z_0 \sinh \gamma l$$
$$I_L = -V_S(\sinh \gamma l)/Z_0 + I_S \cosh \gamma l.$$

Here the coefficients give the b parameters if both signs in the second equation are reversed; thus we find

$$b_{11} = \cosh \gamma l, \; b_{12} = -\sinh \gamma l,$$
$$b_{21} = (\sinh \gamma l)/Z_0, \; b_{22} = -\cosh \gamma l.$$

These are identical to the a parameters, a result which is only to be expected since the transmission line is perfectly symmetrical: input and output can be exchanged with no effect on its behaviour.

The elements of the T and Π equivalents for a line can be found from the above parameters by equating the expressions in either set to the corresponding a or b parameters for the general T and Π forms shown in Fig. 4.4(a) and (b); the resulting equations are solved for the impedances Z_a, Z_b, and Z_c of the T equivalent or for Z_w, Z_x, and Z_y of the Π equivalent. This process yields the equivalent circuits shown in Fig. 10.13. It must be emphasized that these equivalents do not reproduce transmission-line behaviour in its entirety: clearly there is no question of their supporting waves of voltage or current. They are equivalent only in the sense that they yield the correct relationships between voltages and currents at input and output in the steady state.

Many of the results for the behaviour of transmission lines which have been found or discussed in earlier parts of Section 10.3 may seem improbable in the light of everyday experience. We have seen, for example, that the voltage on a line will vary from point to point unless the line is matched with the correct impedance; so the voltage at the load may be markedly different from that at the input, the difference depending on the length of the line. But in a power system, for example, the voltage available to a consumer appears to depend hardly at all on the length of the transmission line or cable which provides the supply, and we certainly do not expect its value to depend appreciably on the

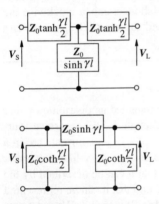

Fig. 10.13
The T and Π equivalents of a transmission line.

value of a load impedance; indeed a power system with such dependence would be almost intolerable. Again, although the wires used to connect various parts of a circuit must in some sense act as transmission lines, it is not usual to worry about standing waves on them: we tend to assume in practice that the impedance presented by a component to a source, for example, will not be appreciably affected by the wires used to connect it, and that has certainly been our assumption in all previous chapters. The apparent divergence here between theory and practical experience turns out to be only a matter of degree, and it hinges on the length of the line. Consider the wavelength of travelling waves on a line at a frequency f typical of a power system, say 50 Hz. Their phase velocity may be taken to be the velocity of light in the surrounding medium (Section 10.3.2), and for a low-loss line in air this is to a good approximation the free-space velocity c, or $3 \times 10^8 \text{ ms}^{-1}$; for different insulation the phase velocity would be somewhat, but not dramatically, less than the free-space value. The wavelength λ is therefore about c/f, which is 6×10^6 m or 6000 km. It follows that even relatively long power lines represent only a small fraction of a wavelength, and hence that for any termination there will be little variation in voltage or current along their length. On the other hand, even a circuit of small dimensions will exhibit wave phenomena at high enough frequencies; this arises in ultra-high-frequency or microwave circuits which are designed to operate at about 1 GHz or more, implying wavelengths in the order of centimetres. In these, printed-circuit connections can take the form of strip (or microstrip) lines which must be designed as transmission lines with characteristic impedances to match the other components.

It appears then that if l is the length of a line or a typical circuit dimension, there is no need to consider wave phenomena so long as $l/\lambda \ll 1$. A line for which this criterion is satisfied is known as a **short line**, whatever its actual length. Further insight can be had by writing the wavelength λ as $2\pi/\beta$; the criterion can then be written as $\beta l/2\pi \ll 1$, or simply as

$$\beta l \ll 1. \tag{10.51}$$

The simplest way in which to see the implications of this form is to apply it to the equivalent circuits of Fig. 10.13. Unless a line is abnormally lossy eqn 10.51 also means that $|\gamma l| \ll 1$ and therefore that

$$\sinh \gamma l \approx \tanh \gamma l \approx \gamma l.$$

It is easily shown from the definitions of γ and of $\mathbf{Z}_0$, eqns 10.28 and 10.33, that

$$\begin{aligned} \mathbf{Z}_0 \gamma l &= \mathbf{Z}l \\ \gamma l / \mathbf{Z}_0 &= \mathbf{Z}l. \end{aligned} \tag{10.52}$$

Hence for a short line the T and Π equivalent circuits of Fig. 10.13

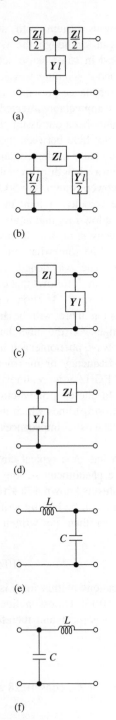

(a)

(b)

(c)

(d)

(e)

(f)

Fig. 10.14

Approximate circuits for a short transmission line.

reduce to the circuits shown in Fig. 10.14(a) and (b). These are lumped circuits in which Zl and Yl are respectively the *total* series impedance and shunt admittance of the line; and they are identical in form to the lumped circuits shown in Fig. 10.1(d) and (e). We may also note that these T and Π versions have identical properties for a given line (as they must if they are correct equivalents) when $|ZY|\,l^2 \ll 1$; this can be written as $(|\gamma|\,l)^2 \ll 1$, a condition which is clearly satisfied by the criterion we have used for a short line. Indeed it can be shown that this condition also means that the asymmetrical forms of Fig. 10.14(c) and (d) are satisfactory representations of a short line.

For a lossless line, Zl becomes $j\omega L$ and Yl becomes $j\omega C$, where L and C now represent total values; the short-line criterion $|\gamma l| \ll 1$ becomes $\omega\sqrt{LC} \ll 1$, and the T and Π equivalents become the simple lumped circuits shown in Fig. 10.14(e) and (f).

10.3.9 Transient behaviour

So far in this chapter we have considered only the steady state, in which the sinusoidal time variation of voltages and currents can be conveniently suppressed by representing them as phasors which vary only with distance at a given frequency. In general, however, transmission lines have a transient response to an arbitrary input; such response can be found by using the Laplace transformation as outlined in Chapter 7, but now the transformation must be applied to functions of both time and distance. In the general case an input waveform will change progressively along the line and at any point will change with time; the velocity of propagation takes on a new significance because an initially quiescent line can carry only a forward wave until the time required for any part of the input to traverse the line has elapsed, but, since no particular frequency is defined, neither a propagation constant nor a velocity can be uniquely specified. After the input signal first reaches the line termination, successive reflections from both ends of the line are superimposed.

Special cases are very much simpler to deal with so far as transient behaviour is concerned, and it can be helpful to consider an input waveform in terms of its spectrum of frequencies. Thus the lossless line propagates any input waveform, unchanged, at the velocity $1/\sqrt{LC}$ because every part of its spectrum travels at that velocity without attenuation; a lossy but distortionless line (Section 10.3.2) propagates the waveform without change of shape but with declining amplitude because all components are equally attenuated. When the waveform first reaches the termination, reflection will occur unless the line is matched; a resistive load differing from Z_0 will reflect a proportion of the waveform without change of shape for both kinds of line, because the reflection coefficient is then finite but independent of frequency. The same happens at the source end (provided that any source impedance is purely resistive) after the time required for two transits of the line, and so on until the superposition of the series of reflected waves on the original converges to a steady state.

Thus a voltage step, for example, applied to a lossless line will produce a succession of steps of voltage and current at any point on the line; but these must converge to a steady-state limit which corresponds simply to the d.c. input voltage applied directly to the load resistance, since such a line has neither voltage drop nor current leakage for d.c. In this case we may arrive at a series expression for the load voltage, say, by an argument on the following lines. Let a step of voltage V be applied to the input at time $t = 0$. If the input source is perfect, with zero impedance and voltage V_S, then $V = V_S$; otherwise, if the source has impedance Z_S (assumed to be resistive), then

$$V = V_S Z_0 / (Z_S + Z_0) \qquad (10.53)$$

by potential division, since initially there is only a forward wave on the line and its input impedance is therefore Z_0. The input step of height V appears at the load at time $t = T$, where T is the transit time for the line (given by l/v_p), and if the voltage reflection coefficient there is ρ_L the load voltage from that instant has the total value $V(1 + \rho_L)$ and may be written

$$V_L = V(1 + \rho_L)u(t - T).$$

The reflected wave, of height $V\rho_L$, arrives back at the input at time $2T$ and if the reflection coefficient there is ρ_S (given by eqn 10.38 with Z_S for Z_L) results in a second forward wave of height $V\rho_L\rho_S$; this reaches the load at time $3T$ and is itself reflected there, so adding to the existing load voltage an amount $V\rho_L\rho_S(1 + \rho_L)$, of which the reflected portion $V\rho_L^2\rho_S$ returns to the input and produces there a third forward wave of height $V\rho_L^2\rho_S^2$. When this reaches the load at $t = 5T$ it adds to the load voltage an amount $V\rho_L{}^2\rho_S{}^2(1 + \rho_L)$. Succeeding reflections continue this pattern, and the total load voltage can therefore be written

$$\begin{aligned} V_L = V\{(1 + \rho_L)u(t - T) + \rho_L\rho_S(1 + \rho_L)u(t - 3T) \\ + \rho_L^2\rho_S^2(1 + \rho_L)u(t - 5T) + ... \qquad (10.54) \\ ... + (\rho_L\rho_S)^{n-1}(1 + \rho_L)u(t - 2nT + T) + ...\}. \end{aligned}$$

If the input voltage source is perfect, with zero impedance, then $\rho_S = -1$; the initial voltage V at the input terminals is simply the voltage V_S of the source itself, and the contribution a backward wave would subsequently make at the input is entirely cancelled by reflection, so that the input voltage maintains the value V_S throughout, as for a perfect source it must. Any other value of ρ_S implies that the source impedance Z_S is non-zero and hence that V_S is greater than V, according to eqn 10.53; in this case the voltage at the input terminals will also vary with time in a series of steps, which must converge to the same steady-state value as the load voltage. The different between the (constant) source voltage and the input voltage must at all times correspond to the drop in Z_S due to the input current, which itself varies with successive reflections.

According to eqn 10.54 the steady-state value of the load voltage must be given by the geometrical series

$$V_L = V\{(1 + \rho_L) + \rho_L\rho_S(1 + \rho_L) + \rho_L^2\rho_S^2(1 + \rho_L) + \dots$$
$$\dots + (\rho_L\rho_S)^{n-1}(1 + \rho_L) + \dots\}$$

which has the sum to infinity

$$V_L = V\frac{1 + \rho_L}{1 - \rho_L\rho_S}. \tag{10.55}$$

For a perfect input source $V = V_S$, $\rho_S = -1$ and eqn 10.55 gives the steady-state value $V_L = V = V_S$ as we should expect.

The load voltage represented by eqns 10.54 and 10.55 for the case $\rho_L = -0.5$ and $\rho_S = -1$ is shown in Fig. 10.15(a).

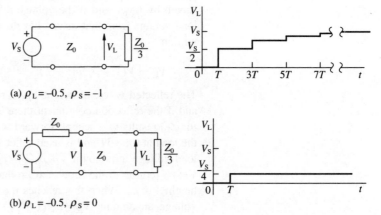

(a) $\rho_L = -0.5$, $\rho_S = -1$

(b) $\rho_L = -0.5$, $\rho_S = 0$

Fig. 10.15
The response of a loaded line to an input step: (a) ideal input source; (b) matched input.

If the source is not perfect but is matched to the line, so that $Z_S = Z_0$ and $\rho_S = 0$, then eqn 10.55 gives V_L as $V(1 + \rho_L)$; since in this case $V = V_S/2$ by eqn 10.53, it follows that now

$$V_L = V_S(1 + \rho_L)/2.$$

When ρ_L is expressed in terms of Z_L and Z_0 this reduces to

$$V_L = V_S Z_L/(Z_L + Z_0) = V_S Z_L/(Z_L + Z_S)$$

which again is what is expected in the steady state by simple potential division. In this case the steady state is reached after a time $2T$, for there is no reflection at the input when the first backward wave arrives, and no further transient; the variation of load voltage is shown in Fig. 10.15(b) for $\rho_L = -0.5$, i.e. for $Z_L = Z_0/3$.

If both load and source are matched to the line so that $Z_L = Z_S = Z_0$ and $\rho_L = \rho_S = 0$ then $V = V_S/2$ again and eqn 10.55 gives $V_L = V = V_S/2$ which is once more the result expected; here the steady state is attained after the transit time T, since there is no reflection at the load.

In the general case, if V, ρ_L and ρ_S are expressed in terms of the source voltage V_S and the impedances Z_L, Z_S, and Z_0, it can be shown that eqn 10.55 reduces to

$$V_L = V_S Z_L / (Z_L + Z_S)$$

as before. These results all confirm that the lossless line has no effect in the d.c. steady state, although in general it affects the transient progression to that state.

The transient response of transmission lines to input waveforms such as pulses is important in practice. We have seen that a waveform applied to a lossless line is delivered unchanged to its load after a delay T, and this fact has led to the use of lines as **delay lines**, whose function is simply to delay an input pulse. The effect of actual lines in distorting and attenuating waveforms is particularly important in communications engineering. It can also be important for the lines used in power systems: although these can be treated as short and therefore as lumped circuits at their normal frequencies (Section 10.3.7) this is not generally so in their response to disturbances such as those caused by switching and, particularly, to lightning strokes which can subject them to relatively large and fast pulses. These appear as surges of voltage and current, and their practical significance gave rise, in the context of power engineering, to the term **surge impedance** for what we have called, and is now usually called, characteristic impedance.

Problems

10.1 A ladder network has unit resistance and conductance for Z and Y respectively. Find the attenuation constant per section, and the fraction of the input amplitude which remains after three sections.

10.2 A ladder network is made from series capacitance of 1 nF and shunt inductance of 1 mH, per section. Find its attenuation and phase constants per section, (a) at 50 Hz and (b) at 500 kHz.

10.3 Find the critical frequency for the ladder in Problem 10.2.

10.4 Suppose that the ladder in Problem 10.2 comprises an infinite series of symmetrical T sections. What is its input impedance at each frequency?

10.5 Find the input impedance at each frequency of the ladder in Problem 10.2 when it comprises an infinite series of Π sections.

10.6 Find the number of sections per wavelength (to the nearest integer) along the ladder of Problem 10.2 at 500 kHz.

10.7 Find the phase and group velocities of the ladder of Problem 10.2 at 500 kHz.

10.8 If the elements making up the ladder of Problem 10.2 are exchanged, each section then having series inductance 1 mH and shunt capacitance 1 nF, find the critical frequency of the new ladder and its phase and group velocities at 100 kHz.

10.9 Find the characteristic impedance of the transmission line of Example 10.5, (a) at 50 Hz and (b) at 500 kHz.

10.10 A transmission line with negligible losses has parameters $L = 0.3\ \mu\text{H m}^{-1}$ and $C = 0.1\ \text{nF m}^{-1}$. Find its characteristic impedance and velocity of propagation.

10.11 Find the voltage reflection coefficient and standing-wave ratio on the line of Problem 10.10 when it is terminated by a resistance of 100 Ω.

10.12 A 30 m length of the line of Problem 10.10 is supplied at 15 MHz while terminated as in Problem 10.11. Find the impedance at the input and at the maxima of voltage and of current, respectively.

10.13 Find the input impedance of the line in Problem 10.12 if the termination is short-circuited.

10.14 Find the characteristic impedance of a quarter-wave transformer which would match the line and load in Problem 10.12.

10.15 Find the length of a short-circuited matching stub, made from the line of Problem 10.10, which would cancel a load reactance of j50 Ω at 15 MHz.

10.16 Find the transmission a parameters of the length of line in Problem 10.12.

10.17 A lossless line of characteristic impedance Z_0 is terminated in a resistance $3Z_0$. Write the series for the voltage at the load following the application of a unit step of input voltage from an ideal source.

11 Diode circuits

11.1 Diodes

11.1.1 The basic diode

The diode is the simplest of those circuit components which may be called electronic. It is a two-terminal device, as its name suggests, and it is passive; it is also non-linear, but its most important property is that it is not bilateral: it allows current to flow freely in one direction only. An early realization of such a device was the vacuum diode; it consisted of two electrodes, an anode and a cathode, in an evacuated envelope, the cathode being heated so as to produce electrons by thermionic emission. If the anode were at a positive potential it attracted a flow of electrons equivalent to a positive current flow from anode to cathode; a negative anode produced no net flow, since electrons simply returned to the cathode as fast as they were emitted.

Nowadays diodes are made of semiconducting material in which a junction is formed between two regions treated with different additives such that one, which acts like an anode, is p-type and the other, the cathode, n-type. Whereas an untreated or **intrinsic** semiconductor has equal numbers of free electrons and of holes, which behave like free positive charges, the free charges in p-type are nearly all holes and those in n-type nearly all electrons. These free charges therefore flow easily across the junction, both kinds contributing to form a current from p to n, when a voltage is applied so as to make the p-type positive and the n-type negative; but a voltage in the other direction will cause almost no current, since only very small numbers of 'minority' charge carriers (electrons in p and holes in n) are available to flow across the junction in response to it. It can be shown that the resulting diode characteristic relating the applied voltage v to the current i, both defined in the 'forward' direction from p to n, can be expressed as

$$i = I_0(e^{ev/nkT} - 1) \tag{11.1}$$

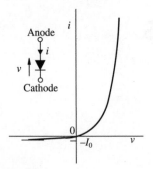

Fig. 11.1
Diode symbol and theoretical characteristic.

for a junction at absolute temperature T, where e is the electronic charge, k Boltzmann's constant, and I_0 a constant for a given diode; the number n is a constant for a given material, and for silicon (the most widely used semiconductor) takes the value 2. The characteristic has the form shown in Fig. 11.1, which also shows the standard symbol for a diode; actual semiconductor diodes follow this theoretical form quite closely, but we shall see below that various approximations to it are often used for practical purposes. The constant I_0 is evidently the limiting value of current in the reverse direction, due to minority carriers; it is sometimes known as the leakage or saturation current, is in the order of a nanoampere for a typical small diode as used in low-power electronic circuits, but may be several microamperes or more for larger diodes designed to carry more than an ampere or so of forward current. The theoretical relationship fails if reverse voltage is applied in excess of some value at which the junction breaks down, for then the reverse current increases sharply; reverse-voltage breakdown is put to practical use in the **Zener diode** (Section 11.1.3).

11.1.2 Approximations to the diode characteristic

As a non-linear device the diode can be treated by methods comparable with those applied to transistors in Chapter 4. Thus the current taken by a diode in series with a resistance from a d.c. voltage source can be found by using a load line to identify an operating point (Section 4.5.2); and at that point the incremental resistance dv/di can be found. However, the analogy really ends there, for there is no question of the diode being an active element with the amplifying properties of the transistor: resistance (or its reciprocal, conductance) is its only incremental parameter; and that is of limited use because most applications of the diode are based on its non-bilateral behaviour, which implies that its voltage and current are subject to large variations rather than small changes about an operating point. It is worth noting, however, that the incremental resistance of a diode corresponds exactly to the parameter h_{ie} of a BJT (Section 4.5.4), for the input characteristic of the latter derives from that of the diode formed by the base-emitter junction.

For many circuits containing diodes it would be far too cumbersome to incorporate eqn 11.1 or its graphical equivalent into analysis or design, and it is common to represent the characteristic by various approximations. First it can be noted that in practice typical values of forward current are much larger than the leakage current I_0 (if it were not so the whole point of the diode would be lost), and it follows that the exponential term in eqn 11.1 is then dominant; hence, for many purposes, the expression for i may be approximated quite well by dropping the term unity. This has the effect of giving i the value I_0 for $v = 0$ and making the leakage zero for large reverse voltages, but is of little account overall. Next we may note that the value of kT/e at room temperature is about 25 mV, so that the approximation can be valid and the forward current i much more than I_0 for values of v less than 1 V.

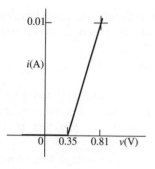

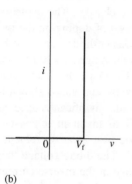

(a)

(b)

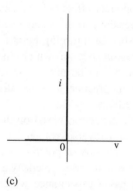

(c)

Fig. 11.2

Approximations to the diode characteristic.

For example, let us suppose that a small diode with $I_0 = 1$ nA, say, in a typical circuit will not have to carry more than 10 mA of forward current, and that a current less than 1 μA can be neglected. Over this range, representing four orders of magnitude, eqn 11.1 shows that for silicon at room temperature v varies between 0.35 and 0.81 V.

This leads to the so-called piece-wise linear approximation shown in Fig. 11.2(a), in which conduction is assumed to be zero below some chosen forward voltage and above that to be linear with a gradient depending on the range of current required. Because the forward current in a diode is almost always determined mainly by other circuit components rather by some precise value of the voltage v applied to it, the only error consequent on this linearization is a relatively small discrepancy in v. For example, with the figures we assumed above the theoretical value of v at a current of 1 mA, say, is 0.69 V while the linear form in Fig.11.2(a) applied to the range chosen gives 0.40 V. Although this difference seems like a large part of v, it is likely to be of little consequence in the circuit as a whole.

For the same reason, the further approximation is often made that the variation of v can be entirely neglected so that v takes a constant value over the chosen range of forward current. Often the range of interest is less than the four decades cited above, and the variation in v correspondingly small: it can easily be shown from the exponential in eqn 11.1 that for silicon at room temperature v changes by only about 115 mV per decade of current. The diode can thus be assumed to have the characteristic shown in Fig. 11.2(b), i.e. to be non-conducting below some fixed voltage V_f which is taken to be the voltage drop across it for any forward current. It is the application of this approximation to the BJT which leads to the common assumption that the base-emitter junction has a voltage drop of about 0.7 V when the transistor is conducting (Section 4.5.1).

Finally, if V_f is small in comparison with other important voltages in a circuit, it may be neglected entirely and the characteristic becomes that shown in (c) which is said to represent the **ideal diode**. It acts as a short circuit in the forward direction and as an open circuit in the reverse direction; just as it cannot carry reverse current, so too it cannot sustain forward voltage. If need be the characteristic of (b) can be regarded as due to an ideal diode in series with an opposing voltage source V_f and that of (a) to the same combination together with a series resistance; such models for the diode are occasionally used but we shall have no further need of them.

Although the characteristics of Fig. 11.2 each consists of two linear regimes, the fact that none is also bilateral means that in general many of the ideas of linear circuits which arose in earlier chapters cannot be applied to a diode circuit taken as a whole. Kirchhoff's laws of course always remain true, but such concepts as superposition and steady-state immittance or transfer function, which depend on the bilateral linearity of passive components, can be applied only to those parts of a diode circuit for which they may be appropriate. It is possible to apply transient or generalized methods to the whole circuit over any interval of time for which the state of each diode in a circuit is known and constant.

11.1.3 Special diodes

By varying its doping (the very small proportions of additives which produce the p- and n-type regions) and physical construction, the behaviour of the basic junction diode can be so modified as to emphasize certain properties for special purposes.

The **Zener diode** makes use of the reverse-voltage breakdown (Section 11.1.1). In the p–n junction there is a thin depletion layer between the two regions where electrons from the n side and holes from the p side recombine to leave virtually no free charges. The depletion layer can be made thinner by increasing the level of doping on the two sides, and this lowers the breakdown voltage. Zener diodes are made by such means to have breakdown voltages V_Z in the range 2 to 200 V, and a given diode has quite a precise value of V_Z. Once breakdown has occurred its voltage increases slightly with reverse current, as shown in Fig. 11.3(a). The rate of this increase is quite significant, since an important application of the Zener diode is to maintain a constant voltage (Section 11.5.1), and can be described best as an incremental resistance; typically it is in the order of 10 Ω. The diode voltage then changes by only 10 mV for every 1 mA change in the reverse current.

A thin depletion layer is also the foundation of the **tunnel diode**. In this the layer is so thin that the diode conducts effectively at any reverse voltage and at very low forward voltages the current is enhanced by the tunnel effect; as a result the characteristic has a turning point as shown in Fig. 11.3(b). There is thus a region AB in which the incremental resistance is negative. This property can be put to use in designing oscillators, for example, although in practice it is usually better to use an active circuit with positive feedback.

The **varactor diode** is used as a variable capacitor, based on the increase in thickness of the depletion layer with reverse voltage in a given junction. The layer, because it separates regions of positive and negative free charges, has intrinsic capacitance which varies inversely as its thickness; hence the reverse-biased diode has a capacitance which decreases with the applied voltage. The values so obtained are typically some tens of picofarads.

In a semiconductor, the recombination of an electron with a hole (each simply neutralizing the other) produces a quantum of radiated energy, and conversely the irradiation of a semiconductor adds to the production of electron-hole pairs over and above the normal rate of production due to thermal energy. In each case the wavelength of the radiation at which the effect is most marked can be selected by choosing an appropriate semiconducting material; this is the origin of the **light-emitting diode** (LED) and the **photodiode**. In the former, the light produced increases with forward current, as electrons pass to the p side and holes to the n side, and it is made available by keeping one

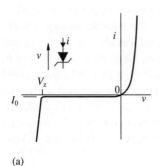

(a)

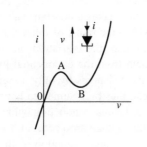

(b)

Fig. 11.3
Symbols and characteristics of (a) the Zener diode, (b) the tunnel diode.

region quite thin and exposing its outer surface. The photodiode, by contrast, operates with reverse bias: the electrons and the holes produced by exposing its surface to radiation then allow a higher than usual leakage current to flow, with a value in proportion to the intensity of light.

Not all diodes are based on the p–n junction. The **Schottky diode**, for example, is a metal-semiconductor junction; it has the useful properties of a low forward voltage drop, around 0.3 V, and a fast response which makes it especially useful in logic circuits.

Several other variations on the basic diode in addition to those mentioned above are used for particular purposes.

11.1.4   The thyristor and the triac

The **thyristor**, known sometimes as the silicon controlled rectifier or SCR, is a device which behaves like a diode with a certain degree of control; it comprises four regions of semiconductor, alternately p-type and n-type, of which the outer two are connected to the main circuit as for an ordinary diode. The control is provided by a third terminal, the gate, which is connected to the layer of p-type adjacent to the cathode; the conventional symbol for the thyristor is that of the diode with the addition of a gate as shown in Fig. 11.4(a).

The control exercised in this case is much more limited than the control of a transistor by its gate or base: following the application of a forward voltage between anode and cathode, provided that voltage is not too high the thyristor does not conduct unless and until it is triggered by a short positive pulse of current applied to the gate; once this happens the thyristor acts like a normal diode and the gate has no effect so long as the circuit can continue to provide forward current. After conduction has ceased it will not restart until it is again triggered while subjected to forward voltage.

In the absence of a gate pulse the thyristor can still be triggered by a high enough applied voltage; it has then the characteristic shown in Fig. 11.4(b). Because of this a thyristor can be designed for use without a gate connection, behaving like a diode except for the critical voltage needed to start conduction. When a gate is used some critical value of voltage is still needed, but it decreases with increasing gate current until it has quite a low value. A gated thyristor is normally so operated that the gate, rather than the applied voltage, determines the beginning of conduction.

The **triac** consists essentially of two thyristors arranged in parallel with opposing polarities and a common gate; it is made as a single semiconductor structure with five regions, and is given the symbol shown in Fig. 11.4(c). The combination of two thyristors gives the device a bilateral characteristic, so that the gate (or the magnitude of the applied voltage) controls the flow of current in either direction. The device can therefore be used in a.c. circuits to control the current taken by a load from a given voltage source, hence its application in, for example, light dimmers.

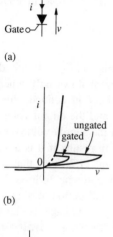

(a)

(b)

(c)

Fig. 11.4

(a), (b) Symbol and characteristics of the thyristor;   (c) triac symbol.

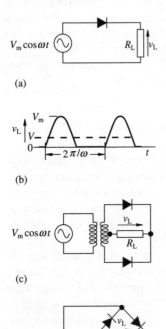

(a)

(b)

(c)

(d)

(e)

Fig. 11.5

Rectification: (a) half-wave rectifier and (b) its waveform; (c), (d) full-wave circuits and (e) their waveform.

11.2 Rectifiers

One of the oldest and most important applications of the diode is the rectification of a.c. voltage and current to d.c. It remains a widespread requirement: although a.c. is virtually universal for public power supplies there is much electrical equipment which needs d.c., including the majority of electronic circuits, certain kinds of motor, and electrolytic processes; there are also d.c. transmission lines in some power systems where rectification and inversion back to a.c. allows an increased power-carrying capacity (due to the absence of reactance) which would otherwise be more expensive, for example in undersea links.

11.2.1 The half-wave rectifier

The simplest rectifier circuit is shown in Fig. 11.5(a). If the diode characteristic is that of Fig. 11.2(b) then current flows only when the supply voltage exceeds V_f in the positive direction defined; this will occur for rather less than half of each period, and the output voltage v_L across the load resistance R_L is then less than the supply voltage by the constant drop V_f. A considerable simplification results if it is assumed that the diode has the ideal characteristic of Fig. 11.2(c), so that V_f can be neglected, and we shall make that assumption in what follows. In this case current flows throughout every alternate half-period, during which time the output voltage v_L is equal to the supply voltage as shown in Fig. 11.5(b), and the arrangement is therefore known as a **half-wave** rectifier. No current flows during the other half-periods and the supply voltage then appears as a reverse voltage across the diode.

The d.c. component of the output waveform in Fig. 11.5(b) is simply its steady-state average and can be found either by integration or from the appropriate Fourier series (Appendix F) to be a fraction $1/\pi$ of the peak value; the d.c. output voltage and current are therefore

$$V = \frac{V_m}{\pi}; \qquad I = \frac{V_m}{\pi R_L}. \qquad (11.2)$$

11.2.2 The full-wave rectifier

The intermittent output of the half-wave rectifier is clearly far removed from pure d.c. It can be improved by one or other of the **full-wave** circuits shown in Fig. 11.5(c) and (d), in both of which the current, while by no means constant, flows throughout each period. Both circuits provide a different current path for each polarity of the a.c. supply: in (c) the two paths are provided by splitting the transformer secondary so as to form two sources, each with its diode, and in (d) the two are provided by a bridge of four diodes of which one pair conducts for each polarity. The output waveform for ideal diodes in both cases is shown in (e), and it follows from eqn 11.2 that its d.c. output is given by

$$V = \frac{2V_m}{\pi}; \qquad I = \frac{2V_m}{\pi R_L}. \qquad\qquad (11.3)$$

11.2.3 Ripple factor

The Fourier series for the waveforms in Fig. 11.5 give, as well as the d.c. component V, a fundamental and a series of odd harmonics. The fundamental and the harmonics can be regarded as together forming the a.c. component or **ripple** of the rectified waveform. It is not difficult to find the r.m.s. value V_{ac} of the voltage ripple in each case by using eqn 9.11, which in this case becomes

$$V_{ac} = \sqrt{V_{rms}^2 - V^2} \qquad\qquad (11.4)$$

when V_{rms} is the r.m.s. value of the total output voltage. In the full-wave case, Fig. 11.5(e), V_{rms} is the same as the r.m.s. supply voltage, since the two voltages have the same magnitude at every instant, and is therefore $V_m/\sqrt{2}$. Since we have $V = 2V_m/\pi$ from eqn 11.3, then

$$V_{ac} = V_m\sqrt{\frac{1}{2} - \frac{4}{\pi^2}} = 0.308\, V_m$$

The size of V_{ac} relative to the d.c. value V is a measure of how well the output approximates to pure d.c., and the **ripple factor** r is a figure of merit defined by the ratio V_{ac}/V. For the full-wave rectifier the ripple factor is therefore

$$r = \frac{0.308V_m}{2V_m/\pi} = 0.483.$$

For the half-wave rectifier the output has a mean-square voltage which must be half that of the full-wave, so its r.m.s. value is $1/\sqrt{2}$ of the full-wave value, that is $V_m/2$. From eqn 11.2 the d.c. voltage is V_m/π; in this case the a.c. component is therefore

$$V_{ac} = V_m\sqrt{\frac{1}{4} - \frac{1}{\pi^2}} = 0.386V_m$$

and the ripple factor is

$$r = \frac{0.386V_m}{V_m/\pi} = 1.21$$

Because the output waveform in these cases has a substantial a.c. component, it follows that any reactance in the load may be significant. However, it must be remembered not only that the ripple contains a range of frequencies but also that the diodes make these circuits non-linear, so that even at a particular frequency ω it is not in general correct

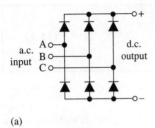

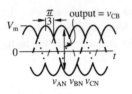

(a)

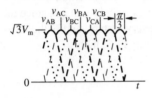

(b)

(c)

Fig. 11.6
Three-phase rectification.

to express the load as an impedance $Z(j\omega)$. Nor is the difficulty merely one of finding the current taken from a voltage waveform like those considered above: both voltage and current are affected by the non-linearity. As it turns out, in practice the most important effect of reactance arises not from the load but from the filters which are often added to rectifier circuits in order to smooth the waveforms and ensure that the voltage and current supplied to the load itself are as near as possible to pure d.c. These filters are considered in Section 11.3.

11.2.4 The three-phase rectifier

Direct current can be produced directly from a three-phase a.c. supply by the circuit shown in Fig. 11.6(a). Here again two diodes conduct at any instant, one from the upper set of three and one from the lower. Since the upper three have common cathodes, connected to the positive d.c. line, the potential of the latter must be that of the most positive a.c. line (for ideal diodes), for if it were less one of these diodes would have to sustain a forward voltage; hence the conducting diode of the three is always that connected to the a.c. line of highest instantaneous potential. Similarly, of the lower set, at any instant the diode connected to the a.c. line of lowest potential is conducting.

Fig. 11.6(b) shows the waveforms for a balanced set of three-phase line-to-neutral voltages, i.e. for the line potentials relative to the neutral point, each having the peak value V_m. By the above argument the output voltage across the load is given by the difference between the upper and lower envelopes. However, since the difference between two line-to-neutral voltages is simply a line voltage (Section 6.2), the output voltage can also be shown as the envelope of the line voltages shown in (c). Its d.c. component is given by its average over an interval of $\pi/3$ such as that shown in the figure and is easily shown to be $3/\pi$ of the peak line voltage, that is $3\sqrt{3}V_m/\pi$. It would be more usual to express this in terms of the r.m.s. line voltage V_ℓ, which is $\sqrt{3}V_m/\sqrt{2}$; the d.c. output then becomes

$$V = \frac{3\sqrt{3}V_m}{\pi} = \frac{3\sqrt{2}V_\ell}{\pi} = 1.35V_\ell. \tag{11.7}$$

Although it appears from the figure that there is appreciable ripple in the potential of each d.c. line, there is much less variation in the difference between the two. Calculation shows that the ripple factor of the waveform shown in (c) is only 0.042, much better than that of the full-wave rectifier supplied from a single phase.

11.2.5 Thyristor rectifiers

In each of the rectifier circuits so far met, the d.c. output voltage has been a function only of the magnitude of the a.c. supply voltage. An effective way of controlling the output is to replace the diode of the half-wave version, or one of the two diodes in each current path in other versions, by a thyristor. For each path, current will then flow only for an interval fixed by the instant when the thyristor gate is pulsed. The angular measure of the delay between the start of the forward voltage and the gate pulse is the **firing angle** α. Figure 11.7 shows a full-wave bridge rectifier with thyristor control, and a typical voltage waveform which it would give across a resistive load. It is easily shown by integration that the average of this waveform gives a d.c. voltage

$$V = \frac{V_m}{\pi}(1 + \cos\ \alpha). \tag{11.8}$$

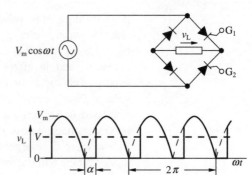

Fig. 11.7
Thyristor control of rectified output.

Clearly the thyristors allow the d.c. output to be varied over the whole range from the normal full-wave value down to zero, by controlling the timing of the gate pulses. This is a relatively easy matter to arrange electronically, and such a rectifier represents a convenient and efficient variable-voltage d.c. supply. The same principle, of course, can be applied to any of the other rectifier circuits already mentioned.

11.3 Smoothing

The ripple in the output of a rectifier circuit can be greatly reduced by interposing between the rectifying device and the load one or more elements which effectively filter out the a.c. component of the output. The process is known as **smoothing**. The arrangements used are often called rectifier filters, but this must not be taken to mean that their performance can be analysed by the methods applied to filters in linear systems (Section 5.4); although the added components are linear, a rectifier circuit as a whole cannot in general be handled by these methods.

The simplest way to smooth a rectified waveform is to use either a large capacitance C in parallel with the load R_L or a large inductance L in series with it: the former tends to keep the load voltage v_L constant, and the latter to keep the load current i_L constant, both having the same effect for a resistive load.

11.3.1 Capacitance smoothing

Figure 11.8(a) shows a half-wave rectifier with a parallel smoothing capacitor added. If the diode is ideal, then while it is conducting the supply voltage v_S is applied directly to the parallel combination of R_L and C and the diode current which flows must therefore be given by

$$i = \frac{v_S}{R_L} + C\frac{dv_S}{dt}. \tag{11.9}$$

Whereas in the simple circuit of Fig. 11.5(a) the diode current falls to zero when v_S reaches zero, eqn 11.9 shows that now i falls to zero when

$$\frac{dv_S}{dt} = -\frac{v_S}{CR_L}. \tag{11.10}$$

If C is large, this value of negative gradient will be reached soon after the supply voltage passes its positive peak, at a point such as A in Fig. 11.8(b). After this the diode stops conducting, has a reverse voltage across it, and in effect isolates the CR_L combination from the a.c. supply. The capacitor therefore discharges through R_L as shown: its voltage (and current) decay exponentially as in the second example discussed in Section 7.2.1, with the time constant CR_L, so long as the reverse voltage across the diode lasts. Two things about this decay follow when C is large: first, the fall in v_L will be slow, so the reverse diode voltage will not return to zero and allow conduction to begin again until some point such as B in the figure, quite close to the next positive peak of the supply voltage; secondly, the time interval AB will be a small part of the time constant CR_L, so the rates of decay in v_L and i_L can be taken to be $1/CR_L$ of their average amplitude between A and B. This enables us to approximate the ripple voltage as a triangular

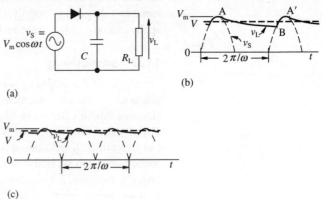

(a)

(b)

Fig. 11.8
Capacitance smoothing.

(c)

'sawtooth' waveform which decays at a rate close to V/CR_L, where V is the d.c. output voltage, for a time which is almost $2\pi/\omega$ or $1/f$, the period of the a.c. supply at frequency f; it then follows the supply voltage v_S for the short time during which the diode conducts before the next decay starts at A'. The sawtooth therefore has a peak-to-peak amplitude V/fCR_L. The table of waveforms in Appendix F shows that the r.m.s. value of such a waveform is $1/2\sqrt{3}$ of its peak-to-peak amplitude, giving in this case

$$V_{ac} = V/2\sqrt{3}fCR_L.$$

The ripple factor of the output is then the ratio of·this to the d.c. output V, namely

$$r = 1/2\sqrt{3}fCR_L, \tag{11.11}$$

which is clearly small for large CR_L.

Equation 11.11 shows that if the capacitance C is large enough the ripple is negligible and the output is virtually pure d.c.; and it is clear from Fig. 11.8(b) that the output voltage then takes the *peak* value of the a.c. supply. This illustrates the need to avoid applying linear analysis indiscriminately to a diode circuit: if we were to imagine that the diode simply applied the unsmoothed half-wave of Fig. 11.5(b) to the CR_L combination, and considered the response of the latter to each Fourier component as in a linear circuit, we should deduce wrongly that the d.c. voltage across the load must be the average V_m/π whatever the value of C. In reality the addition of C changes the waveform of the rectifier output voltage in a way which can be found only by some such argument as has been used above.

The smoothing of a full-wave output by a parallel capacitor follows exactly the same principle, but now the ripple factor is improved by a factor of about two for a given CR_L, because the decay of v_L extends only for about one half-period, as shown in Fig. 11.8(c).

Because there can be no d.c. current in a capacitance, it follows that the average diode current, in either the half-wave or the full-wave circuits, must be equal to the d.c. current in the load. But with increased smoothing the period of diode conduction, such as the interval BA′ in Fig. 11.8(b), becomes shorter and in the limit of large C tends to zero; the current in the diodes, and hence also in the supply, therefore becomes correspondingly larger to maintain the required average. Effective capacitive smoothing implies that both diodes and supply must be able to operate with short high-current pulses.

11.3.2 Inductance smoothing

There is little point in applying smoothing by series inductance alone to a simple half-wave rectifier, as indicated in Fig. 11.9(a). For one thing, the current and therefore the load voltage v_L must be zero for some part of every cycle however large the inductance L: if the diode were always conducting it could be omitted, implying that an a.c. source can provide d.c. current without rectification. Also, the current, whose waveform is the transient response of the circuit following the onset of conduction, tends to lag the supply voltage and therefore falls to zero only after the supply voltage has reversed as shown; this evidently lowers the average of the voltage v across the L–R_L combination, which must be the same as the d.c. output voltage V since the average voltage across an inductance is (ideally) zero. It follows that the use of a large inductance in this circuit simply depresses the d.c. output while providing ineffective smoothing.

This performance can be considerably improved by the addition of a second diode, known as a **commutating diode** and shown in

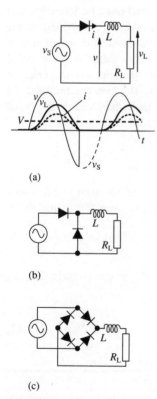

(a)

(b)

(c)

Fig. 11.9
Inductance smoothing.

Fig. 11.9(b), connected across L and R_L in such a direction that no appreciable reverse voltage can appear across the combination. The output voltage V, the average of v_L, therefore remains at the half-wave value of V_m/π. In addition, the commutating diode provides a path by which the load current can continue to flow until the supply voltage is again in the forward direction; hence current can flow continuously and can be much smoother.

A full-wave rectifier, by contrast, can benefit directly from the smoothing effect of a series inductor, as shown for example in the bridge circuit of Fig. 11.9(c). It can be shown formally that now the load current does not fall to zero; but it can be more simply deduced by noting as before that in one half-period of the supply the tendency of current to lag voltage means that some current must still be flowing when the next half-period begins. In that case one or other of the diode pairs is always carrying current and the L–R_L combination is therefore permanently connected to the supply, but with a polarity change in each half period: in other words it is connected to the full-wave rectified voltage of Fig. 11.5(e). For this reason it is possible to analyse the output by the response of the series combination to the appropriate Fourier series. The d.c. output voltage V across R_L is simply the average value $2V_m/\pi$ for any L, since again there can be no d.c. voltage drop in the inductance. The a.c. voltage across the load must be found for one frequency at a time; it cannot be found from the a.c. component of the voltage across the combination, i.e. of the unsmoothed waveform. In fact only the fundamental component of the latter (having twice the supply frequency) is significant, and the appropriate series in Appendix F shows that its peak amplitude is $4V_m/3\pi$; the amplitude of the corresponding component of load voltage is therefore, by potential division,

$$V_{2m} = \frac{4V_m}{3\pi} \times \frac{R_L}{\sqrt{R_L^2 + 4\omega^2 L^2}}.$$

If L is large enough the square root in this can be taken to be simply $2\omega L$. In that case the r.m.s. value of the ripple is given approximately by

$$V_{ac} \approx \frac{V_{2m}}{\sqrt{2}} \approx \frac{\sqrt{2}V_m R_L}{3\pi\omega L}$$

and the ripple factor by the ratio of this to the d.c. voltage $2V_m/\pi$, namely

$$r = \frac{R_L}{3\sqrt{2}\omega L} = \frac{R_L}{6\sqrt{2}\pi f L}. \tag{11.12}$$

Example 11.1 Find the values of capacitance and inductance respectively which would reduce to 1% the ripple factor of a full-wave rectifier supplying a 1 kΩ load from a 50 Hz source.

From eqn 11.11 the capacitance required for a ripple factor r in the case of a half-wave rectifier has the value

$$C = 1/2\sqrt{3}fR_L r$$

which with the above figures gives

$$C = 1/(2 \times \sqrt{3} \times 50 \times 1000 \times 0.01) = 577 \ \mu F.$$

Since the ripple factor is twice as good for the full-wave case, the value required here is half this, namely

$$C = \textbf{289} \ \boldsymbol{\mu}\textbf{F}.$$

With inductance smoothing the value of L needed with full-wave rectification is given, from eqn 11.12, by

$$L = R_L/6\sqrt{2}\pi f r$$
$$= 1000/(6 \times 1.414 \times \pi \times 50 \times 0.01) = \textbf{75 H}.$$

Both of these answers indicate that in general quite large values of inductance or capacitance are needed to obtain a highly smoothed output.

11.3.3 Other smoothing circuits

The smoothing of rectified waveforms can be improved, without improbably high values of series L or parallel C, by using combinations of the two; as with filters generally, the more elements used the more effective their performance. Figure 11.10 shows circuits which can be used: (a) shows an LC, or L-section, filter. If L is large enough, at every

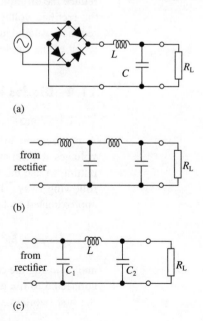

(a)

(b)

(c)

Fig. 11.10
Smoothing filters for rectifiers.

instant one or other pair of diodes carries current, as in full-wave inductance smoothing, and the unsmoothed full-wave voltage is therefore applied to the filter. The low reactance of the capacitance and the high reactance of the inductance then ensure that, by potential division, the ripple voltage across the load is readily made smaller than for either of these on its own. Figure 11.10(b) shows a similar circuit with two *LC* sections in cascade, serving further to attenuate the ripple voltage. In the Π-section filter shown in (c), on the other hand, the capacitance C_1 (often called **reservoir** capacitor) acts like that used in capacitance smoothing, causing the load to be isolated from the supply voltage for part of each half-period. As in Section 11.3.1, the resulting ripple voltage across C_1 is approximately a saw-tooth waveform, but now potential division between *L* and the C_2R_L combination ensures that only a small fraction of it appears across the load itself. For simplicity and economy the inductance is sometimes replaced by a resistance; the resulting circuit is still quite effective at reducing the ripple voltage but, unlike purely reactive filters, it somewhat reduces the d.c. output also and introduces power loss.

In practice, whatever combination of elements is used, passive filters for rectifiers cannot provide the stable, ripple-free d.c. supplies which many electronic circuits need. Apart from ripple, the d.c. output voltage tends to vary with the current drawn as well as with any variation in the a.c. supply. A typical rectified power supply therefore uses a **voltage regulator** in a feedback circuit, as shown schematically in Fig. 11.11: in essence a chosen fraction of the final d.c. output voltage is compared with a constant reference voltage (obtained from a Zener diode: Section 11.5); any difference is amplified and applied to the base of a transistor, interposed between rectifier and load, in such a way as to reduce the difference. As a result the transistor takes up the variations in the rectified voltage, leaving the load voltage accurately constant. The rectifier usually includes at least a reservoir capacitor to remove the worst of the ripple from the voltage to be regulated.

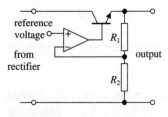

Fig. 11.11
Stabilizing circuit for rectifier output.

11.4 Diodes in modulation, detection, and clamping

11.4.1 Amplitude modulation by diode

A requirement in amplitude modulation is the multiplication of two voltages at different frequencies (Section 9.3.5). This can be achieved by putting to use the non-linearity of an actual diode characteristic, in the following way. The characteristic expressed by eqn 11.1 can be approximated by taking as many terms as we please of the power series

$$i = av + bv^2 + \ldots$$

and choosing the constants *a, b, ...* which best fit the curve. If a diode is connected to the two voltages $V_1 \sin \omega_1 t$ and $V_2 \sin \omega_2 t$ in series, then the first two terms of this series give a current

$$i = a(V_1 \sin \omega_1 t + V_2 \sin \omega_2 t)$$
$$+ b(V_1^2 \sin^2 \omega_1 t + 2V_1 V_2 \sin \omega_1 t \sin \omega_2 t + V_2^2 \sin^2 \omega_2 t)$$
$$= a(V_1 \sin \omega_1 t + V_2 \sin \omega_2 t)$$
$$+ \frac{b}{2}\{ V_1^2 + V_2^2 - V_1^2 \cos 2\omega_1 t - V_2^2 \cos 2\omega_2 t$$
$$+ 2V_1 V_2[\cos(\omega_1 - \omega_2)t - \cos(\omega_1 + \omega_2)t] \}.$$

This current can be passed through a resistance, small enough not to affect its waveform, to produce a corresponding voltage v_{out}, say. Now if ω_2 represents a signal frequency and ω_1 its carrier, then $\omega_1 \gg \omega_2$ and a narrow-band filter centred on ω_1 will pass only those components of v_{out} which have the frequencies ω_1, $\omega_1 + \omega_2$, and $\omega_1 - \omega_2$. We may therefore write

$$v_{out} = K\{aV_1 \sin \omega_1 t + bV_1 V_2[\cos(\omega_1 - \omega_2)t - \cos(\omega_1 + \omega_2)t]\}$$
$$= K\{aV_1 \sin \omega_1 t + 2bV_1 V_2 \sin \omega_1 t \sin \omega_2 t\}$$

where K is a constant which depends on the value of the resistance used and the gain of the filter. The output evidently has the same form as the general expression for amplitude modulation given by eqn 9.29.

11.4.2 The diode detector

A diode application which is closely related to rectification is the recovery of a signal $x(t)$ from an amplitude-modulated carrier (Section 9.3.5) by applying the latter to the circuit shown in Fig. 11.12(a). If the modulation has the form given by eqn 9.29, with $| bx(t) | < a$, then the resulting waveform has the form shown in Fig. 9.19 and its envelope does not change sign. The circuit acts as a half-wave rectifier with capacitance smoothing; if the time constant CR is chosen to be much longer than a carrier period but much less than the shortest period of the modulating signal, then the output voltage will follow the peaks of the rectified waveform, i.e. its envelope, as indicated in Fig. 11.12(b), and the circuit is known as an **envelope detector**. There will be some degree of ripple at the carrier frequency, but that can be removed by a low-pass filter to leave an output which closely represents the original modulating signal $x(t)$.

If the time constant of the same circuit is made very large, either by omitting R or by choosing large values, the output will decay very little after each peak; it will therefore retain the value of the highest peak reached by the input waveform and so acts as a **peak detector**.

11.4.3 Voltage clamping

For some purposes in communication circuits it is necessary to add to an a.c. waveform a d.c. component of sufficient value to keep the resulting waveform always above (or below) some chosen value; a circuit to achieve this is known variously as a **voltage clamp** or a **d.c. restorer**. (The term 'clamp' is also sometimes used for the voltage

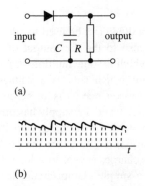

(a)

(b)

Fig. 11.12
Envelope detection of amplitude modulation.

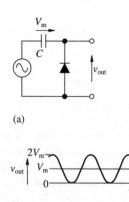

(a)

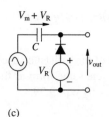

(b)

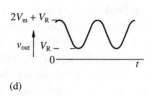

(c)

(d)

Fig. 11.13
Voltage clamping.

limiters described in the following section.) In essence a clamp is based, like the detectors of the previous section, on a diode and a capacitor; but now the circuit is arranged as shown in Fig. 11.13(a). When the a.c. input voltage is in the direction which causes the diode to conduct, the capacitance charges up to nearly the peak value of the input voltage, V_m, say; once charged, it remains so because the diode prevents reverse current flow. Consequently, the output voltage across the diode is the input waveform superimposed on a voltage which approximates to V_m; it is therefore virtually unidirectional, as shown in (b). The polarity of the output can of course be changed by reversing the diode. In theory the diode never conducts once the steady state has been reached. In practice the capacitance would discharge slightly in the course of a cycle, especially if the output is connected to some load rather than an open circuit as shown, and the diode would conduct for a short time to top up the charge.

In the case just considered the output voltage had the lower (or upper) limit of zero. If some other limit is required, this can be achieved by including in the circuit a voltage source, or reference voltage, to give an additional d.c. bias V_R in either direction. Figure 10.13(c) shows an example, for which the resulting output waveform is shown in (d).

11.4.4 Voltage multiplication

The clamping action of a capacitance with a diode can be used to add a d.c. voltage not only to an a.c. input but also to the d.c. output of a rectifying circuit, giving a total d.c. output which is some multiple of the peak input. Figure 11.14(a) shows a circuit which acts as a voltage doubler based on a half-wave rectifier. The capacitance C_1 is, as before, charged up to virtually the peak supply voltage when the supply has one polarity; when the polarity changes C_1 is unable to discharge and its voltage is added to that of the supply, charging C_2 to twice the peak value. Because C_2 is likewise unable to discharge, except to a load, it tends to hold the d.c. voltage $2V_m$. As in a simple clamp circuit, any discharge to a load is replenished during the following half-period; in this respect C_2 behaves like a smoothing capacitance.

Higher multiples can be obtained by adding further diodes and capacitances: the circuit in Fig. 11.14(b), for example, has four of each and acts as a voltage quadrupler.

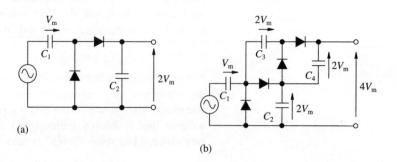

Fig. 11.14
Voltage multiplication:
(a) doubler; (b) quadrupler.

11.5 Voltage regulation and voltage limiting

The fact that there are practical limits to the voltage across a diode in each direction — the forward voltage at normal currents is small and often negligible, and the backward voltage is effectively restricted to the breakdown value — gives rise to a variety of applications involving the control of circuit voltages.

11.5.1 The Zener diode regulator

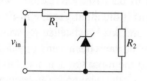

Fig. 11.15
A Zener regulator.

If a Zener diode is so connected as to remain in its breakdown region, which means that it must always carry a reverse current in excess of the value I_B at which breakdown occurs, then the reverse voltage across it will change very little from the breakdown value V_Z (Section 11.1.3). Figure 11.15 shows a circuit which can achieve this. To ensure that the diode is always in breakdown the voltage v_{in} and the resistances R_1 and R_2 must satisfy the condition

$$v_{in} > V_Z + R_1 \left(I_B + \frac{V_Z}{R_2} \right)$$

or

$$v_{in} \frac{R_2}{R_1 + R_2} > V_Z + I_B \frac{R_1 R_2}{R_1 + R_2}. \tag{11.13}$$

Since I_B is in the order of the diode leakage current I_0, the last term may be negligible provided that the resistances are not too large; in that case the condition reduces simply to a statement of potential division. At the same time the need to limit the diode current to some maximum value I_m, say, evidently requires that

$$\frac{v_{in} - V_Z}{R_1} - \frac{V_Z}{R_2} \leq I_m$$

or

$$v_{in} \frac{R_2}{R_1 + R_2} \leq V_Z + I_m \frac{R_1 R_2}{R_1 + R_2} \tag{11.14}$$

on the assumption that the Zener voltage remains at the breakdown value for any current up to I_m. Provided that these two conditions are always satisfied, the output provides a reference voltage which remains at the value V_Z whatever the value of v_{in} and the load resistance R_2, and the circuit can be used to cope with changes in either or both of these quantities. In practice any change in R_2 may be specified in terms of a range of the output current V_Z/R_2. The small increase in voltage with current indicated by the Zener characteristic can be ignored for many purposes but if necessary can be taken into account by an incremental resistance (Section 11.1.3).

Equations 11.13 and 11.14 are simply formal expressions of limits which are evident on inspection of the circuit in Fig. 11.15: for a given

(a) $v_{AB} > -V_f$

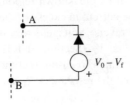

(b) $v_{AB} > -V_0$

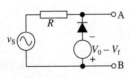

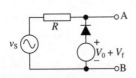

(c)

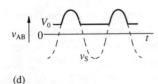

(d)

Fig. 11.16
Voltage limiting: lower limit only.

diode, v_{in} and R_1 set the total current, of which the amount V_Z/R_2 is the output and the balance must flow in the diode. It follows that the maximum-current limit I_m is approached by increasing v_{in} or decreasing R_1, so increasing the total current, or by increasing R_2, which then takes less of the total and causes more to flow in the diode; the opposite trends cause the diode current to approach its lower limit I_B.

Example 11.2   A Zener regulator is to be based on a diode with $V_Z = 20$ V, $I_B = 1$ mA and $I_m = 100$ mA. (a) If the input voltage is 40 V, find the values of the input resistance R_1 and load resistance R_2 which will allow the maximum range of output current from zero upwards. (b) If the output is to be fixed at 50 mA and the input voltage may fall to 30 V, find the appropriate value of R_1 and the maximum acceptable value of v_{in}.

(a) Since the output current may fall to zero, the total current in R_1 must not exceed I_m; it should be set at this value to allow the maximum output current, which is then $I_m - I_B$ or 99 mA. The required value of R_1 is then given by

$$R_1 = (v_{in} - V_Z)/I_m = 20/0.1 = \mathbf{200\ \Omega},$$

and the load resistance may evidently vary between an open circuit and the value 20/0.099 Ω; so the range of R_2 is given by

$$R_2 > \mathbf{202\ \Omega}.$$

(b) Since $I_B = 1$ mA, an output current of 50 mA implies that R_1 must be small enough to pass a total of not less that 51 mA at the minimum input voltage of 30 V; but R_1 should be set to the maximum value which this allows in order to cope with large input voltages. It follows that

$$R_1 = (30 - 20)/0.051 = \mathbf{196\ \Omega}.$$

The maximum of v_{in} is limited by I_m, and must give a total current in R_1 not exceeding $I_m + 50$ mA, i.e. 150 mA. The upper limit is therefore given by

$$v_{in} = 0.15R_1 + V_Z = \mathbf{49.4\ V}.$$

The Zener diode regulator cannot of course act on its own as a rectifier, since any a.c. input voltage would certainly violate eqn 11.13. Nor is it usually connected directly to a rectifier output: for one thing, that would restrict the final output to some chosen voltage V_Z; for another, even the small variation of the Zener voltage with current may not be acceptable for a good d.c. supply. Instead, the use of a regulator circuit with feedback (Section 11.3.3) allows any chosen fraction of the final output to be kept equal to the Zener voltage; and since the latter is applied only to the input of a feedback amplifier its variation with current does not arise. As a result, the output voltage can be made both highly stable and variable at will.

11.5.2 Limiters or clippers

It is clear from the previous section that the Zener regulator can maintain a constant output voltage only if the input is within certain limits. However, there are many applications in which the requirement is not that the output voltage should be constant but that it should remain within certain limits, and there are several kinds of diode circuit which can achieve this for any input. As a simple example, suppose it is required that the voltage v_{AB} between points A and B in a circuit should under no circumstances take an appreciable negative value; this can be achieved by connecting a diode between A and B as shown in Fig. 11.16(a). Evidently the diode allows v_{AB} to take any positive value while having no effect on the circuit so long as it does not reach breakdown. But whatever the state of the rest of the circuit it limits the negative amplitude of v_{AB} to the forward drop V_f of the diode, which is a fraction of a volt (provided only that abnormally large diode currents are not allowed to flow: any circuit designer would have to ensure that this could not happen, if only to protect the diode against damage).

This idea can easily be extended. Thus, if v_{AB} is allowed negative values up to some magnitude V_0, say, i.e. if it is restricted to the range $v_{AB} > -V_0$, that can be achieved by inserting in series with the diode a d.c. reference voltage of value $V_0 - V_f$ with the polarity shown in Fig. 11.16(b). Then, for example, a sinusoidal voltage source of amplitude exceeding V_0 applied to the circuit in (c) would result in the waveform shown for v_{AB} provided that the resistance R was enough to take up the 'lost' voltage without excessive current in the diode. Any load connected between A and B would affect the general form of v_{AB} but not its limiting value.

If, on the other hand, v_{AB} was required not to fall below some *positive* limit V_0, keeping it to the range $v_{AB} > +V_0$, that can be achieved in a similar manner simply by connecting the reference voltage $V_0 + V_f$ with the reverse polarity shown in Fig. 11.16(d); for a sinusoidal input as before this would give v_{AB} as shown.

The same principle can readily be applied to put both upper and lower limits on a voltage by using a diode and source combination for each: the circuit shown in Fig. 11.17(a), for example, would ensure that the voltage v_{AB} was always in the range bounded by $V_1 + V_f$ and $-(V_2 + V_f)$. A sinusoidal input so limited is shown in (b).

The d.c. reference voltages to set the limits in these examples are often obtained from Zener diodes; now, however, in contrast to the Zener regulators discussed in Section 11.5.1, each diode breaks down only when the appropriate voltage limit is reached. Fig. 11.17(c) shows a circuit whose output voltage v_{out} lies within the limits V_{Z1} and $-V_{Z2}$.

Because these circuits effectively remove part of the input voltage waveform, they are often referred to as **clippers**. They are commonly

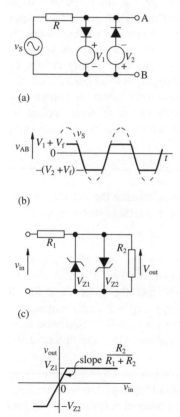

(a)

(b)

(c)

(d)

Fig. 11.17

Voltage limiting: upper and lower limits.

used either to modify a waveform to a required shape or simply to protect parts of a circuit against temporary voltages which might be large enough to be damaging in some way. It can be useful to represent such a circuit by an input/output characteristic: the behaviour of the circuit shown in Fig. 11.17(c), for example, can be represented as in (d). Since the circuits considered here have no frequency dependence the characteristic applies to any waveform, and it should not be confused with the kinds of transfer function encountered in previous chapters for purely linear systems.

11.6 Function forming

The characteristic of a clipper circuit as in Fig. 11.17(d) is a special case of a general class of characteristics which can be achieved by combinations of diodes and resistances, together with d.c. reference sources. If voltage is applied to a circuit containing several diodes as well as resistances, then in general some diodes will act virtually as short circuits and others as open circuits, the numbers in each category depending on the instantaneous input voltage. For any range of voltage over which no diode changes its state the characteristic will be a straight line, since the circuit is then a fixed arrangement of resistances; it follows that the overall characteristic comprises a sequence of linear segments (that is, it is 'piecewise linear'), in which every change of slope means that at least one diode has changed from an open circuit to a short circuit or vice versa. A point at which this occurs is known as a **breakpoint**, not to be confused with the same term applied to frequency response in Chapter 5, and in general there is one such point for every diode in the circuit.

Example 11.3 Find the characteristic relating the current i to the input voltage v_{in} in the circuit shown in Fig. 11.18(a), on the assumption that the diodes are ideal.

In this case it is immediately clear that the current i, given by $(i_1 + i_2 + i_3)$, is zero for all negative input voltages. It is also obvious that, in milliamperes, the current i_1 will take the value $v_{in}/2$ for all positive v_{in}; that i_2 is given by $(v_{in} - 2)/4$ for $v_{in} > 2$ V and is otherwise zero; and similarly that i_3 is $(v_{in} - 4)/0.8$ for $v_{in} > 4$ V, otherwise zero. The summation of these three linear relations gives the characteristic shown in Fig. 11.18(b).

In order to produce a particular functional relationship between input and output, the desired function may be approximated to any required degree of accuracy by linear segments; a combination of diodes and resistances can then be designed to realize it. As usual, it is rather easier to analyse a given circuit than to design one in this way. Even so, for such non-linear systems analysis is, like design, a matter of trial and error. It is not generally possible to predict by inspection which of

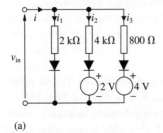

(a)

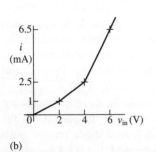

(b)

Fig. 11.18
Circuit and characteristic for Example 11.3.

several diodes in a circuit will conduct for a given value of input voltage. The usual approach is to assume (or guess) that certain diodes will conduct and analyse the resulting circuit as for any resistive combination with a d.c. input; if this leads to the conclusion that any diode must support either an appreciable forward voltage or a reverse current then the assumed configuration is wrong.

11.7 Diodes in logic circuits

Although the logic gates which are common in digital electronics are based on transistors, almost invariably in integrated circuits, the essential logic functions in some versions depend on diodes or on diode action in the junction transistor. Indeed, gates can be made with only diodes as the logic elements. For example, the circuit shown in Fig. 11.19(a) acts as an AND gate: because the diodes ensure that the output voltage can be only slightly greater than the lower of the two inputs, it follows that the output can assume a high voltage (of a value depending on the supply voltage V^+) only if *both* inputs are at a comparably high voltage. The circuit shown in (b) acts as an OR gate, because only if both inputs are at earth potential (or at any rate at a small voltage below the diode drop V_f) can the output also be at a low voltage; if either input is at a significantly higher voltage, or if both are, the output must be almost as high. There is no need, in either of these circuits, to restrict the number of inputs to two: further diodes can be added as required.

Better gates can be made by using diodes in a similar way but in combination with transistors; the generality of such gates are said to comprise **diode–transistor logic** (DTL). Fig. 11.19(c) shows an elementary NAND gate based on the same two diodes as in (a) but now connected to the base of a BJT; in this case the latter simply acts as an inverter, changing the AND function to NAND. If either input is low, then the base potential is also low and the transistor therefore tends to cut-off, giving the maximum output voltage V^+; otherwise, if both inputs are significantly above zero potential, neither diode conducts and the d.c. supply provides ample base current to saturate the transistor, giving a low output voltage. However, the gate as shown in (c) would be unsatisfactory in practice, because with one or both inputs at zero, making at least one diode conduct, the base would actually be at the potential V_f; this by no means ensures that the transistor would be completely cut off, since some base current could flow. It is imperative in logic circuits, where transistors act simply as switches, that each should be unambiguously in one or other of the two appropriate states, either cut off or fully conducting. The difficulty is easily overcome in this case by inserting another diode as shown in (d); so long as one of the inputs is zero, no appreciable current can now flow to the base because the potential V_f at the point A cannot provide the total voltage drop which would be needed for the new diode and the base-emitter junction to conduct in series. The use of a diode in this way to offset a voltage drop in the order of V_f which may be troublesome is quite

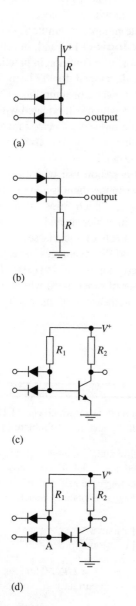

(a)

(b)

(c)

(d)

Fig. 11.19
Logic gates: (a) AND; (b) OR; (c), (d) NAND.

common in circuit design, even in those logic circuits, mentioned below, where the essential function is not provided by diodes.

The combination of three diodes in the circuit of Fig. 11.19(d) is equivalent to an n-p-n transistor with a double emitter giving two base-to-emitter junctions in parallel. Such transistors are in fact used for logic gates; they are better for the purpose than discrete diodes, partly because they have a faster response (diodes, like other devices, have a bandwidth beyond which their behaviour deteriorates from that at low frequencies). Here too more than two inputs can be accommodated if necessary, by fabricating the transistor with the appropriate number of emitters. Gates in this class form **transistor–transistor logic** (TTL) and, in rather more refined versions than described here, are quite common in practice. Particular variants of the BJT have been developed for TTL in the interests of still faster response and low power consumption. They include the so-called **Schottky transistor**, which is fabricated with a Schottky diode connected so as to bypass the base-to-collector junction and prevent it from reaching saturation when it is in the conducting state; the transistor can then be turned off faster.

There are other kinds of logic gates in common use which do not depend directly on diode behaviour; notable among them is a large class based on FETs: it includes NMOS logic, which uses n-channel MOSFETs as switches, and CMOS logic in which n and p channel complement each other in matched pairs. Each of the classes which remain in common use has its advantages and disadvantages in respect of speed, power consumption, and a property known as **fanout**. The latter, for a given gate, is simply the number of other gates which may safely be connected to its output and is a measure of its ability to maintain its correct function when loaded.

Problems

11.1 Find the incremental resistance at a current of 1 mA and a temperature of 300 K of a silicon diode for which the saturation current is 1 nA. What is the actual ratio of voltage to current at this point?

11.2 Estimate the firing angle needed to produce an average voltage of 120 V from a full-wave thyristor rectifier supplied at 240 V r.m.s.

11.3 A half-wave rectifier supplies a resistive load of 250 Ω from a 240 V, 50 Hz supply with a smoothing capacitance of 1 μF. For how long does the diode continue to conduct after each positive peak in voltage?

11.4 Show that the assumptions on which eqn 11.11 is based cannot be valid for the rectifier of Problem 11.3.

11.5 By solving the differential equation for a series RL circuit with an applied sinusoidal voltage, show that the current which flows during each half-period in a full-wave rectifier with load R_L and smoothing inductance L is given by

$$i = \frac{V_m}{\sqrt{R^2 + \omega^2 L^2}} \left\{ \frac{2\sin\phi}{1 - e^{-\pi/\omega\tau}} e^{-t/\tau} + \sin(\omega t - \phi) \right\}$$

where $\phi = \tan^{-1}(\omega L/R_L)$ and $\tau = L/R_L$. [Hint: set the final current equal to the initial current.]

11.6 The diodes in the circuit shown may be assumed to have zero voltage drop when conducting. Show that all three diodes conduct when $V = 0$ and that only D1 conducts when $V = 5$ V.

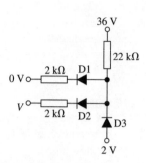

11.7   Find the current in each diode of Problem 11.6 when $V = 0$.

11.8   In the circuit shown, V_1 and V_2 can take only the values 1 V or 4 V, and the diodes may be assumed to have a forward voltage drop of 0.6 V. Find the currents in each resistor (a) when both inputs are 4 V, (b) when they differ, and (c) when both are 1 V.

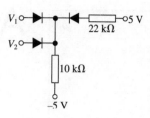

11.9   Find the range of V_{in} for which each diode, taken to be ideal, conducts in the circuit shown.

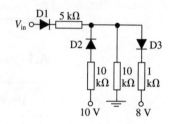

11.10   Find the current in each diode of Problem 11.9 when $V_{in} = 16$ V.

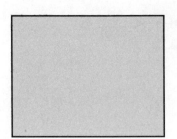

Appendices

Appendix A: SI units

A.1   Base units

Quantity	Unit	Symbol
Length	metre	m
Mass	kilogram	kg
Time	second	s
Electric current	ampere	A
Temperature	kelvin	K
Luminous intensity	candela	cd
Amount of substance	mole	mol

A.2   Some derived units

Quantity	Unit	Symbol	Equivalent
Frequency	hertz	Hz	s^{-1}
Force	newton	N	$kg\ m\ s^{-2}$
Pressure, stress	pascal	Pa	$N\ m^{-2}$
Work, energy, heat	joule	J	$N\ m$
Power	watt	W	$J\ s^{-1}$
Electric charge	coulomb	C	$A\ s$
Electric potential, e.m.f.	volt	V	$J\ C^{-1}$
Electric capacitance	farad	F	$C\ V^{-1}$
Electric resistance	ohm	Ω	$V\ A^{-1}$
Electric conductance	siemens	S	$A\ V^{-1}$
Magnetic flux	weber	Wb	$V\ s$
Magnetic flux density	tesla	T	$Wb\ m^{-2}$
Inductance	henry	H	$Wb\ A^{-1}$

A.3 Multiples and sub-multiples

Multiplier	Prefix	Symbol	Multiplier	Prefix	Symbol
10^{18}	exa	E	10^{-1}	deci	d
10^{15}	peta	P	10^{-2}	centi	c
10^{12}	tera	T	10^{-3}	milli	m
10^{9}	giga	G	10^{-6}	micro	μ
10^{6}	mega	M	10^{-9}	nano	n
10^{3}	kilo	k	10^{-12}	pico	p
10^{2}	hecto	h	10^{-15}	femto	f
10	deca	da	10^{-18}	atto	a

Appendix B: Passive circuit elements

B.1   Resistance

B.1.1   Ohm's law

The basis of Ohm's law is the experimental fact that the current I which flows in a conducting material due to a voltage V is often, under constant conditions, closely proportional to V. A common form of the law is therefore

$$V = IR,$$

in which the constant of proportionality R is called the resistance, and depends not only on the material which carries the current but also on the size and shape of the specimen concerned. (The law can of course be expressed equally well in terms of conductance G, defined for this purpose simply as the reciprocal of resistance.) A more fundamental statement of the law relates the electric field strength E at any point in the material to the current density J at that point by the equation

$$E = \rho J \tag{B.1}$$

in which ρ is the **resistivity**, is a constant of the material for fixed conditions, and is measured in ohm-metres. An equivalent alternative statement is

$$J = \sigma E \tag{B.2}$$

in which σ is the **conductivity**, is the reciprocal of resistivity and is measured in siemens per metre. Because electric field is the gradient of potential (that is, voltage per unit length) and current density is current per unit area, the resistivity and conductivity of a material can be defined as the resistance and conductance respectively between opposite faces of a unit cube.

Both E and J are vector quantities and if, as we might expect, current always flows in the direction of the voltage gradient, then ρ and σ must be scalar quantities. This is the case for any homogeneous isotropic material, and therefore for most conducting materials, in most circumstances and we shall take it to be so in what follows; but it is not variably true, and to be perfectly general ρ and σ should be expressed as tensors of the second order. Materials in which ρ and σ are scalar and constant over a significant range of applied field for fixed conditions are said to be linear or **ohmic**, and the assumption of constant R in linear circuit theory is based on the common availability of such materials. For the most part we use σ rather than ρ in what follows, since its significance mirrors that of ρ and it is slightly the more convenient of the two in some ways.

B.1.2 The nature of ohmic conduction

Under given physical conditions, the conductivity of a material depends on the number density and mobility of those of its charge-carrying particles which are free to move (electrons in a metal, positive holes as well as electrons in semiconductors generally, electrons and ionized atoms in conducting fluids) in response to an applied electric field. Particles of number density n each having charge q and travelling with average velocity v in a given direction constitute a current density

$$J = nqv. \tag{B.3}$$

The average velocity of a population of particles in a given direction is sometimes called a **drift** velocity; it is in contrast to, and superimposed upon, the random velocities due to thermal motion which of course give no resultant current. In a conducting material v represents a kind of terminal velocity at which the force qE exerted by the electric field on each of the moving charges is on average balanced by the resistance of the material to their motion. In ohmic conduction the resisting force is on average proportional to the drift velocity; it follows that in equilibrium the velocity v is proportional to the applied electric field E, and the ratio v/E is defined as the **mobility** μ of the charged particles. We may therefore write eqn B.3 as

$$J = nq\mu E. \tag{B.4}$$

It is the virtual constancy of the number and mobility of the free charges in many materials, for given conditions and over a considerable range of E, which lies at the root of Ohm's law. The value of μ itself depends on the charge q, because that governs the force acting on each charged particle in the field E; on the effective mass m of the particle, which fixes its acceleration for a given force; and on the interactions between the moving particles and the atomic structure of the material. It is, of course, these interactions which resist the motion and restrict the charged particles to a terminal velocity: without them the mobile charges would continue to accelerate. It is convenient for most purposes to describe the interactions in terms of an average time interval τ between what may be loosely called 'collisions', variously known also as a relaxation time or an average time for momentum transfer. It can then be shown that the mobility is given by

$$\mu = q\tau/m. \tag{B.5}$$

It follows from eqns B.2, B.4, and B.5 that the conductivity due to a single species of charge can be written as

$$\sigma = J/E = nq\mu = nq^2\tau/m. \tag{B.6}$$

In materials with more than one species of free charges, such as semiconductors, each kind gives rise to a conductivity having the form of eqn B.6 and hence contributes to the total current according to eqn B.4.

In the foregoing we have assumed a steady state in which forces are on average in equilibrium. There is fortunately no need in the present context to consider the question of a transient period before the steady state is reached (for example, on first applying the field) since an overall equilibrium is attained very quickly, in times much shorter than those of interest in most electrical circuits. Similarly, although the inertia due to the mass of the charged particles means that their response to an alternating field must be frequency dependent (clearly there will be frequencies so high that the particles scarcely have time to move between reversals of the field), this usually becomes important only at frequencies much higher than are relevant to conventional circuits. For the frequencies relevant to most a.c. circuits the mobilities of charged particles are virtually the same as at d.c. and values of conductivity may therefore be taken to be independent of frequency. (Note that the same cannot be said of the resistance of a given body, because of the skin effect mentioned below.)

The value of σ for a particular material depends somewhat on its physical conditions and of these one of the most important is the temperature T. In general both n and τ may vary with T. In metallic conductors, however, n is virtually constant and the principal effect of temperature is on τ, which decreases with rising T because of the increased levels of atomic vibration; hence σ also falls, and the result is the familiar increase in resistance with temperature. In semiconductors, on the other hand, the populations of free charges, both holes and electrons, increase rapidly with temperature to the extent that the effect on τ is much less significant; the result is that σ rises and resistance values fall with temperature, giving a negative temperature coefficient of resistance.

B.1.3 The calculation of resistance

The resistance of a conducting body between any two points, or equipotential surfaces, can be obtained in terms of σ or ρ by relating the voltage to the field E and the total current to the current density J. In general, the voltage between points P and Q, say, is given by the line integral

$$V_{QP} = -\int_{P}^{Q} E \cdot dl \tag{B.7}$$

along a path of length l, and the current flowing is the area integral

$$I = \int_{A} J \cdot dA \tag{B.8}$$

over a cross-section of area A. In the absence of magnetic induction the field E is conservative, which means that between any two points the integral of eqn B.7 has the same value whatever the path chosen; if, as now, we are not considering inductance then V can be evaluated over any convenient path joining P and Q. Similarly, the continuity of current means that the integral of eqn B.8 has the same value for any area through which the entire current flows, so it can be evaluated for any convenient choice of A. The evaluation of either integral requires a knowledge of the pattern of the field E and hence that of J; in general the field pattern in a conducting body is found by solving Laplace's equation and for a body of arbitrary shape this must be done by some numerical method. For simple geometries, however, the field pattern is obvious, so the two integrals are easily evaluated and hence the resistance calculated in terms of σ or ρ. Since the resistance of a passive conductor is always a positive quantity the negative signs which may arise (depending on the assumed polarity of P relative to Q in the line integral B.7) can be discounted.

The simplest case is the familiar one of a uniform conductor of length l and cross-section A. If it is assumed that the current I is uniformly distributed over the cross-section, then both E and J are in the axial direction with the same value everywhere and the integrals, evaluated respectively along a straight path between the ends of the conductor and over any plane cross-section, become simply

$$V = El$$

and

$$I = JA.$$

Combining these with Ohm's law in the form of eqn B.1 or eqn B.2 gives the resistance of the conductor as

$$R = \frac{V}{I} = \frac{El}{JA} = \frac{l}{\sigma A} = \frac{\rho l}{A}. \tag{B.9}$$

The assumption that J is uniform over the cross-section of a homogeneous uniform conductor is adequate for many practical purposes, and is correct for d.c., but fails at high frequencies because of the skin effect, which tends to make the current flow near the surface because of magnetic induction inside the conductor due to its own magnetic field (Section B.3.4).

Another simple case is the resistance or conductance in the medium between two coaxial cylinders, which might represent, for example, the leakage parameter G of a transmission line. Here the current density J must be purely radial and if we consider a length l over which the voltage between the cylinders can be considered constant then J will not vary with axial distance and must therefore depend on radius only. If then the total current between the cylinders over a length l is I, the (radial) current density through the cylindrical surface of radius r and length l shown in Fig. B.1, whose area A is $2\pi rl$, must have the value

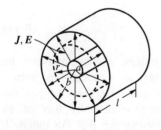

Fig. B.1
The electric field between coaxial cylinders.

$$J = I/A = I/2\pi rl$$

and the (radial) electric field at radius r has therefore the value

$$E = J/\sigma = I/2\pi\sigma rl.$$

From this the voltage between the cylinders can be found by integrating in the radial direction; if their radii are a and b as in the figure, then the voltage between them has the magnitude

$$V = \int_a^b E dl = \frac{I}{2\pi\sigma l} \ln \frac{b}{a}$$

and the resistance of the length l is then

$$R = \frac{V}{I} = \frac{\ln(b/a)}{2\pi\sigma l}. \tag{B.10}$$

It may be noted that in this case the resistance decreases with l, as expected, because for radial flow the area A corresponding to a given radius increases with length. (The conductance correspondingly *increases* with l, of course, which leads to the idea of conductance per unit length as used in Chapter 10.)

A comparable calculation gives the resistance between two concentric hemispheres (or spheres if required). Here again the current flow is clearly radial, but now the area A through which the current I passes at radius r is that of a hemisphere, $2\pi r^2$; thus the current density there has the value

$$J = I/2\pi r^2$$

and the electric field is therefore

$$E = J/\sigma = I/2\pi\sigma r^2.$$

Integrating this along a radius gives the voltage between two radii a and b, say, as

$$V = \int_a^b E dr = \left[-\frac{I}{2\pi\sigma r} \right]_a^b = \frac{I(b-a)}{2\pi\sigma ab}$$

and the resistance is therefore

$$R = \frac{V}{I} = \frac{b-a}{2\pi\sigma ab}. \tag{B.11}$$

A special case of this arises when b is infinite: R then represents the resistance of a semi-infinite medium acting as a current sink from a hemisphere of radius a; practically this could be, for example, the resistance presented to a current flowing to earth by way of an embedded hemispherical electrode as shown in Fig. B.2. For infinite b the resistance becomes

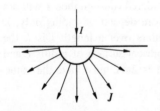

Fig. B.2
Current flow from a hemisphere.

$$R = 1/2\pi\sigma a.$$

B.1.4 Non-ohmic conduction

There are materials in which Ohm's law fails, for a variety of reasons, apart from the common situation when the heat of power dissipation causes an appreciable temperature rise. Conduction in gases, for example, is highly non-linear: here there are normally very few free electrons and ions, and the numbers needed for appreciable conduction are to a large extent produced by ionization due to the applied electric field itself (because of the energy it imparts, in the first instance to such electrons as are available) when that is high enough. The processes by which this happens, and the numbers produced, depend on the pressure of the gas and its boundary conditions; the conducting gas is also highly inhomogeneous, with regions of net space charge, and the effective area of conduction may be ill-defined and variable.

In vacuum, or at very low pressure, current can flow provided that there is a source of mobile charges, such as an emitting electrode. Here, however, their velocity in a given field is not limited by a mobility, since collisions or interactions are rare, and the idea of a conductivity is not appropriate; under these conditions the motion of charged particles is determined by their acceleration in an electric field whose distribution depends on the accumulation of space charge, and the result is a non-linear voltage–current characteristic. It is also quite possible under these conditions for current to flow in a region of no electric field by virtue of charged particles which have already acquired a directed velocity; the ordinary cathode-ray tube in which electrons are first accelerated in a 'gun' and then flow to the screen as a beam is such an example. Current flow of this nature is sometimes called **convection current**.

The drift velocity of charges which constitutes a flow of current need not even be caused by an electric field in the first place. In general, particles having a number density n which decreases with distance x, say, acquire a drift velocity in the x direction by diffusion due to their concentration gradient dn/dx. If the particles have a diffusion coefficient D their drift velocity from this cause has magnitude

$$v = -\frac{D}{n}\frac{dn}{dx}.$$

In general vector notation this becomes

$$v = -(D/n)\nabla n$$

and if the particles have charge q then by eqn B.3 their drift velocity constitutes a current density

$$J = nqv = -qD\nabla n. \tag{B.12}$$

This is called a **diffusion current**, and is independent of any electric field although it often arises in conjunction with such a field; in that case the total current is the algebraic sum of the contribution due to diffusion and that due to electrical conductivity. If two different charged species are present, then as for conductivity each may contribute to a resultant diffusion current. In that case the separation of charge caused by different diffusion rates tends to produce an electric field even in an otherwise neutral material. In a situation where there can be no net current flow these various contributions may still arise, but must then have such relative proportions as to cancel exactly.

Diffusion currents are especially important where inhomogeneity in the density of free charges arises at the junction of two different materials: the behaviour of such junctions is the basis of an important range of devices, not least the various forms of transistor.

All of the forms of current flow described above involve the actual transport of charge, and are sometimes known collectively as **conduction current** in order to distinguish them from **displacement current**, which does not involve the conduction of charge. Displacement current is that form which, for example, allows us to speak of the current flowing in a capacitance despite the fact that an ideal dielectric cannot conduct; it turns out that continuity of the current flow through a capacitance is maintained by virtue of displacement current in its dielectric. We discuss this further in the following section.

B.2 Capacitance

B.2.1 The electric field

Capacitance defines the relationship between electric charge and its associated potential in given circumstances. The capacitance C of two conducting bodies (one of which may be earth, or may be infinitely remote) is by definition the ratio Q/V, where Q is the magnitude of their electrical charges, when these are equal and opposite, and V is the potential difference which in consequence exists between them. This ratio depends not only on their geometrical arrangement but also on the material medium between and around them. If that medium should be conducting the definition is not changed, but there would then also be a finite resistance (and conductance) between the same bodies. What we know as a capacitor is a device designed to have capacitance and, so far as is practicable, only that; in that case the medium must be a dielectric which ideally has zero conductivity, and for the present we assume such a dielectric.

The dependence of capacitance on the dielectric, for a given geometry, stems from the fact that the electric field strength E around a given charge depends on the medium. In dealing with electric fields it is useful to introduce a vector quantity which at a given point depends only on the charge producing the field and not on the medium; it is called the electric **flux density** D. The flux density is related to the charge density ρ at any point by the Maxwell equation

$$\nabla \cdot D = \rho$$

and, what is mathematically equivalent, it is related to the total charge Q

within any closed surface A by **Gauss's law**

$$\int_A \boldsymbol{D} \cdot \mathrm{d}\boldsymbol{A} = \int_V \rho \mathrm{d}V = Q \qquad (B.13)$$

in which V is the volume enclosed by A. Equation B.13 simply states that the total electric flux leaving any closed volume is equal to the charge it contains. The flux density $\boldsymbol{D}$ is related to the field strength $\boldsymbol{E}$ by the **constitutive equation**

$$\boldsymbol{D} = \epsilon \boldsymbol{E} \qquad (B.14)$$

in which ϵ is the *permittivity* of the medium. We should note that although both $\boldsymbol{D}$ and $\boldsymbol{E}$ are due to charge, they are dimensionally very different: the former is measured in charge per unit area, the latter in voltage per unit length or force per unit charge; hence ϵ has dimensions coulombs per volt metre, or farads per metre.

There is an obvious correspondence between eqn B.14 and Ohm's law in the form of eqn B.2; both relate $\boldsymbol{E}$ to a kind of flux density: in the one case to $\boldsymbol{D}$, which we have defined as electric flux density, and in the other to $\boldsymbol{J}$, the current density, which is really the density of a flux of charge. Because $\boldsymbol{D}$ is charge per unit area and $\boldsymbol{E}$ is voltage per unit length, it follows that the permittivity ϵ is the capacitance of a unit cube, just as the conductivity σ is the conductance of a unit cube. Permittivity, like conductivity, should in general be a tensor quantity but for homogeneous isotropic materials is a scalar, in which case $\boldsymbol{D}$, like $\boldsymbol{J}$, is in the same direction as $\boldsymbol{E}$. There is, however, an important difference between these two properties of media: whereas σ covers a very wide range of values, is virtually zero for many materials, and is by definition zero for free space, ϵ is finite in all media, free space and conductors included. Hence, the electric field due to any charge inevitably has both $\boldsymbol{D}$ and $\boldsymbol{E}$, while in a conducting medium these are accompanied by a current density $\boldsymbol{J}$; in a homogeneous isotropic medium all three vector fields have identical distributions.

The energy stored in an electric field is equivalent to an amount

$$e = \int_0^D \boldsymbol{E} \cdot \mathrm{d}\boldsymbol{D}$$

per unit volume, and for constant scalar ϵ this becomes

$$e = \frac{1}{2}ED = \frac{1}{2}\epsilon E^2 = \frac{1}{2}D^2/\epsilon. \qquad \textbf{(B.15)}$$

B.2.2. Displacement current

Recalling that the permittivity ϵ is the capacitance of a unit cube, and noting that the voltage across such a cube is the electric field E, we may use the usual current–voltage relation for capacitance (eqn 1.3) to write the current through it as

$$i = \epsilon\, \mathrm{d}E/\mathrm{d}t$$

which by eqn B.14 is $\mathrm{d}D/\mathrm{d}t$. But the current through a unit cube is simply the current density; hence the current density in a dielectric, according to our view of capacitance as a circuit element, has the value $\mathrm{d}D/\mathrm{d}t$. However, an ideal dielectric is non-conducting and in it there can be no current density J or σE in the sense of charge transport as described in Section B.1; the quantity $\mathrm{d}D/\mathrm{d}t$ is therefore referred to as **displacement current**. We have tacitly assumed that the unit cube has voltage applied to it between two of its faces from some external circuit; in that case a conduction current J which flows from the circuit to these faces, but cannot flow in the dielectric, becomes $\mathrm{d}D/\mathrm{d}t$ in the latter and the continuity of current which is a fundamental principle in circuit theory is assured. The plausibility of this is apparent if we note that current flowing to and from conducting bodies separated by a non-conducting dielectric must cause these to accumulate charge at a corresponding rate, and this in turn causes a changing flux density throughout the dielectric.

Although we have arrived at this conclusion by considering a unit cube of ideal dielectric, the density of displacement current is always $\mathrm{d}D/\mathrm{d}t$ at any point, whatever the material and however non-uniform the current flow may be; the *total* density of current can be written in general as $J + \mathrm{d}D/\mathrm{d}t$ and the two contributions are always so related as to preserve current continuity.

In non-dielectrics displacement current is usually negligible except at very high frequencies; but in dielectrics it is essential, for it is, ideally, the only possible form of current flow.

B.2.3 Permittivity

The **permittivity of free space** ϵ_0 has the value 8.854×10^{-12} Fm^{-1}. The permittivity of matter can be greater or less than this but for the insulating materials known as dielectrics, with which alone we are concerned here, it exceeds ϵ_0. The permittivity of a dielectric material is often expressed as

$$\epsilon = \epsilon_r \epsilon_0 \tag{B.16}$$

where ϵ_r is the **relative permittivity** or **dielectric constant**. The increase in ϵ over its free-space value arises from the fact that an electric field causes **polarization** of its atoms or molecules, that is the small separation of positive and negative charge which gives each particle a **dipole moment**, defined for point charges as the product of charge and separation. In some materials, said to be **polar**, the molecules have permanent dipole moments; in the absence of an applied field these are randomly directed and their effects cancel, but they tend to align with an applied field to produce **orientational polarization**. Whether or not a material is polar, an applied field causes **electronic** or **ionic**

polarization by inducing charge separation and hence a dipole moment in each particle: in a non-conductor the electrons are strongly bound to the atoms and a small displacement of the one relative to the other sets up a restoring force sufficient to balance the force due to the applied field; if the molecular structure has positive and negative ions they will react in a similar way. The **polarizability** of a material is the total dipole moment per unit volume, due to any kind of polarization, produced by unit electric field.

The effect of polarization is to reduce the electric field strength which the original charges would otherwise produce, in other words to reduce E for given D everywhere and thereby, as we shall see, to reduce V for given Q on charged conductors. Because an absence of polarization (as in free space) corresponds to $\epsilon_r = 1$, it is convenient to introduce an **electric susceptibility** χ, defined as $\epsilon_r - 1$, which can be shown to be proportional to the polarizability. For many dielectrics the polarizability and hence also χ, ϵ_r and ϵ are virtually constant over a considerable range of conditions, but in polar materials ϵ depends quite strongly on temperature (because that affects the degree of alignment on average). In addition, ϵ always depends on frequency (and sometimes, unlike conductivity, to a significant extent at frequencies which arise in conventional circuits) because the movement of charge in polarization is subject to the inertial effect of its mass as well as to restoring forces.

B.2.4 Dielectric loss

Although actual dielectric materials typically have very low values of conductivity (which in a dielectric represents unwanted leakage of conduction current and consequent energy loss), for some practical purposes losses may have to be considered. The effective conductivity of a dielectric is frequency dependent and at certain frequencies shows peaks of dissipation which are connected with resonance of the oscillating charges rather than the ordinary processes of conduction discussed in Section B.1 above. For this reason it is common to specify the energy loss in a dielectric at a given frequency without reference to a conductivity σ. There are two related quantities which are used to do so: one is the **loss tangent**, the other the **complex permittivity**, and we discuss these next.

Consider again a unit cube of dielectric, across which is applied a voltage given by a sinusoidal electric field at frequency ω and represented by a phasor E. If its permittivity is ϵ', there will be a displacement current density dD/dt or $\epsilon' dE/dt$ which in phasor terms becomes $j\omega\epsilon' E$ (the j indicating, as it should, that this is a capacitive current which leads the applied voltage). If there is also a leakage or conduction current due to an effective conductivity σ, we could write that as σE, representing, as required, a current in phase with E; instead, suppose that the permittivity, in addition to ϵ', has an imaginary part $-\epsilon''$ so that its complete value is

$$\epsilon = \epsilon' - j\epsilon''$$

and in consequence the total current in the cube is

$$i = j\omega\epsilon E = j\omega(\epsilon' - j\epsilon'')E$$
$$= E(\omega\epsilon'' + j\omega\epsilon'). \qquad (B.17)$$

Hence the current in phase with E has become $\omega\epsilon''E$ instead of σE, and the effective conductivity at frequency ω is $\omega\epsilon''$. Instead of quoting a normal (real) permittivity ϵ and conductivity σ, it is therefore possible to specify for an imperfect dielectric a complex permittivity $\epsilon' - j\epsilon''$. Both parts of ϵ should properly be assumed to be frequency dependent, although for the practical purposes of circuits operating at frequencies up to a few megahertz or so it is common to assume that ϵ' at least, and hence the values of capacitance which depend on it, are constant.

Equation B.17 indicates that the angle between current and voltage, ideally $\pi/2$, is in fact $\tan^{-1}(\omega\epsilon'/\omega\epsilon'')$ or $\tan^{-1}(\epsilon'/\epsilon'')$. The difference between this and $\pi/2$, which can be attributed to the effective conductivity, is known as the loss angle δ and is given by

$$\tan\delta = \epsilon''/\epsilon'. \qquad (B.18)$$

This too should strictly be regarded as frequency dependent. The actual value of the losses per unit volume can be expressed either as σE^2 (or its equivalent $J^2\rho$) as for a conducting material, or as $\omega\epsilon''E^2$. The stored energy of the electric field, given by eqn B.15, becomes in the present notation $\frac{1}{2}\epsilon'E^2$ per unit volume (or $\frac{1}{2}D^2/\epsilon'$).

A lossy capacitor can also be represented at a given frequency, in circuit terms, by a perfect capacitance and an associated resistance such that the phase angle of the resultant impedance is $\pi/2 - \delta$ at the frequency in question. Although this can be done with either series or parallel resistance the latter is a more natural choice since capacitance is defined in terms of voltage, which in that case is common to both elements.

The conductivity of a dielectric material can increase catastrophically if a sufficiently high electric field is applied to it, a phenomenon known as breakdown which occurs when the field is large enough to produce significant numbers of free charges by ionization of atoms. The value of the field at which this happens is the **dielectric strength** of the material (not to be confused with its dielectric constant) and it depends on several conditions including temperature and, for a gas, pressure. In many devices, including most solid-state components, breakdown amounts to failure since it is likely to cause irreversible damage; but in others it is reproducible and can be put to practical use, as for example in the Zener diode (Section 11.1.3).

B.2.5 The calculation of capacitance

The capacitance between two conductors is calculated, as the ratio of the charge on each to the voltage between them, by a process quite similar to the calculation of resistance. The voltage can be related to the electric

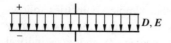

Fig. B.3

The electric field between parallel plates.

field strength E by eqn B.7, as before; now however we obtain it not in terms of current density and hence, by eqn B.8, total current, but in terms of flux density and hence, by Gauss's law eqn B.13, total charge. For this purpose we simply write E as D/ϵ instead of J/σ or $J\rho$. The simplest case is that of the parallel-plate capacitor, which corresponds to the uniform straight conductor in having, ideally, a uniform field. To apply Gauss's law, we must assume (without justification at this point) that flux passes directly from one plate to the other, as shown in Fig. B.3, and has uniform density between the plates and zero density elsewhere. In that case, applying the law to a surface around one plate shows that the flux density is given by

$$D = Q/A$$

for plates of area A. Hence the electric field strength E is $Q/\epsilon A$ and the voltage between plates separated by a distance d is

$$V = Ed = Qd/\epsilon A;$$

from this the capacitance is

$$C = Q/V = \epsilon A/d \qquad \qquad (B.19)$$

which is the familiar expression for a parallel plate capacitor. It illustrates the general feature that capacitance is maximized by using conductors of large area and close spacing. Hence a compact capacitor requires a very small separation between its conductors; this in turn implies maximizing the electric field in its dielectric for a given voltage, with consequent risk of breakdown (see below). It follows that the design of a capacitor must balance voltage rating against compactness.

The assumption that the field exists only between the plates and is uniform there needs a little justification. By symmetry a single infinite sheet of charge would emit flux uniformly on both sides; from this it follows that the fluxes from two infinite parallel plates carrying equal and opposite charges would cancel everywhere except for the space between them, where it would be uniform as has been assumed. But if the plates have finite area, there is an edge effect because even if the dielectric is confined to the region between the two the material outside, whatever it may be, can carry flux and some (called **fringing flux**) will pass that way as shown in Fig. B.4. To neglect this requires either that the dielectric has a much higher permittivity than the surrounding region (which is unlikely, since few materials have ϵ_r greater than about 10) or that the distance d between the plates is very small compared with their extent; in the latter case the fringing flux makes a negligible contribution to the total. The reason why no such constraint was considered in deducing the resistance of a uniform conductor (eqn B.9) is the dramatic difference between the conductivities of good conductors and those of insulators (i.e. dielectric materials) under normal conditions: it covers many orders of magnitude, and allows us to assume as a matter of course throughout circuit theory that current flow is confined to conducting paths and is zero elsewhere.

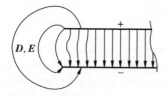

Fig. B.4

Fringing flux at an edge.

The calculation of capacitance for other simple geometries follows the same line. Between coaxial cylinders, for example, the flux is purely radial (if fringing at each end can be neglected) and a cylindrical Gaussian surface between the two gives the flux density at radius r between cylinders of length l carrying charges Q as

$$D = Q/2\pi rl$$

and so the field strength is

$$E = Q/2\pi \epsilon rl$$

from which the voltage between the cylinders of radii a and b, say, follows by integration exactly as in the resistance calculation, to give the result

$$C = \frac{Q}{V} = \frac{2\pi \epsilon l}{\ln(b/a)}. \tag{B.20}$$

This result is frequently quoted with the length l omitted, to give the capacitance per unit length of, for example, a coaxial transmission line. A point to note about eqn B.20 and the corresponding expression for resistance, eqn B.10, is that multiplication of the two gives

$$CR = \epsilon/\sigma = \epsilon\rho. \tag{B.21}$$

This useful relationship arises from the similarity of the two calculations, which pertain to the same field pattern, and is true for any two equipotential surfaces separated by a homogeneous isotropic medium.

Another important case is that of twin wires. If their separation d between centres is much greater than the radius a of each, then the charge on each may be assumed to produce a symmetrical radial flux pattern because the small radius allows only a negligible displacement of the charges by mutual attraction. Hence the flux density produced by each at radius r from its axis is again

$$D = Q/2\pi rl$$

if the wires carry charges $\pm Q$ and have length l. To deduce the voltage, however, we must this time integrate the resulting E field from a to $d-a$; and since each conductor must contribute equally to the overall field pattern (unlike the coaxial case, where the outer cylinder makes no contribution to the field between the two) the voltage due to one only must be doubled. The resulting expression for capacitance is

$$C = \frac{\pi \epsilon l}{\ln\left(\dfrac{d-a}{a}\right)} \approx \frac{\pi \epsilon l}{\ln(d/a)}. \tag{B.22}$$

This too is often quoted without the length l, to give a value per unit length.

An important practical case which does not lend itself to simple calculation is that of the strip lines or microstrip lines which are used to connect components in electronic circuits for ultra-high or microwave frequencies (Section 10.3.8). Because of the need to design these as compact transmission lines with a specified characteristic impedance, their width-to-spacing ratio is too small to allow fringing to be neglected as in the parallel-plate case described above. Their parameters can be calculated by more refined methods.

It is possible to define mutual capacitance, in a way which is analogous to mutual inductance (Section 3.10). It arises when several conducting bodies can have different potentials. With respect to earth, taken to be at zero potential, the potential of each conductor is a linear function of the charges on all, and conversely the charges on each are therefore a function of all the potentials. It is therefore possible to regard the charge induced on conductor i per unit potential of conductor j, when all except j are earthed, as a mutual capacitance C_{ij}. However, it can be shown that the overall effect can be simply accounted for by allocating a capacitance between each conductor and earth, and between every pair of conductors. Any of these capacitances can be calculated in the usual way when all the others are ignored.

B.2.6   Stored energy

It would be possible to find the total energy stored in the electric field of a capacitor by integrating eqn B.15, giving the energy per unit volume, over the volume of the dielectric. However it is better to start from the basic fact that moving a small charge dQ through a potential difference V requires energy VdQ. From the definition of capacitance as Q/V this can be written as $CVdV$, so long as we take C to be constant, and it follows that to charge a capacitor to a potential V needs an input of energy given by

$$\mathscr{E} = \int_0^V CVdV = \frac{1}{2}CV^2$$

which can also be written $\frac{1}{2}QV$ or $\frac{1}{2}Q^2/C$.

B.3   Inductance

B.3.1   The magnetic field

As capacitance expresses the effectiveness of a given geometrical arrangement in accumulating charge (and therefore electric flux) from a potential difference, so inductance describes the effectiveness of a given arrangement in producing a magnetic field and hence a total magnetic flux from a current. Since current necessarily flows in one or more closed loops the flux in question is that which passes through, or 'links', such a loop. Inductance is defined in general as the ratio

$$L_{jk} = \Phi_{jk}/i_j \qquad (B.23)$$

where Φ_{jk} is the flux linking a loop k due to a current[†] i_j in loop j. If $k = j$ then the ratio is L_{jj}, the self-inductance of the jth loop; if $k \neq j$ then the ratio defines the mutual inductance of the jth and kth loops which is often given the symbol M (as in Chapter 3).

Both self- and mutual inductance depend on the medium which carries the magnetic flux as well as on the geometrical arrangement of the current paths; in this case, by contrast with the electric field, it is the vector quantity defined as the flux density B which depends on the medium, and the corresponding magnetic field strength H which at any point depends only on the configuration of current responsible for it. These two magnetic field properties are related by the constitutive equation

$$B = \mu H \qquad (B.24)$$

in which μ is the **permeability** of the medium.

Magnetic field strength is related to the current density at any point by the Maxwell equation

$$\nabla \times H = J + \mathrm{d}D/\mathrm{d}t \qquad (B.25)$$

and to the total current I passing through any area A by the mathematically equivalent relation, usually known as **Ampere's law**,

$$\int_l H \cdot \mathrm{d}l = I \qquad \textbf{(B.26)}$$

where l is the closed path which bounds A. The current I, which must include displacement current if need be, can be found by integrating current density over A exactly as in eqn B.8 (including any contribution from $\mathrm{d}D/\mathrm{d}t$). In practice I is often simply the current flowing in a conductor and l is a convenient path in the medium surrounding it; for simple geometries the path can usually be chosen so that H is everywhere parallel to it, and the vector notation can then be dropped. In cases where the integral in eqn B.26 cannot be easily evaluated because the variation of H along an appropriate path l cannot be predicted, an alternative is to use the **Biot–Savart equation**. This gives the contribution to H at any point due to a current element $I\mathrm{d}l$ as the vector

---

[†] In this Appendix, the use of lower and upper case symbols for voltage and current varies according to the usage common in each context, rather than the conventions of previous chapters.

$$dH = Idl \times r/4\pi r^3 \qquad\qquad\qquad (\textbf{B.27})$$

where r is the distance vector from the element to the point considered. In this dl is now an element of a current loop (not, as in eqn B.26, a path over which H is to be integrated), and eqn B.27 can be integrated to find H at any point due to the whole loop.

The total magnetic flux Φ through any area is found from the flux density B by a surface integral corresponding exactly to eqn B.8 for current (and to eqn B.13 for electric flux, although that equation, being a statement of Gauss's law, referred to a closed surface in particular); that is, the flux through an area A is given by

$$\Phi = \int_A B \cdot dA. \qquad\qquad\qquad (B.28)$$

In many simple cases an area can be so chosen that B is everywhere normal to it, so that the directions of B and of the vector A are the same and the vector notation becomes redundant.

The unit of magnetic flux density is the tesla and that of flux is the weber (equivalent to one volt-second); the field strength H is measured in units of amperes per metre. In consequence the permeability μ has units henries per metre; like conductivity and permittivity, it is a scalar quantity for a homogeneous isotropic medium. The assumption of constant inductance in linear circuit theory implies constant μ, and is of only limited validity for those many devices in which magnetic materials (discussed below) are used.

The energy stored in a magnetic field is equivalent to an amount

$$e = \int_0^B H \cdot dB \qquad\qquad\qquad (B.29)$$

per unit volume; if the permeability is scalar and constant this becomes

$$e = \frac{1}{2}BH = \frac{1}{2}\mu H^2 = \frac{1}{2}B^2/\mu. \qquad\qquad\qquad (\textbf{B.30})$$

That there is a kind of duality between electric and magnetic fields, which lies behind the duality between capacitance and inductance, is apparent from a comparison of eqns B.24 and B.29 with B.14 and B.15, for example. However it must be remembered that the two fields are different in fundamental character: the magnetic field can be shown to be **solenoidal**, which is to say that it has zero divergence. It follows that the magnetic counterpart to Gauss's law, eqn B.13, states only that the total magnetic flux through any *closed* area is zero; in other words exactly as much flux leaves a closed volume as enters it. On the other hand, the electric field which originates from charge (the so-called electrostatic field) and leads to the idea of capacitance is **conservative**, which is to say that it has zero curl. From this it follows that for such an electric field the counterpart to Ampere's law, eqn B.26, states only that the integral of the field strength around any closed path is zero. (But this is *not* true of all electric fields: those which are magnetically induced do have curl; and in general the integral of an electric field around a closed path constitutes an e.m.f., arising from magnetic induction or otherwise, which is the basic source of Kirchhoff's second law.)

B.3.2 Permeability

Like permittivity, permeability has a finite value for all materials and for free space; according to the definitions adopted for SI units its value μ_0 for free space is $4\pi \times 10^{-7}$ Hm^{-1}. In general the permeability of a material can be written as

$$\mu = \mu_r \mu_0 \tag{B.31}$$

where μ_r is the **relative permeability**. The value of μ_r can be greater or less than unity, but for most materials the difference is very small (typically much less than one per cent). For a few metallic elements (principally iron, nickel, and cobalt) and their alloys, however, μ_r is much greater than unity; these are what are loosely known as magnetic materials but are better called **ferromagnetic**.

The reason why μ for any material differs, however slightly, from its free-space value lies in the polarization of its particles in an applied magnetic field. Whereas electric dipoles consist of pairs of separated charges, magnetic dipoles are formed by loops of current which on the atomic scale consist of orbiting electrons having a dipole moment proportional to their angular momentum. (In the simple case of a loop of current I bounding a plane area A, the dipole moment is defined as IA.) As a magnetic field is applied the e.m.f. of magnetic induction imparts a dipole moment to each atom which by Lenz's law opposes the applied field. This induced polarization therefore tends to make μ less than μ_0, in which case μ_r is less than unity and the **magnetic susceptibility** χ_m, defined as $\mu_r - 1$ (corresponding to the electric susceptibility χ), is negative. Materials for which this is true are said to be **diamagnetic**, but the values of χ_m which result are so small that for nearly all engineering purposes they can be neglected, so such materials are assumed to have $\mu = \mu_0$.

In some materials the atoms have permanent magnetic dipole moments; an applied field tends to align these in such a way that their effect on average tends to increase μ. This effect, which is comparable to orientational polarization in dielectrics, can outweigh that of the induced dipoles, making the resultant μ_r greater than unity and χ_m positive. It is temperature dependent and shows **saturation**: because the alignment becomes virtually complete if the applied field is strong enough, the average dipole moment per unit volume tends to a maximum and the value of χ_m falls as the field increases further.

Materials of this kind, for which χ_m is positive and $\mu_r > 1$, are in general said to be **paramagnetic**; but they too have in most cases negligibly small values of χ_m which allow the assumption that $\mu = \mu_0$. Materials with values of permeability close to μ_0, whether diamagnetic or paramagnetic, are often loosely said to be non-magnetic.

Ferromagnetic materials, on the other hand, are those in which the paramagnetic effect due to permanent dipole moments is greatly enhanced because their atoms tend to form groups known as **domains** in which the dipoles are all parallel. In an applied field the boundaries of these domains change in such ways as to give, in effect, very strong alignment and hence high values of μ_r and χ_m; these can be in the order of 10^3–10^4 for ordinary ferrous metals and can reach values of around 10^6 in special alloys. It is the phenomenon of ferromagnetism, and in particular the ready availability of iron, which allows relatively large values of flux density B (greatly in excess of that in the earth's magnetic field, for example) to be produced from modest values of current. (Since it is this in turn which allows the large-scale generation of electrical energy, it can be argued that ferromagnetism is uniquely fundamental to modern industrial civilization.)

B.3.3 Maximizing inductance

Although any current-carrying circuit produces a magnetic field and therefore has inductance, the effect is normally quite small and to form a useful inductor steps must be taken to maximize it, that is to produce the largest practicable flux for a given current. There are two ways of doing so, according to the expression B.28 for flux: one is to increase the area of the loop of current, and the other to increase the flux density B.

Increased area is most easily achieved by winding the current-carrying conductor in the form of a close-pitched helix, that is to say in a coil; the area of the surface so bounded is large in relation to the overall dimensions of the helix. Although it is a single surface, its area is very nearly equal to the number of turns times the projected area of the helix on a plane normal to its axis (loosely called the area of one turn). Nearly all of the flux produced by such a coil passes through the helix along its whole length; thus, if the flux density is integrated over the area of one turn to give what we may call the flux per turn, Φ_t say, then that much flux passes N times through the area bounded by a coil of N turns. It therefore represents a total contribution to the flux, given formally by the integral in eqn B.28, which to a good approximation is

$$\Phi = \Phi_t N, \tag{B.32}$$

a product which is known as the **flux linkages** of the coil but is in fact a value of total flux no different in principle from any other except that it is most easily calculated in the product form. A helical coil also has the effect of increasing the magnetic field strength in the cylindrical region it encloses (which governs the value of Φ_t) to a value much greater than that outside. As a result, the path by which the current loop is completed by joining the ends of a coil is of little importance, since it is likely to add only a small fraction to the total area, and that in a region of low field; hence we are justified in referring to the inductance of a coil even when we cannot specify exactly how its circuit is completed.

While the use of a helical coil increases not only the effective area for flux but also the field strength H and hence B, the latter can clearly be increased still further by means of a ferromagnetic material. For this reason inductors and any device, like a transformer (Section 4.9) or rotating electrical machine, which needs as much magnetic flux as possible for economic operation are often made with cores of such material. The core may be simply a straight bar round which the coil is wound, but it is more effective if it forms a magnetic circuit: that is, a more or less continuous high-permeability path linking the coil, or linking two coils if a high mutual inductance is the aim. Figure 4.32 showed typical arrangements for transformers, but the same idea can be applied to any inductance; rotating machines too are conventionally designed around a magnetic circuit, which in their case must include an air gap. Evidently there must be a tendency for the flux to be confined to the magnetic core, because a given value of H will produce a much higher flux density there. There is an obvious analogy with the conductors in an electrical circuit, and we consider this in the next section.

B.3.4 Magnetic circuits

The analogy between a magnetic circuit and an electrical circuit of conducting material is not perfect because, as pointed out earlier, the conductivity of an insulating medium is very many orders of magnitude less than that of a good conductor and can often be neglected; whereas the finite permeability of any medium, like its permittivity, is more likely to be significant. Thus, even in the presence of a magnetic circuit some flux will flow in the surrounding medium; it may be called **leakage flux**, although that term is most often applied in connection with mutual inductance in particular (Section 3.8.2). Despite this, magnetic circuits are sometimes assumed to have a high enough permeability for leakage to be negligible; the analogy with linear electric circuits is then close.

Magnetic flux is analogous to electric current; both can be expressed as the integral of a flux density over an area, as in eqns B.8 and B.28. Both are continuous and flow in closed paths; analogous to Kirchhoff's first law, therefore, is the fact that the fluxes at any junction in a magnetic circuit add to zero. Analogous to Kirchhoff's second law is Ampere's law, eqn B.26, in which the current I or, for a coil of N turns each carrying I, the **ampere-turn** product NI is equivalent to an e.m.f. and is known as the **magnetomotive force** or m.m.f. The line integral of H between two points can be regarded as a magnetic potential difference, analogous to the line integral of E in eqn B.7.

These ideas lead to the magnetic equivalent of resistance, known as **reluctance** and defined as the ratio of magnetic potential difference to flux. In general it is therefore given by the ratio of the appropriate integrals, but for a uniform length l of magnetic material having cross-sectional area A it is evidently given by

$$\Re = Hl/BA = l/\mu A \tag{B.33}$$

which may be compared with eqn B.9 for resistance. When a magnetic circuit is made up of various uniform branches in series or in parallel, their reluctances can be combined by exactly the same rules as resistance. (The inverse of reluctance, corresponding to conductance, is known as **permeance** and may equally well be used, according to convenience.) So far as any one coil on a magnetic circuit is concerned, therefore, the complete circuit can be reduced to a single reluctance which gives the ratio of ampere-turns to flux (per turn) for that coil; that is

$$\Re = NI/\Phi_t. \tag{B.34}$$

This result in turn leads to a useful expression for the self-inductance of the coil; for, by the definition of inductance, eqn B.23, we have

$$L = \Phi/I$$

where Φ is the flux linkages, given by $\Phi_t N$ according to eqn B.32. Combining this with eqn B.34 gives

$$L = N^2/\Re. \tag{B.35}$$

We should note here that whereas L is a property of a coil with its magnetic circuit, $\Re$ is characteristic of the circuit alone. These relationships enable a magnetic circuit to be solved in the sense of deducing fluxes in every branch for given ampere-turns in one or more coils, or of finding the ampere-turns needed for a given flux in some branch. Their usefulness is limited, however, by the fact that μ is far from constant for ferromagnetic materials. The alternative approach is to combine Ampere's law, around one or more closed paths, and continuity of flux with the $B–H$ relationship described in the next section.

B.3.5 Ferromagnetic saturation and losses

The use of a magnetic circuit, while it allows maximization of flux, has some attendant disadvantages as outlined in Section 4.9. One is that the permeability of a ferromagnetic material is very far from constant. Its variation is most conveniently represented by a **magnetization curve** of B against H; an example is shown in Fig. B.5. The curve is highly non-linear, and shows saturation, indicated by a much reduced gradient, for reasons similar to those which apply to paramagnetism: at those values of the applied field where the domains approach complete alignment, further increase in the field strength H produces a relatively small increase in the flux density B. The values at which this occurs vary but typically saturation begins at a flux density in the order of one tesla. The curve also shows **hysteresis**: because the changes in domain

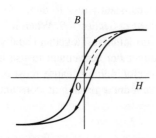

Fig. B.5
A magnetization curve, or $B–H$ loop.

boundaries due to an applied field are irreversible (in the strict sense) the B–H relation differs according as the field is increasing or decreasing; only in the saturation region do the two curves coincide. If H varies between positive and negative saturation (for example if it is produced by a large enough alternating current) then the field will trace out the loop shown; if H varies between lower limits then it will trace out a smaller loop. The area of such a hysteresis loop represents a loss of energy per unit volume for each cycle of the field, during which the stored energy fluctuates between zero and maxima given by eqn B.29. The consequent dissipation of power in a ferromagnetic material carrying an a.c. field is known as **hysteresis loss**. Its value is evidently proportional to frequency, volume, and the area of the B–H loop.

An alternating magnetic field always induces an electric field which can cause current to flow in a conducting medium. Currents due to this cause can arise in any conductor and may be superimposed on any current which flows from other sources. They are known as **eddy currents**; in some situations they are put to good use (as in eddy-current heating), in others they are undesirable but inevitable (as in the skin effect, mentioned in Section B.1). In ferromagnetic materials, which are mostly reasonably good conductors but are nearly always intended (in electrical applications) for carrying magnetic flux rather than current, these eddy currents are a source of power loss. It can be shown that the loss per unit volume is proportional to the square of frequency and to the square of the peak (or r.m.s.) flux density. Eddy-current loss can be minimized by laminating the material with thin layers of insulation to block the current paths, but remains significant in many cases.

Apart from hysteresis and eddy-current losses in magnetic circuits energized by a.c., the conducting wire from which the winding of an inductor is made has inevitable resistance, giving rise to a further loss of power. As for a capacitor, all the losses can be represented in circuit terms either by a series resistance or by an equivalent parallel resistance; in each case the value must depend on the applied voltage, or current, and the frequency. Because the current which produces the flux is necessarily also that which flows in the actual resistance of the winding, it is natural to represent the latter as a series element having that resistance (which can be assumed to be constant); this is done, for example, in the transformer circuits described in Section 4.9. When it is required to represent hysteresis and eddy-current losses, together usually known simply as iron losses, it is normally more convenient to use a parallel element such as R_m (or G_m) in Section 4.9.3, because most of the devices for which these are important are operated at constant voltage.

B.3.6 The calculation of inductance

The inductance of a given geometrical arrangement can be found by a method analogous to that used for capacitance. The magnetic field

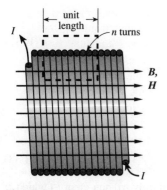

Fig. B.6
The magnetic field in a long solenoid.

strength H due to a postulated current I is found as a function of position either from Ampere's law, eqn B.26, or from the Biot–Savart law, B.27; the flux density B follows as μH and is then integrated over the effective area of the current loop (for self-inductance) or of the other loop in question (for mutual inductance) to give the required total flux (or flux linkages) Φ. The inductance is finally Φ/I.

The simplest case in theory is that of a helical coil of infinite length, a section of which is shown in Fig. B.6. Such a coil is usually known as a long solenoid; the stipulation of infinite length means that the form of the field at the ends can be ignored, so that the field outside can be taken to be zero everywhere and that inside may be assumed uniform. If the latter has strength H when the solenoid carries current I then Ampere's law applied around a path enclosing unit length as shown gives

$$H = nI$$

if there are n turns per unit length, and the flux density inside the solenoid is therefore

$$B = \mu nI$$

for a medium of permeability μ, now taken to be constant. (It is this dependence of B and H on the product nI which gives rise to the idea of ampere-turns, rather than current alone, as the quantity responsible for magnetic field strength; a similar dependence applies to any coil.) If the area of a turn (that is, of the cross-section of the coil) is A then the flux linkages per unit length are

$$\Phi = \Phi_t n = BAn = \mu An^2 I$$

from which the self-inductance *per unit length* is

$$L = \Phi/I = \mu An^2. \tag{B.36}$$

If the proportions of a coil make its length comparable with its diameter it can no longer be assumed to approximate to the results found for infinite length: the distribution of the field cannot be predicted with any precision, so the integral in Ampere's law cannot be evaluated and the field at any point must be found by the Biot–Savart law, eqn B.27. Since a varying flux density must then be integrated over the area of a turn to find a total flux, the calculation of inductance is not in general a simple one.

If a coil is wound on a magnetic circuit (Section B.3.4) the calculation of its inductance may be greatly simplified by the assumption that the flux flows uniformly in the magnetic material, and only in that, whatever its shape. For example, if a coil of N turns is wound on a continuous core of average length l and constant cross-section A with permeability μ, then a current I will produce a field strength H given according to Ampere's law by

$$Hl = NI$$

from which the flux density is

$$B = \mu H = \mu N I / l; \tag{B.37}$$

the flux per turn is BA and the total linkages are therefore given by

$$\Phi = BAN = \mu N^2 I A / l$$

from which the self-inductance is then

$$L = \Phi / I = \mu N^2 A / l. \tag{B.38}$$

Strictly this result is a constant of the coil only to the extent to which μ can be taken to be constant; some account can be taken of varying μ by evaluating the inductance, as above, for a particular value of current.

The similarity with eqn B.36 for the long solenoid is clear. Both expressions show that L depends on the square of the number of turns; this is true of coils generally, because both the ampere-turns and the flux linkages increase in proportion to the turns. It is also clear that eqn B.38 follows immediately from eqn B.35 because the reluctance of the magnetic circuit in this simple case of a single uniform path for flux is given directly by eqn B.33. For a less simple magnetic circuit there is a choice between applying Ampere's law and flux continuity, and deducing an overall reluctance (Section B.3.4). For example, if the uniform core of length l and area A already considered has in addition a short air gap of length g, then the gap can be accounted for either by adding to Hl the amount $(B/\mu_0)g$ and proceeding as before, or by adding the reluctance of the air gap to that of the core (since they are in series) to give a total reluctance

$$\mathfrak{R} = (l/\mu + g/\mu_0)/A.$$

Although inductors are commonly made in the form of coils, inductance can be an important parameter in other configurations, such as transmission lines of which the coaxial line is a familiar example. To find the inductance of coaxial cylinders carrying equal and opposite currents (not to be confused with forward and backward *waves*, as discussed in Chapter 10) we need first to find the flux Φ which encircles the gap between the two, that is between an inner radius a and an outer radius b as shown in Fig. B.7, since that is the flux which links the circuit in this case. Because of the circular symmetry H is purely azimuthal and must have a constant value around any path concentric with the cylinders; by Ampere's law, for such a path having radius r as shown the value of H due to a current I in the inner cylinder is given by

$$2\pi r H = I.$$

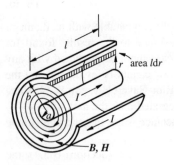

Fig. B.7

The magnetic field between coaxial cylinders.

It may be noted that the outer cylinder contributes nothing to the field in the gap, and that its contribution to the field outside both cylinders cancels that from the inner to give zero external field, in agreement with the fact that any external path must encircle a net current of zero. From the above relation the flux density follows as

$$B = \mu H = \mu I / 2\pi r.$$

This can be easily integrated over the area of a radial plane between a and b. If the cylinders have length l integration gives a flux

$$\Phi = \int_a^b Bl \, dr = \frac{\mu Il}{2\pi} \int_a^b \frac{dr}{r} = \frac{\mu Il}{2\pi} \ln \frac{b}{a}$$

and therefore a self-inductance

$$L = \frac{\Phi}{I} = \frac{\mu l}{2\pi} \ln \frac{b}{a}. \tag{B.39}$$

Since the medium in a coaxial system would usually be non-magnetic, and for a transmission line it would be normal to quote the inductance per unit length, the expression B.39 is often seen with l omitted and μ written as μ_0. This result may be compared with the corresponding expression for the capacitance of coaxial cylinders, eqn B.20. Their product gives

$$LC = \mu\epsilon$$

and confirms the result which was pointed out in Section 10.3.2. Its simplicity rests on the fact that in finding both C and L we have considered electric and magnetic flux only in the medium between the cylinders. But if the inner cylinder has its current distributed over a finite area (as in a solid wire at low frequencies, say) then it must support some internal magnetic flux, because any concentric path in its cross-section will enclose a proportion of its current. This adds to the inductance found above an amount which is not easy to calculate by the same means but can be shown on the basis of stored energy to be

$$L' = \mu / 8\pi$$

per unit length, if the current is uniformly distributed. Comparing this with eqn B.39 shows that the internal self-inductance becomes less important as the ratio b/a increases; L' also becomes unimportant at frequencies high enough to make the skin depth less than the radius a.

The inductance of uniform parallel wires can be found in a comparable way, by integrating the flux density due to both across the plane joining their centres. If internal flux is ignored this gives

$$L = \frac{\mu}{\pi} \ln \left(\frac{d - a}{a} \right) \tag{B.40}$$

for wires of radius a with centres d apart. This result taken with eqn B.22 again confirms that $LC = \mu\epsilon$ for uniform conductors.

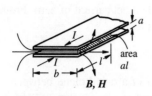

Fig. B.8
The magnetic field between
parallel strips.

A further example where Ampere's law is easily applied is that of a uniform line comprising closely spaced parallel strips. If the strips have width b and spacing a, as shown in Fig. B.8, and carry equal and opposite currents I, then it may be assumed that the field H is uniform in the gap and zero elsewhere (compare the assumptions made for the long solenoid, above). In that case an amperian path enclosing one conductor gives

$$Hb = I$$

and B is therefore $\mu I/b$. For length l the flux in the gap is that crossing the area al, which is

$$\Phi = Bal = \mu Ial/b$$

and the inductance is therefore

$$L = \Phi/I = \mu al/b \tag{B.41}$$

or $\mu a/b$ per unit length. However, it should be noted that this simple result is not applicable to the strip lines and microstrip lines used in printed circuits for high frequencies (Section 10.3.8); in these the spacing between the strip and the ground plane(s) is large enough, relative to the strip width, to invalidate the above assumptions.

Mutual inductance is calculated by similar procedures to those given above for self-inductance. A common case is that of two coils wound on the same magnetic circuit. Suppose that this has average length l, a uniform cross-section of area A and a (constant) permeability μ, and that the coils have N_1 and N_2 turns respectively. From eqn B.37 the flux density in the core produced by a current I_1 in coil 1 is given by $\mu N_1 I_1/l$, and the flux through the core, which is the flux per turn for each coil, is therefore

$$\Phi_t = \mu N_1 I_1 A/l.$$

If we assume that the magnetic circuit assures negligible leakage, so that the coils are perfectly coupled and ϕ_t links both completely whatever their relative positions on the core, then the flux linkages in coil 2 which result from I_1 are

$$\Phi_{12} = \Phi_t N_2 = \mu N_1 N_2 I_1 A/l.$$

From the definition of inductance, eqn B.23, it follows that the mutual inductance of the coils is given by

$$M_{12} = \Phi_{12}/I_1 = \mu N_1 N_2 A/l. \tag{B.42}$$

An exactly equivalent argument starting from current I_2 in coil 2 would obviously give an identical expression for M_{21}. This would also confirm the reciprocity relationship

$$M_{12} = M_{21}$$

for the particular case of perfect coupling; a more general proof would show it to hold in any linear system, i.e. for any two loops in a homogeneous medium of constant μ.

Equations B.38 and B.42 together also show that for perfectly coupled coils

$$M^2 = L_1 L_2,$$

confirming eqn 3.45, and

$$L_2 N_1 / N_2 = M = L_1 N_2 / N_1$$

in agreement with eqn 4.37.

B.3.7 Stored energy

The total energy stored in the magnetic field of an inductor can be found from eqn B.30 but as with capacitance the energy can be more simply found in terms of quantities which apply to the inductance as a whole. Since the e.m.f. in an inductor with flux Φ is $d\Phi/dt$, the power needed to change its flux when it carries current I would be $Id\Phi/dt$ and its energy input can therefore be expressed as $Id\Phi$. From the definition of L, assumed to be a constant, we may write $d\Phi$ as LdI and the total energy stored becomes

$$\mathscr{E} = \int_0^I IL dI = \frac{1}{2} L I^2 \qquad \textbf{(B.43)}$$

which can also be written $\frac{1}{2}\Phi^2/L$ or $\frac{1}{2}\Phi I$.

The energy stored in mutual inductance is additional to that stored in its constituent self-inductances, given by eqn B.43 for each. It can be shown by a comparable argument that the mutual inductance M between self-inductances carrying currents I_1 and I_2 stores additional energy given by $MI_1 I_2$. (This can be regarded as the sum of the two contributions $\frac{1}{2}M_{12}I_1 I_2$ and $\frac{1}{2}M_{21}I_2 I_1$, which in a linear circuit are equal.) In the case of several coupled inductances there is energy stored in each self-inductance and additional energy in the mutual inductance of every pair.

Appendix C: Transistor symbols and characteristics

Junction FET or JFET

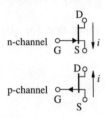
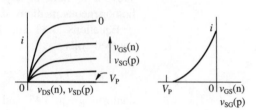

Metal-oxide-semiconductor FET or MOSFET

(a) Depletion mode or D-MOST

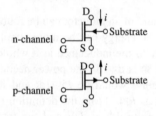
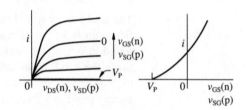

(b) Enhancement mode or E-MOST

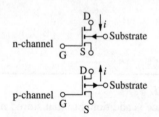
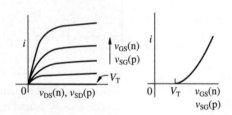

Bipolar Junction Transistor or BJT

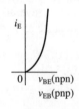
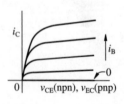

Appendix D: The economy of three-phase transmission

Suppose that a three-phase three-wire transmission line, with balanced line voltages and currents of magnitudes V_3 and I_3 respectively, is required to transmit the same total volt-amperes as a single-phase transmission line with voltage V_1 and current I_1. Then

$$V_1 I_1 = \sqrt{3} V_3 I_3$$

and if the two lines are allowed the same voltage between conductors we have $V_1 = V_3$ and

$$I_1 = \sqrt{3} I_3.$$

If now the respective conductors have resistance R_1 and R_3 per unit length and the total power losses are required to be the same for both lines, then

$$2I_1^2 R_1 = 3I_3^2 R_3 = I_1^2 R_3$$

from which

$$R_3 = 2R_1.$$

Since the resistance of a conductor is in inverse proportion to its cross-sectional area A it follows that

$$A_1 = 2A_3$$

and the ratio of the total volumes of conductor material per unit length of the two lines is therefore

$$r = 3A_3/2A_1 = 3/4$$

in favour of the three-phase line. A four-wire system would of course reduce the advantage, which would be lost entirely in the unlikely case of a neutral conductor identical to the other three.

Alternatively, if the conductors of the three-wire system use in total the same amount of material as those of the single-phase then $2A_1 = 3A_3$ and therefore $R_3 = 3R_1/2$. For the same total volt-amperes, that is for $I_1 = \sqrt{3} I_3$ as before, the total loss in the three-phase case is

$$3I_3^2 R_3 = I_1^2 R_3 = 3I_1^2 R_1/2$$

as against $2I_1^2 R_1$ in the single-phase; the improvement is again a factor of $3/4$.

If the single-phase voltage is limited to the line-to-neutral voltage of the three-phase line, the advantage of the latter is increased.

Appendix E: The Laplace transformation

E.1   General relations

Function	Transform
$Af(t)$	$AF(s)$
$Af_1(t) + Bf_2(t)$	$AF_1(s) + BF_2(s)$
$f(t - T)u(t - T)$	$e^{-sT}F(s)$
$e^{-at}f(t)$	$F(s+a)$
$f'(t)$	$sF(s) - f(0+)$
$f''(t)$	$s^2F(s) - sf(0+) - f'(0+)$
$f^{(n)}(t)$	$s^nF(s) - s^{n-1}f(0+) - s^{n-2}f'(0+)$
	$\quad - ... - f^{(n-1)}(0+)$
$\int_0^t f(\tau)d\tau$	$F(s)/s$
$\int_0^t f_1(t-\tau)f_2(\tau)d\tau$	$F_1(s)F_2(s)$

E.2 Transforms of functions

Functions	Transform
$\delta(t)$	1
A	A/s
$u(t)$	$1/s$
$t^{n-1}/(n-1)!$	$1/s^n$
e^{-at}	$1/(s+a)$
$(1 - e^{-at})/a$	$1/\{s(s+a)\}$
$\cos at$	$s/(s^2 + a^2)$
$\cosh at$	$s/(s^2 - a^2)$
$\sin at$	$a/(s^2 + a^2)$
$\sinh at$	$a/(s^2 - a^2)$
$(1 - \cos at)/a^2$	$1/\{s(s^2 + a^2)\}$
$(at - \sin at)/a^3$	$1/\{s^2(s^2 + a^2)\}$
te^{-at}	$1/(s+a)^2$
$e^{-at}(1 - at)$	$s/(s+a)^2$
$(\sin at - at \cos at)/2a^3$	$1/(s^2 + a^2)^2$
$t \sin at/2a$	$s/(s^2 + a^2)^2$
$t \cos at$	$(s^2 - a^2)/(s^2 + a^2)^2$
$e^{-at} \cos bt$	$(s+a)/\{(s+a)^2 + b^2\}$
$e^{-at} \sin bt$	$b/\{(s+a)^2 + b^2\}$
$(e^{-bt} - e^{-at})/(a - b)$	$1/(s+a)(s+b)$
$(ae^{-at} - be^{-bt})/(a - b)$	$s/(s+a)(s+b)$
$e^{-\alpha t} \sin \beta t/\beta$	$1/(s^2 + 2\zeta\omega_0 s + \omega_0{}^2)*$
$e^{-\alpha t}(\cos \beta t - \dfrac{\alpha}{\beta} \sin \beta t)$	$s/(s^2 + 2\zeta\omega_0 s + \omega_0{}^2)*$

*In these $\alpha = \zeta\omega_0$, $\beta = \omega_0\sqrt{1 - \zeta^2}$, and $\zeta < 1$.

E.3 Partial fractions

The ratio of two polynomials is s which can be expressed in the form

$$F(s) = \frac{P(s)}{(s + p_1)(s + p_2)...(s + p_n)} \tag{E.1}$$

where the degree of $P(s)$ is less than n, can be written in partial fractions as

$$F(s) = \frac{A_1}{s + p_1} + \frac{A_2}{s + p_2} + ... + \frac{A_n}{s + p_n} \tag{E.2}$$

where the A are constants. The two expressions for $F(s)$ are identically equal, so may be equated for any value of s; equating them for $s = -p_r$ yields

$$A_r = |\ (s + p_r)F(s)\ |_{s=-p_r} \tag{E.3}$$

Example Find the partial fractions for

$$F(s) = \frac{s + 1}{(s + 2)(s + 3)}.$$

Let

$$F(s) = \frac{A_1}{s + 2} + \frac{A_2}{s + 3};$$

then

$$A_1 = |\ (s + 2)F(s)\ |_{s=-2} = \left.\left|\frac{s + 1}{s + 3}\right|\right|_{s=-2} = -1$$

and

$$A_2 = |\ (s + 3)F(s)\ |_{s=-3} = \left.\left|\frac{s + 1}{s + 2}\right|\right|_{s=-3} = 2$$

so

$$F(s) = -\frac{1}{s + 2} + \frac{2}{s + 3}.$$

which can readily be confirmed to be equivalent to the given form.

The condition that the degree m, say, of the numerator $P(s)$ is less than n, the degree of the denominator, is often satisfied for physical systems, but it is possible in principle that $m \geq n$. In that case $F(s)$ can be written

$$F(s) = C + F'(s)$$

where C is a polynomial of degree $m - n$ (i.e. a constant for $m = n$) and $F'(s)$ is a ratio of polynomials with $m < n$ as before.

If the denominator of $F(s)$ contains a repeated factor $(s + p_r)^q$ then the partial fractions must include

$$\frac{B_1}{s + p_r} + \frac{B_2}{(s + p_r)^2} + \ldots + \frac{B_q}{(s + p_r)^q}.$$

Here it can be shown that

$$B_q = | (s + p_r)^q F(s) |_{s=-p_r}$$

$$B_{q-1} = | \frac{d}{ds} \{(s + p_r)^q F(s)\}|_{s=-p_r} \tag{E.4}$$

$$B_{q-j} = | \frac{1}{j!} \frac{d^j}{ds^j} \{(s + p_r)^q F(s)\} |_{s=-p_r}.$$

Example Find the partial fractions for

$$F(s) = \frac{(s + 1)(s + 2)}{(s + 3)^3}.$$

Let

$$F(s) = \frac{B_1}{s + 3} + \frac{B_2}{(s + 3)^2} + \frac{B_3}{(s + 3)^3};$$

then

$$B_3 = | (s + 3)^3 F(s) |_{s=-3} = | (s + 1)(s + 2) |_{s=-3} = 2$$

$$B_2 = | \frac{d}{ds} \{(s + 1)(s + 2)\} |_{s=-3} = | 2s + 3 |_{s=-3} = -3$$

$$B_1 = | \frac{1}{2} \frac{d^2}{ds^2} \{(s + 1)(s + 2)\} |_{s=-3} = 1$$

and

$$F(s) = \frac{1}{s + 3} - \frac{3}{(s + 3)^2} + \frac{2}{(s + 3)^3}.$$

A quadratic factor $s^2 + as + b$ in the denominator of $F(s)$ which does not have two real roots is best dealt with by including a partial fraction of the form

$$\frac{As + B}{s^2 + as + b}.$$

The constants A and B can be determined by equating powers of s in the equivalent numerators.

Example Find the partial fractions for

$$F(s) = \frac{s + 3}{(s + 1)(s^2 + s + 2)}.$$

Let

$$F(s) = \frac{As + B}{s^2 + s + 2} + \frac{C}{s + 1},$$

which can be written

$$F(s) = \frac{(As + B)(s + 1) + C(s^2 + s + 2)}{(s + 1)(s^2 + s + 2)}$$

Equating the numerators gives

$$(A + C)s^2 + (A + B + C)s + B + 2C = s + 3$$

in which equating coefficients yields

$$A + C = 0$$
$$A + B + C = 1$$
$$B + 2C = 3.$$

These give

$$A = -1$$
$$B = 1$$
$$C = 1$$

so that

$$F(s) = \frac{1 - s}{s^2 + s + 2} + \frac{1}{s + 1}.$$

Appendix F: Fourier series

The series here are expressed in terms of an angle θ and have period 2π. Similar waveforms in the variable t with period T have the same series with the substitution of $2\pi t/T$ for θ. The origin is wherever possible chosen to make the waveforms even functions ($b_n = 0$). The parameters α and β are angles and k is an integer.

	Series	Mean square value
	$\dfrac{4A}{\pi}\{\cos\theta - \tfrac{1}{3}\cos 3\theta + \tfrac{1}{5}\cos 5\theta - ...\}$	A^2
	$A\left\{\dfrac{\alpha}{\pi} + \dfrac{2}{\pi}(\sin\alpha\,\cos\theta + \tfrac{1}{2}\sin 2\alpha\,\cos 2\theta \right.$ $\left. + \tfrac{1}{3}\sin 3\alpha\,\cos 3\theta + ...)\right\}$	$\dfrac{A^2\alpha}{\pi}$
	$\dfrac{2A}{\pi}\{1 + \tfrac{2}{3}\cos 2\theta - \tfrac{2}{15}\cos 4\theta + \tfrac{2}{35}\cos 6\theta - ...\}$	$\dfrac{A^2}{2}$
	$\dfrac{A}{\pi}\left\{1 + \dfrac{\pi}{2}\cos\theta + \tfrac{2}{3}\cos 2\theta - \tfrac{2}{15}\cos 4\theta + ...\right\}$	$\dfrac{A^2}{4}$
	$\dfrac{A}{\pi}\{(\sin\alpha - \alpha\cos\alpha) + (\alpha - \tfrac{1}{2}\sin 2\alpha)\cos\theta$ $+ (\sin\alpha + \tfrac{1}{3}\sin 3\alpha - \cos\alpha\sin 2\alpha)\cos 2\theta$ $+ (\tfrac{1}{2}\sin 2\alpha + \tfrac{1}{4}\sin 4\alpha - \tfrac{2}{3}\cos\alpha\sin 3\alpha)\cos 3\theta + ...\}$	$\dfrac{A^2}{2\pi}\{\alpha - \tfrac{3}{2}\sin 2\alpha$ $+ 2\alpha\cos^2\alpha\}$
	$\dfrac{8A}{\pi^2}\{\cos\theta + \tfrac{1}{9}\cos 3\theta + \tfrac{1}{25}\cos 5\theta + ...\}$	$\dfrac{A^2}{3}$
	$\dfrac{A}{\pi\alpha}\left\{\dfrac{\alpha^2}{2} + 4\sin^2\tfrac{\alpha}{2}\cos\theta + \sin^2\alpha\cos 2\theta \right.$ $\left. + \tfrac{4}{9}\sin^2\tfrac{3\alpha}{2}\cos 3\theta + ...\right\}$	$\dfrac{A^2\alpha}{3\pi}$
	$\dfrac{4A}{\pi\beta}\{\sin\beta\,\cos\theta - \tfrac{1}{9}\sin 3\beta\,\cos 3\theta$ $+ \tfrac{1}{25}\sin 5\beta\,\cos 5\theta - ...\}$	$A^2\left(1 - \dfrac{4}{3}\dfrac{\beta}{\pi}\right)$
	$\dfrac{2A}{\pi}\{\sin\theta + \tfrac{1}{2}\sin 2\theta + \tfrac{1}{3}\sin 3\theta + ...\}$	$\dfrac{A^2}{3}$
	$\dfrac{Ak}{\pi}\sin\dfrac{\pi}{k}\left\{1 + \dfrac{2}{k^2 - 1}\cos k\theta - \dfrac{2}{4k^2 - 1}\cos 2k\theta \right.$ $\left. + \dfrac{2}{9k^2 - 1}\cos 3k\theta - ...\right\}$	

Appendix G: The Fourier transformation

In the following the transform of $x(t)$ is given in the form $X(\omega)$. The same transform in the form $X(f)$ is obtained by the substitution of $2\pi f$ for ω and $\delta(f \pm f_0)/2\pi$ for $\delta(\omega \pm \omega_0)$. By convention the same symbol is used for both forms; that is, $X(f)$ is taken to mean $X(2\pi f)$, which is the exact equivalent to $X(\omega)$.

G.I. General relations

Function	Transform
$Ax(t)$	$AX(\omega)$
$x(t-T)$	$e^{-j\omega T}X(\omega)$
$e^{j\omega_0 t}x(t)$	$X(\omega - \omega_0)$
$x(t)\cos\omega_0(t)$	$\frac{1}{2}\{X(\omega + \omega_0) + X(\omega - \omega_0)\}$
$x^{(n)}(t)$	$(j\omega)^n X(\omega)$
$\int_{-\infty}^{t} x(\tau)d\tau$	$\frac{1}{j\omega}X(\omega) + \pi X(0)\delta(\omega)$
$\int_{-\infty}^{\infty} x_1(t-\tau)x_2(\tau)d\tau$	$X_1(\omega)X_2(\omega)$

G.2 Transforms of functions

Function	Transform
$\delta(t)$	1
A	$2\pi A\delta(\omega)$
$u(t)$	$\pi\delta(\omega) + 1/j\omega$
$e^{j\omega_0 t}$	$2\pi\delta(\omega - \omega_0)$
$\cos\omega_0 t$	$\pi\{\delta(\omega + \omega_0) + \delta(\omega - \omega_0)\}$
$\sin\omega_0 t$	$j\pi\{\delta(\omega + \omega_0) - \delta(\omega - \omega_0)\}$

Function	Transform
Sign function, $\epsilon(t)$	

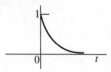

$2/j\omega$

Rectangular pulse, $\Pi(t/\tau)$

$\tau\,\text{sinc}(\omega\tau/2\pi)$

Triangular pulse, $\Lambda(t/\tau)$

$\tau\,\text{sinc}^2(\omega\tau/2\pi)$

Carrier pulse, $\Pi(t/\tau)\cos\omega_0 t$

$$\frac{\tau}{2}\left\{\text{sinc}\frac{(\omega-\omega_0)\tau}{2\pi}+\text{sinc}\frac{(\omega+\omega_0)\tau}{2\pi}\right\}$$

Exponential pulse, $u(t)e^{-at}$
$(a > 0)$

$$\frac{1}{(a+j\omega)}$$

Double-sided exponential pulse, $e^{-a|t|}$
$(a > 0)$

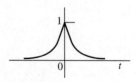

$$\frac{2a}{a^2+\omega^2}$$

Function	Transform
Gaussian pulse, $e^{-\pi(t/\tau)^2}$ 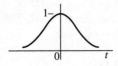	$\tau e^{-(\omega\tau)^2/4\pi}$
Cosine-squared pulse, $\Pi\left(\dfrac{t}{\tau}\right)\cos^2\left(\dfrac{\pi t}{2\tau}\right)$ 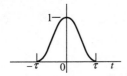	$\dfrac{\pi^2 \sin \omega\tau}{\omega(\pi^2 - \omega^2\tau^2)}$
Impulse train, $\displaystyle\sum_{-\infty}^{+\infty} \delta(t - nT)$ 	$\dfrac{\pi}{T}\displaystyle\sum_{-\infty}^{+\infty} \delta\left(\omega - \dfrac{2\pi n}{T}\right)$

Appendix H: Answers to problems

1.1 $10 \cos 1000t$ mA; $10^{-5} \sin 1000t$ C.

1.2 constant $+ 100t$ V.

1.3 constant $+ 10t$ kA.

1.4 $10 \, \mu$V.

1.5 $10 + 1000t$ V.

1.6 $0.5 \, \Omega$.

1.7 6.05 V.

1.8 12 V; $2 \, \Omega$.

1.9 0–$111 \, \Omega$.

1.10 0–20 V.

2.1 -1 A.

2.2 2.5 A.

2.3 8.1 A; 3.1 A.

2.4 5.71 A; 0.71 A.

2.5 3.33 V; 4 V.

2.6 7.48 V; 7.23 V.

2.7 342 kΩ.

2.8 $1818 \, \Omega$; $202 \, \Omega$.

2.9 $1.875 \, \Omega$; $1.875 \, \Omega$; $0.563 \, \Omega$.

2.10 0.3 V; $1.95 \, \Omega$.

2.11 11.75 V; $0.05 \, \Omega$.

2.12 2.5 A; $4 \, \Omega$.

2.13 7.03 V; 0.45 V.

2.14 3 A; -4 A.

2.15

$$\begin{pmatrix} 4 & -3 & 0 \\ -3 & 16 & -3 \\ 0 & -3 & 5 \end{pmatrix}.$$

2.16 5.

2.17 53.8 W.

2.18 3.33 W; 8 W; 5.56 W; 0.44 W; 5.33 W.

2.19 $5 \, \Omega$; 0.8 W.

3.1 (a) $31.8 \, \mu$F; 318 mH. (b) 3.18 pF; 31.8 nH.

3.2 (a) 7.07 V; 14.14 V. (b) 50 Hz. (c) $\pi/3$. (d) 18.71 V. (e) 0.714 rad.

3.3 $106 \, \lfloor 58° \,$ mA.

3.4 $5.7 \, \lfloor -15° \,$ V; $2.5 - j2.5 \, \Omega$.

3.5 $R/(1 + \omega^2 C^2 R^2) - j\omega C R^2/(1 + \omega^2 C^2 R^2)$.

3.6 $8.8 - j1.6 \, \Omega$.

3.7 $6.72 \, \lfloor 35° \, \Omega$.

3.10 $I(j\omega C R - \omega^2 L C)/(1 + j\omega C R - \omega^2 L C)$.

3.11 -0.00402 S; 0.00108 S.

3.12 2 mH; 1.98 kΩ.

3.15 72 W; 71.6 W; 0.29 W.

3.16 4.87 W; 0.0635.

3.17 $8.5 \, \mu$W; $35 \, \mu$W.

3.18 55 W; -10 VAR.

3.19 $-j544 \, \Omega$.

3.20 $\sqrt{3}\omega$; 25 W.

4.1 $R_1 + R_2$; R_2; R_2.

4.2 $5/3$, $2/3$, $1/3 \, \Omega$; $2/3$, $5/3$, $-1/3$ S.

4.3 5, -2, $-3 \, \Omega$, 3 S; 2, -5, $-3 \, \Omega$, 3 S.

4.4 $1 - \omega^2 LC$, -1, $-j\omega L$, $j\omega C$.

4.5 $j\omega L$, $j\omega C$, 1, -1.

4.7 $3 \, \Omega$, $3/4 \, \Omega$, $3 \, \Omega$.

4.8 $(z_{11}z_{22} - z_{12}^2)/z_{22}$, $1/z_{22}$, z_{12}/z_{22}, $-z_{12}/z_{22}$.

4.9 $500 \, \Omega$, $500 \, \Omega$, 0, 10.

4.10 $175 \, \mu$A.

4.11 -9.29.

4.12 -250.

4.13 33.3; -167.

4.14 -125; -38.5.

4.15 $147 \, \Omega$; -6.8.

4.17 $h_{ie} + (h_{fe} + 1)R_e$.

4.21 37.5%.

4.22 1390; 3.33 kΩ.

4.23 1 kΩ; 0.75.

4.24 52 kΩ, $19.2 \, \Omega$; 0.98.

4.25 $2 \, \Omega$, $1 \, \Omega$; $1.91 \, \Omega$, $1.04 \, \Omega$.

4.26 (a) 16 V, $j1.13 \, \Omega$; (b) 20 V, 0.

4.27 $0.0795 \, \lfloor -6.4° \,$; $0.1 \, \lfloor 0° \,$.

4.28 $-2.40 \, \lfloor 72.6° \,$; $-3.00 \, \lfloor 72.6° \,$.

4.29 394 mH, 65.7 mH.

4.30 $25.8°$.

4.31 0.0822; 3.

4.32 0.0909; 2.90.

4.33 $0.122 \, \Omega$, $1.13 \, \Omega$.

4.34 $4.5 \, \lfloor -77.2° \,$ mS.

4.35 256 V; 0.969.

4.36 57.9 kVA; 0.77.

5.2 10^6, 1.5×10^6 rad s^{-1}; -9.54 dB, -6.02 dB.

5.3 -15%, $+ 17.7\%$; ± 1.41 dB.

5.4 $11.5°$.

5.5 2.13 krad s^{-1}, 204 krad s^{-1}; 13.2 dB, 52.9 dB.

5.6 $78.3°$.

5.7 0.0576 J, 0.584 J; 159 Hz; 0.0576 J.

5.8 $2(1)$ kΩ; $0.632(0.5)$ kΩ; $0.632(0.5)$ kΩ.

5.9 $0.5(1)$ mS; $0.4(0.5)$ mS; $0.4(0.5)$ mS.

5.10 1 mS; 0.555 mS; 0.555 mS.

5.11 2.5 kΩ.

5.12 0.4 mS, 0.5 mS.

5.13 (a) $\pm 56°$, (b) $\pm 31°$.

5.14 0.996; 5 kΩ.

5.15 655 Ω.

5.16 2, 0.5.

5.17 11.2; 7.09 kHz; 634 Hz.

5.18 796 Hz; 0.125; 4; 4000.

5.19 27.6 kHz; $1/\sqrt{3}$; 15.9 kHz, $1/\sqrt{2}$; 2/3.

5.20 6.44 kHz, 15.5; 10^4, 44.2.

5.21 152 kHz, 168 kHz; 2.12, 2.00, 2.36.

5.22 0.673, 0.637, 0.601.

5.23 0.876, 0.815, 0.703; 0.964, 0.942, 0.892.

6.1 416 V; 159 kV.

6.2 $286 \underline{|-45°}$ A.

6.3 $410 \underline{|-40°}$ A.

6.4 10 μF.

6.5 16 A, 20.4 A, 13.9 A.

6.6 23 A.

6.7 9.27 A.

6.8 12 kW.

6.9 $\pm jV/\sqrt{3}$, 0.

6.10 19.3 A, 9.75 A, 7.66 A.

7.1 4.6 μs; 3 μs.

7.2 $-16.7e^{-10^6 t}$ mA.

7.3 4.68 μs; 19.3 mA.

7.4 200 Ω; 10 μs; 73.6 mA.

7.5 25000 s^{-1}; 15.4 kHz; 0.2.

7.7 $10^{-3}\delta(t) + 10^3 + 10^7 t$ V.

7.8 $(2/L)e^{-tR/L}$.

7.9 $(10^7/s^2)\left(1 - e^{-10^{-6}s}\right) - (10/s)e^{-10^{-6}s}$.

7.10 $(10^7/s^2) - \{10/s(e^{10^{-6}s} - 1)\}$.

7.11 $100(s + 10^4)/\{s(s + 2 \times 10^4)\}$.

7.13 $e^{-2t} - e^{-3t}$.

7.14 $\frac{1}{6}\left(1 - 3e^{-2t} + 2e^{-3t}\right)$.

7.15 $\frac{1}{\sqrt{2}} e^{-2t} \sin \sqrt{2} t$.

7.16 $\delta(t) - 2e^{-3t}$.

7.17 $\omega_n = \sqrt{91}$ rad s^{-1}, $\alpha = 3$ s^{-1}.

7.18 $20(1 - e^{-10^6 t})$ mA.

7.19 Zeros: $-10^4(1 \pm j3)$ s^{-1}; pole: -2×10^4 s^{-1}.

8.1 $A_0/(1 + \frac{1}{2}A_0 + j\omega T)$; $(1 + \frac{1}{2}A_0)/T$ rad s^{-1}.

8.2 $A_0 > 2$.

8.3 5 rad s^{-1}, 0.25; 7.07 rad s^{-1}, 0.177.

8.4 1110.

8.5 29; $1/\sqrt{6}RC$.

8.6 0.09; 10.

8.7 0.111%.

8.8 0.066 Ω.

8.9 15.9 kHz; 796 mV.

8.10 1 mV; 8000.

8.11 0.213 V.

9.1 $0.0862 \cos 2\pi \times 10^4 t + 0.00958 \cos 6\pi \times 10^4 t + 0.00345 \cos \pi \times 10^5 t$ V; 0.036%.

9.2 4.11 mA.

9.3 0.614 mA.

9.4 0.986.

9.5 $2.95/\tau$.

9.6 1.80×10^{-3} V^2s; 5%.

9.7 10^5 s^{-1}; 50 kHz.

9.8 0.95, 0.97, 0.99, 1, 1.01, 1.03, 1.05 MHz.

9.9 3.20×10^{-14} V^2Hz^{-1}; 1.66×10^{-16} V^2Hz^{-1}.

9.10 72.4 K.

9.12 1.07×10^{-15} W.

10.1 0.962; 0.0557.

10.2 16.1, $\pm \pi$; 0, 0.320.

10.3 79.6 kHz.

10.4 $-j1.59$ MΩ, 987 Ω.

10.5 $-j0.628$ Ω, 1.013 kΩ.

10.6 20.

10.7 -9.83×10^6, 9.74×10^6 sections s^{-1}.

10.8 318 kHz; 9.83×10^5, 9.49×10^5 sections s^{-1}.

10.9 $556 \underline{|-41.4°}$ Ω, $200 \underline{|-0.02°}$ Ω.

10.10 54.8 Ω; 1.83×10^8 ms^{-1}.

10.11 0.292; 1.83.

10.12 $93.2 \underline{|15.3°}$ Ω, $100 \underline{|0°}$ Ω, $30 \underline{|0°}$ Ω.

10.13 $-j12.3$ Ω.

10.14 74.0 Ω.

10.15 4.65 m.

10.16 -0.976, $-j12.0$ Ω, j4.01 mS, 0.976.

10.17 $1 + 1/2 - 1/2 - 1/4 + 1/4 + 1/8 -1/8-\dots$.

11.1 51.7 Ω; 714 Ω.

11.2 83.6°.

11.3 4.75 ms.

11.7 D1: 1 mA; D2: 1 mA; D3: 455 μA.

11.8 10 kΩ: 840 μA, 840 μA, 540 μA; 22 kΩ: 45.5 μA, 45.5 μA, 182 μA.

11.9 D1: $V_{in} > 5$ V; D2: $V_{in} < 25$ V; D3: $V_{in} > 11$ V.

11.10 D1: 1.46 mA; D2: 129 μA; D3: 714 μA.

Index